Stream Ciphers

Andreas Klein

Stream Ciphers

 Springer

Andreas Klein
Dept. of Pure Mathem. & Computer Algebra
State University of Ghent
Ghent, Belgium

ISBN 978-1-4471-5078-7   ISBN 978-1-4471-5079-4 (eBook)
DOI 10.1007/978-1-4471-5079-4
Springer London Heidelberg New York Dordrecht

Library of Congress Control Number: 2013936538

Mathematics Subject Classification: 94A60, 68P25, 11T71

Printed on acid-free paper

Springer is part of Springer Science+Business Media (www.springer.com)

Preface

Cryptographic ciphers come in two flavours: symmetric (AES, etc.) and asymmetric (RSA, etc.). The symmetric ciphers are further divided into block ciphers and stream ciphers. Block ciphers work on large blocks simultaneously (typically comprising 128 or 256 bits) and have no internal state (at least not in their basic version). Stream ciphers work on single bits or single words and need to maintain an internal state to change the cipher at each step.

Typically stream ciphers can reach higher speeds than block ciphers, but their theory is less developed. This is why stream ciphers are often skipped in books on cryptography.

This does not reflect the real importance of stream ciphers. They are used in several everyday applications (for example RC4 is used in wireless LAN and mobile telephones use A5). This book should fill the gap and provide a detailed introduction to stream ciphers.

I wrote this book in the years 2008–2010 when I had a research position at Ghent University.

I want to thank all my colleagues in Ghent for the pleasant time I had there, but especially Prof. Leo Storme who first gave me the opportunity to come to Ghent. We did some nice research together.

I also thank the team of the Springer Verlag who did a great job in improving this book. In addition I want to thank the anonymous referee, without whom the chapter on the Blum-Blum-Shub generator would be missing and there would be no exercises.

Wettenberg, Germany
Andreas Klein

Contents

1   Introduction to Stream Ciphers   1
   1.1   History I: Antique Ciphers   1
   1.2   Lessons from History: The Classification of Ciphers   3
   1.3   History II: The Golden Age of Stream Ciphers   8
   1.4   Lessons from the Enigma   8
   1.5   History III: Towards Modern Cryptography   10
   1.6   When to Use Stream Ciphers?   11
   1.7   Outline of the Book   11

Part I   Shift Register-Based Stream Ciphers

2   Linear Feedback Shift Registers   17
   2.1   Basic Definitions   17
   2.2   Algebraic Description of LFSR Sequences   18
      2.2.1   Generating Functions   19
      2.2.2   Feedback Polynomials Without Multiple Roots   20
      2.2.3   Feedback Polynomials with Multiple Roots   21
      2.2.4   LFSR Sequences as Cyclic Linear Codes   23
   2.3   Properties of m-Sequences   24
      2.3.1   Golomb's Axioms   24
      2.3.2   Sequences with Two Level Auto-Correlation   27
      2.3.3   Cross-Correlation of m-Sequences   29
   2.4   Linear Complexity   30
      2.4.1   Definition and Basic Properties   30
      2.4.2   The Berlekamp-Massey Algorithm   33
      2.4.3   Asymptotic Fast Computation of Linear Complexity   37
      2.4.4   Linear Complexity of Random Sequences   42
   2.5   The Linear Complexity Profile of Pseudo-random Sequences . . .   44
      2.5.1   Basic Properties   44
      2.5.2   Continued Fractions   46

        2.5.3   Classification of Sequences with a Perfect Linear
               Complexity Profile 48
  2.6 Implementation of LFSRs 50
        2.6.1   Hardware Realization of LFSRs 51
        2.6.2   Software Realization of LFSRs 52

3   **Non-linear Combinations of LFSRs** 59
  3.1 De Bruijn Sequences 59
  3.2 A Simple Example of a Non-linear Combination of LFSRs 64
  3.3 Different Attack Classes 65
        3.3.1   Time-Memory Trade-off Attacks 65
        3.3.2   Algebraic Attacks 65
        3.3.3   Correlation Attacks 66
  3.4 Non-linear Combinations of Several LFSR Sequences 66
        3.4.1   The Product of Two LFSRs 67
        3.4.2   General Combinations 70
  3.5 Non-linear Filters 72
  3.6 Correlation Immune Functions 75
        3.6.1   Definition and Alternative Characterizations 75
        3.6.2   Siegenthaler's Inequality 78
        3.6.3   Asymptotic Enumeration of Correlation Immune Functions 80

4   **Correlation Attacks** 91
  4.1 CJS-Attacks 91
        4.1.1   The Basic Version 91
        4.1.2   Using Relations of Different Size 94
        4.1.3   How to Search Relations 96
        4.1.4   Extended Relation Classes 98
        4.1.5   Twice Step Decoding 101
        4.1.6   Evaluation of the Relations 103
  4.2 Attacks Based on Convolutional Codes 105
        4.2.1   Introduction to Convolutional Codes 105
        4.2.2   Decoding Convolutional Codes 107
        4.2.3   Application to Cryptography 111
  4.3 Attacking LFSRs with Sparse Feedback Polynomials 114

5   **BDD-Based Attacks** 117
  5.1 Binary Decision Diagrams 117
        5.1.1   Ordered BDDs 118
        5.1.2   Free BDDs 124
  5.2 An Example of a BDD-Based Attack 126
        5.2.1   The Cipher E_0 126
        5.2.2   Attacking E_0 127

6   **Algebraic Attacks** 131
  6.1 Tools for Solving Non-linear Equations 131
        6.1.1   Gröbner Bases 131

6.1.2 Linearization 143
6.2 Pre-processing Techniques for Algebraic Attacks 147
6.2.1 Reducing the Degree 147
6.2.2 Dealing with Combiners with Memory 149
6.3 Real World Examples 151
6.3.1 LILI-128 151
6.3.2 E_0 153

7 Irregular Clocked Shift Registers 155
7.1 The Stop-and-Go Generator and the Step-Once-Twice Generator . 155
7.2 The Alternating Step Generator 157
7.3 The Shrinking Generator 158
7.3.1 Description of the Cipher 159
7.3.2 Linear Complexity of the Shrinking Generator 159
7.3.3 Correlation Attacks Against the Shrinking Generator . . . 161
7.4 Side Channel Attacks 163

Part II Some Special Ciphers

8 The Security of Mobile Phones (GSM) 169
8.1 The GSM Protocol 169
8.2 A5/2 170
8.2.1 Description of A5/2 170
8.2.2 An Instance of a Ciphertext-Only Attack 172
8.2.3 Other Attacks Against A5/2 175
8.3 A5/1 176
8.3.1 Description of A5/1 176
8.3.2 Time-Memory Trade-off Attacks 176
8.3.3 Correlation Attacks 179

9 RC4 and Related Ciphers 183
9.1 Description of RC4 183
9.2 Application of RC4 in WLAN Security 184
9.2.1 The WEP Protocol 184
9.2.2 The WPA Protocol 185
9.2.3 A Weakness Common to Both Protocols 187
9.3 Analysis of the RC4 Key Scheduling 190
9.3.1 The Most Likely and Least Likely RC4 Permutation . . . 191
9.3.2 Discarding the First RC4 Bytes 196
9.4 Chosen IV Attacks 199
9.4.1 Initialization Vector Precedes the Main Key 199
9.4.2 Variants of the Attack 200
9.4.3 Initialization Vector Follows the Main Key 202
9.5 Attacks Based on Golić's Correlation 202
9.5.1 Initialization Vector Follows the Main Key 204
9.5.2 Initialization Vector Precedes the Main Key 205
9.5.3 Attacking RC4 with the First n Bytes Discarded 207

          9.5.4   A Ciphertext-Only Attack 209
    9.6   State Recovering Attacks 209
    9.7   Other Attacks on RC4 212
          9.7.1   Digraph Probabilities 213
          9.7.2   Fortuitous States 218
    9.8   RC4 Variants 222
          9.8.1   An RC4 Variant for 32-Bit Processors 222
          9.8.2   RC4A 224
          9.8.3   Modifications to Avoid Known Attacks 227

10 The eStream Project 229
    10.1 Trivium 229
    10.2 Rabbit 232
    10.3 Mosquito and Moustique 235

11 The Blum-Blum-Shub Generator and Related Ciphers 241
    11.1 Cryptographically Secure Pseudo-random Generators 241
    11.2 The Blum-Blum-Shub Generator 244
    11.3 Implementation Aspects 247
    11.4 Extracting Several Bits per Step 251
    11.5 The RSA Generator and the Power Generator 253
    11.6 Generators Based on Other Hard Problems 254
    11.7 Unconditionally Secure Pseudo-random Sequences 256

Part III Mathematical Background

12 Computational Aspects 261
    12.1 Bit Tricks 261
          12.1.1   Infinite 2-adic Expansions 261
          12.1.2   Sideway Addition 262
          12.1.3   Sideway Addition for Arrays 263
    12.2 Binary Decision Diagrams, Implementation Aspects 264
          12.2.1   Memory Management 264
          12.2.2   Implementation of the Basic Operations 266
          12.2.3   Implementation of Reordering Algorithms 267
          12.2.4   Emulating a BDD Base 271
    12.3 The O-Notation 272
    12.4 The Complexity Classes $\mathcal{P}$ and $\mathcal{NP}$ 273
    12.5 Fast Linear Algebra 278
          12.5.1   Matrix Multiplication 278
          12.5.2   Other Matrix Operations 289
          12.5.3   Wiedmann's Algorithm and Black Box Linear Algebra . . 291

13 Number Theory 293
    13.1 Basic Results 293
    13.2 The Group $(\mathbb{Z}/n\mathbb{Z})^{\times}$ 294
    13.3 The Prime Number Theorem and Its Consequences 295

13.4 Zsigmondy's Theorem 297
13.5 Quadratic Residues 299
13.6 Lattice Reduction 301

14 Finite Fields 305
14.1 Basic Properties 305
14.2 Irreducible Polynomials 305
14.3 Primitive Polynomials 307
14.4 Trinomials 308
14.5 The Algebraic Normal Form 309

15 Statistics 311
15.1 Measure Theory 311
15.2 Simple Tests 312
15.2.1 The Variation Distance 312
15.2.2 The Test Problem 313
15.2.3 Optimal Tests 314
15.2.4 Bayesian Statistics 315
15.3 Sequential Tests 316
15.3.1 Introduction to Sequential Analysis 316
15.3.2 Martingales 316
15.3.3 Wald's Sequential Likelihood Ratio Test 319
15.3.4 Brownian Motion 322
15.3.5 The Functional Central Limit Theorem 326

16 Combinatorics 329
16.1 Asymptotic Calculations 329
16.2 Permutations 332
16.3 Trees 334

Part IV Exercises with Solutions

17 Exercises 339
17.1 Proposals for Programming Projects 344

18 Solutions 347

Part V Programs

19 An Overview of the Programs 365

20 Literate Programming 371
20.1 Introduction to Literate Programming 371
20.2 Pweb Design Goals 371
20.3 Pweb Manual 372
20.3.1 Structure of a WEB-Document 372
20.3.2 Text Sections 372
20.3.3 Code Sections and Modules 373
20.3.4 Macros 374

20.3.5 Special Variable Names 375
20.3.6 Include Files 375
20.3.7 Conditional Compilation 375
20.3.8 More pweb Commands 376
20.3.9 Compatibility Features 376
20.3.10 Common Errors 376
20.3.11 Editing pweb Documents 377
20.3.12 Extending pweb 377

Notations 379

References 381

Index 395

List of Figures

Fig. 1.1 Encrypting a text with a Vigenère cipher 3
Fig. 1.2 Stream-oriented block cipher modes 5
Fig. 1.3 Encrypting a text with an auto key cipher 5
Fig. 1.4 A synchronous stream cipher 6
Fig. 1.5 A self-synchronizing stream cipher 7

Fig. 2.1 A feedback shift register 18
Fig. 2.2 The sum of two LFSRs 32
Fig. 2.3 Construction for the Berlekamp-Massey algorithm 35
Fig. 2.4 Combination of the two LFSRs of Fig. 2.3 35
Fig. 2.5 The linear complexity profile of 10101111000100110101011100 . 45
Fig. 2.6 A typical linear complexity profile 45
Fig. 2.7 The Fibonacci implementation of an LFSR 51
Fig. 2.8 The Galois implementation of an LFSR 51
Fig. 2.9 Right shift over several words 53

Fig. 3.1 The smallest de Bruijn graphs 60
Fig. 3.2 The Geffe generator 64
Fig. 3.3 A simple non-linear filter 72
Fig. 3.4 $1 - \cos(x) \geq \frac{2}{\pi^2} x^2$ 86

Fig. 4.1 A simple convolutional code 105
Fig. 4.2 Three different encoders of the same code 106
Fig. 4.3 An example for the Viterbi algorithm 108
Fig. 4.4 A tree diagram for a (2, 1) encoder 109

Fig. 5.1 A non-reduced binary decision diagram 118
Fig. 5.2 A reduced binary decision diagram 118
Fig. 5.3 Reducing a binary decision diagram 118
Fig. 5.4 Algorithm 5.2 applied to the diagram of Fig. 5.1 122
Fig. 5.5 The melt of two BDDs 123
Fig. 5.6 A free binary decision diagram 125
Fig. 5.7 The control graph of the free BDD in Fig. 5.6 125

Fig. 5.8   The cipher E_0 127
Fig. 5.9   Basic BDDs for attacking E_0 129

Fig. 6.1   A combiner with memory 149
Fig. 6.2   The LILI-128 keystream generator 152

Fig. 7.1   The stop-and-go generator 156
Fig. 7.2   The alternating step generator 158
Fig. 7.3   The shrinking generator 159

Fig. 8.1   Outline of the GSM protocol 170
Fig. 8.2   Diagram of A5/2 171
Fig. 8.3   Diagram of A5/1 177

Fig. 9.1   The S-box of the Temporal Key Hash (Part 1) 188
Fig. 9.2   Temporal Key Hash (Part 1) 188
Fig. 9.3   Temporal Key Hash (Part 2) 188
Fig. 9.4   The graph representation of $S = (0\ 1)(2\ 3) = (3\ 1)(2\ 3)(1\ 2)(0\ 1)$ 193
Fig. 9.5   The FMS-attack key scheduling 200
Fig. 9.6   Digraph repetition 217
Fig. 9.7   Example of a 3-fortuitous state 218

Fig. 10.1   The cipher Trivium 230
Fig. 10.2   The cipher Rabbit 233
Fig. 10.3   The cipher Moustique 236
Fig. 10.4   Mapping between q_j^i and $a_k^{(0)}$ 237

Fig. 12.1   Memory layout 265
Fig. 12.2   A BDD node in memory 265
Fig. 12.3   Variable swapping 268
Fig. 12.4   The variable 3 jumps up 269
Fig. 12.5   Moving a variable from the top to the bottom 270
Fig. 12.6   Sifting down 270
Fig. 12.7   A Turing machine 274

Fig. 15.1   A Brownian motion path 323

Fig. 16.1   Comparison of $\int_a^b f(x)\mathrm{d}x$ and $1/2 f(a) + f(a+1) + \cdots +$
            $f(b-1) + 1/2 f(b)$ 330
Fig. 16.2   A labeled tree 335

Fig. 18.1   Binomial coefficients modulo 2 352
Fig. 18.2   Basic BDD for attacking the self-shrinking generator 355
Fig. 18.3   The densities of $\mathcal{N}(0, 1)$ and $\mathcal{N}(0, 2)$ 362

Fig. 19.1   An example of the Doxygen documentation 366
Fig. 19.2   An example of the pweb documentation 366

Fig. 20.1   The literate programming environment 372

List of Tables

Table 1.1   The Vigenère tableau   4

Table 2.1   Tests for the algorithms   42
Table 2.2   Speed of different LFSR implementations (128 bit LFSR)   57
Table 2.3   Speed of an LFSR with feedback polynomial $z^{127} + z + 1$ 57

Table 4.1   A Fano metric for a (2, 1) convolutional code and a BSC with
             $p = 0.25$ 110
Table 4.2   Example of the sequential decoding algorithm   111

Table 9.1   A 4-order, 7-generative pattern   213
Table 9.2   Digraph probabilities of RC4   216
Table 9.3   The number of fortuitous states and their expected occurrence . .   222

Table 10.1   Number of bits per cell in the CCSR of Moustique   237
Table 10.2   Bit updating in the CCSR of Moustique   238

Table 12.1   Comparing sideway addition algorithms for arrays   264
Table 12.2   Speed of Pan's algorithm   287

Table 14.1   Primitive and irreducible polynomials over $\mathbb{F}_2$ of low weight . .   310

Table 18.1   Comparison of block cipher modes   347

Table 20.1   Files needed by pweb   378

List of Algorithms

Algorithm 2.1   The Berlekamp-Massey algorithm   34
Algorithm 2.2   Massey(i, i')   38
Algorithm 2.3   feedback(i')   42
Algorithm 2.4   Right shift over several words   53
Algorithm 2.5   Sideway addition mod 2 (32 bit version)   53
Algorithm 2.6   Sideway addition mod 2 (without multiplication)   54
Algorithm 2.7   LFSR byte-oriented implementation (table look-ups)   55
Algorithm 2.8   Parallel sideway addition mod 2   56
Algorithm 2.9   LFSR update with parallel sideway addition mod 2   57
Algorithm 2.10 Generating an LFSR sequence with the feedback
                polynomial $z^n + z^k + 1$ 58

Algorithm 4.1   Simple fast correlation attack (CJS)   93
Algorithm 4.2   Twice step decoding 102
Algorithm 4.3   Fast Fourier transform over the group $\mathbb{Z}_2$ 105
Algorithm 4.4   Viterbi decoding 108
Algorithm 4.5   Sequential decoding 110
Algorithm 4.6   Meier's and Staffelbach's attack against LFSRs with sparse
                feedback polynomials 115

Algorithm 5.1   Counting solutions of an ordered BDD 119
Algorithm 5.2   Reducing an ordered BDD 120
Algorithm 5.3   Check that a given BDD is free 124

Algorithm 6.1   Multivariate division with remainder 135
Algorithm 6.2   Buchberger's algorithm 140
Algorithm 6.3   The Gröbner walk 142
Algorithm 6.4   XL-algorithm 146
Algorithm 6.5 F_4 algorithm (simplified) 147

Algorithm 7.1   The alternating step generator 158
Algorithm 7.2   The alternating step generator (alternative form) 159

Algorithm 8.1   A5/2 initialization 174
Algorithm 8.2   A5/1 initialization 176
Algorithm 8.3   Enumerating special states of A5/1 178

Algorithm 9.1   RC4 key scheduling 184
Algorithm 9.2   RC4 pseudo-random generator 184
Algorithm 9.3   Temporal Key Hash 186
Algorithm 9.4   Temporal Key Hash S-box 187
Algorithm 9.5   CRC encoding 189
Algorithm 9.6   CRC decoding 189
Algorithm 9.7   Idealized RC4 key scheduling 191
Algorithm 9.8   Computing the key from an early permutation state 203
Algorithm 9.9   A simple internal state recovering attack 211
Algorithm 9.10   Computing the digraph probabilities 213
Algorithm 9.11   Computing the digraph probabilities (1. Transformation) . . 215
Algorithm 9.12   Computing the digraph probabilities (2. Transformation,
                 inner loops) 215
Algorithm 9.13   Searching fortuitous states 219
Algorithm 9.14   Enumerating fortuitous states (fast) 220
Algorithm 9.15   RC4(n,m) key scheduling 223
Algorithm 9.16   RC4(n,m) pseudo-random generator 223
Algorithm 9.17   RC4(n,m) pseudo-random generator, old version 224
Algorithm 9.18   RC4A pseudo-random generator 225
Algorithm 9.19   RC4 key scheduling 227
Algorithm 9.20   Paul's suggestion for key scheduling 228

Algorithm 10.1   Trivium key stream generation 230
Algorithm 10.2   Trivium key scheduling 231

Algorithm 11.1   The Blum-Micali generator 243
Algorithm 11.2   Discrete logarithm generator 243
Algorithm 11.3   The Blum-Blum-Shub generator 244
Algorithm 11.4   Enhancing the success probability 245
Algorithm 11.5   Montgomery reduction 250
Algorithm 11.6   Variation of the Blum-Blum-Shub generator for use with
                 Montgomery reduction 251
Algorithm 11.7   A variation of the Blum-Micali generator that outputs j
                 bits per step 252
Algorithm 11.8   The RSA generator 253
Algorithm 11.9   The Fisher-Stern generator 255
Algorithm 11.10 The QUAD cipher 256

Algorithm 12.1   Sideway addition based on table look-up 262
Algorithm 12.2   Sideway addition (64 bit words) 263
Algorithm 12.3   Sideway addition Harley-Seal method 264
Algorithm 12.4   Winograd's algorithm for multiplying small matrices 279

Algorithm 12.5   Strassen's algorithm to multiply 2×2 matrices 280
Algorithm 12.6   Pan's matrix multiplication 286
Algorithm 12.7   Multiplication of 64×64 binary matrices 287
Algorithm 12.8   Multiplication of 8×8 binary matrices (MXOR) 288

Algorithm 13.1   Evaluating the Jacobi symbol 301
Algorithm 13.2   LLL basis reduction 302
Algorithm 13.3   Coppersmith's method (univariate case) 303

Algorithm 14.1   Choosing a random primitive element of $\mathbb{F}_q$ 307

Algorithm 15.1   Wald's sequential test 320

Algorithm 17.1   Sideway addition mod 2 340
Algorithm 17.2   A weak variation of the RC4 pseudo-random generator . . . 342

Algorithm 18.1   Choosing a random de Bruijn sequence 351
Algorithm 18.2   Binary Decision Diagrams: The ternary-and operator 353
Algorithm 18.3   Binary Decision Diagrams: The constrain operator 355
Algorithm 18.4   Sideway addition for sparse words 358

Chapter 1
Introduction to Stream Ciphers

1.1 History I: Antique Ciphers

The art of writing secret messages is very old. In the early days few people could write, so effectively every text was encrypted. It took a millennium for true cryptosystems to appear. An early example was the Scytale ($\sigma\kappa\upsilon\tau\alpha\lambda\eta$) which was used by the Spartanians in the Persian wars (3rd century BC). Cryptography has been reinvented many times independently. For an extensive history of the subject, see [141].

Another cipher was used by the Roman emperor Gaius Julius Caesar. Sueton writes:

Exstant et [epistolae] ad Ciceronem, item ad familiares de rebus, in quibus, si qua occultius perferenda erant, id est sic structo litterarum ordine, ut nullum verbum effici poset; quae si qui investigare et persequi velit, quartam elementorum litteram, id est D pro A et perinde reliquas commutet.

In English this reads:

There exist [letters from Caesar] to Cicero and his friends in which he uses a cipher, when something has to be transmitted confidentially, i.e. he changed the order of the letters in such a way that no word could be recognized. If one wants to read the content, he must convert the fourth letter, i.e. D, into an A and must proceed with the other letters in the same way.

Ancient cryptology did not distinguish between the algorithm used for encryption (the cipher) and the secret key. It took more than a millennium for the modern distinction between cipher and key to be introduced. In 1883 Kerckhoffs [146] stated his famous principle: The security of an encrypted message must not rely on the security of the encryption algorithm, but only on the security of the secret key.

History has proved *Kerckhoffs' principle* to be true many times. In modern cryptography we always require that the cipher has to be public and that there is public research about its security. Many people have thought that they could violate the principle and use a secret cipher. The result has always been the same: sooner or

A. Klein, *Stream Ciphers*, DOI 10.1007/978-1-4471-5079-4_1,
© Springer-Verlag London 2013

later (most times sooner) the cipher was leaked to the public and usually the cipher had some serious flaws.

We transform Caesar's cipher into a "modern" cipher with a key by declaring that the cipher is the substitution of each letter by another and that the key should be a permutation of the alphabet which the sender and the receiver have to agree on. This class of ciphers is called *monoalphabetic*. The key space has size $26! \approx 2^{88}$ which is, even for modern computers, too big to do an exhaustive search.

However, with the development of statistics it became clear that simple monoalphabetic ciphers can be broken by analyzing letter frequencies. At first, this was only known to some experts in the military and the secret service, but in time the approach became publicly known. In the 19th century attacks against monoalphabetic ciphers had become a popular theme in adventure literature (see Edgar Allan Poe [213] or Arthur Conan Doyle [81]).

So the simple idea of the monoalphabetic cipher needs an extension. There are three ways to obfuscate the letter frequency.

- In a *homophone cipher* we assign several ciphertext symbols to each letter. Common letters like e are assigned many different ciphertext symbols and rare letters like z get only a few. Each time we want to encode a letter we choose one of the associated ciphertext symbols at random.

  The Beale cipher [275], which is probably the most famous cryptogram in history, is of this type. The oldest known usage of a homophone cipher is dated at 1401 (see [141]).
- In a *polyalphabetic cipher* one uses very simple substitutions for each letter (normally cyclic shifts or involutions), but the substitution is changed for every letter in a previously agreed way. Changing the substitution masks the redundancy in the plaintext.

  The oldest use of a polyalphabetic cipher is dated at 1568 (see [141]). The Enigma machine, which is famous for its role in the second world war (see Sect. 1.3), is a sophisticated example of a polyalphabetic cipher.
- In a *polygraphic cipher* one groups the letters in blocks and uses a substitution on the block. This masks the letter frequency and, if the block size is large enough, blocks will almost never repeat, which is a good defense against attacks based on the redundancy in the plaintext.

  Polygraphic ciphers are relatively new. The Playfair cipher, which was invented in 1854 by Charles Wheatstone (see [141]), is the oldest known example.

All three approaches provide security against simple attacks based on letter frequency. Homophone ciphers have the disadvantage that the ciphertext is longer than the plaintext, which is unacceptable in many applications. Furthermore they are not well suited for automatic encryption, which is the reason that homophone ciphers do not play a role in modern cryptography.

The other two approaches work well. The modern descendants of polyalphabetic ciphers are known as *stream ciphers* while the polygraphic ciphers are the ancestors of the modern *block ciphers*.

```
SECRETSECRETSECRETSECRETSECRETSECRETSECRETSECRETSECRET
MANYYEARSAGOICONTRACTEDANINTIMACYWITHAMRWILLIAMLEGRAND
EEPPCWETJEYSKTSFXTRGLIFRRARVZQSGANMLLCDVOMNCMSQNVKJEPU
```

Fig. 1.1 Encrypting a text with a Vigenère cipher

1.2 Lessons from History: The Classification of Ciphers

To get a feeling for the modern ciphers, it helps to understand their historic counterparts. The most famous polygraphic cipher is the Vigenère cipher. It is named after the French cryptologist BLAISE DE VIGENÈRE, but it is older (see [141]).

The idea is simply to change the width of the cyclic-shift used in the Caesar cipher for every letter. One selects a keyword, for example "SECRET". On the first letter of the plaintext we apply the cyclic shift that would move A to the first letter of the keyword. (In the example we would apply the shift $A \mapsto S$, $B \mapsto T$, ..., $Z \mapsto R$.) On the second letter we apply the shift that maps A to the second letter of the keyword and so on. After we reach the last letter of the keyword we then go back and use the first letter again.

Example 1.1 Let us encrypt the first sentence of Edgar Allan Poe's novel "The Gold-Bug". The keyword is "SECRET". In Fig. 1.1 you see in the first row the repeated keyword, in the second row the plaintext, and in the third row the ciphertext.

The first few characters already demonstrate that the Vigenère cipher can map different plaintext characters to the same character in the ciphertext and that different characters in the ciphertext may encode the same plaintext character. Thus it prevents the simple cryptanalysis that works against the monoalphabetic ciphers.

The decryption is similar: one must simply apply the reverse shift. As an aid for carrying out the encryption and decryption one can use the Vigenère tableau (Table 1.1) which shows the results of all possible shifts.

The Vigenère cipher is easily susceptible to cryptographic attacks that first recover the length of the keyword (such as the Kasiski test and Friedman's coincidence index, see for example [259]). Here the attacker must solve several Caesar ciphers, which can be done by searching for the most frequent letter (usually corresponding to the letter E).

Nevertheless the Vigenère cipher has some interesting features:

- If the keyword is completely random and has the same length as the plaintext, one obtain the 'one-time pad', which is unconditionally secure.[1] One can think

---

[1]This is an important point that is often missed. Cryptography and proofs have a very special relation.

What does it mean to say that a one-time pad is provably unconditionally secure? It is of course pointless to try to guess a pattern in a truly random sequence. This is exactly what the proof says. However, there are some rare examples where people try to use a one time-pad, but use a (weakly)

Table 1.1 The Vigenère tableau

```
     a b c d e f g h i j k l m n o p q r s t u v w x y z

a    A B C D E F G H I J K L M N O P Q R S T U V W X Y Z
b    B C D E F G H I J K L M N O P Q R S T U V W X Y Z A
c    C D E F G H I J K L M N O P Q R S T U V W X Y Z A B
d    D E F G H I J K L M N O P Q R S T U V W X Y Z A B C
e    E F G H I J K L M N O P Q R S T U V W X Y Z A B C D
f    F G H I J K L M N O P Q R S T U V W X Y Z A B C D E
g    G H I J K L M N O P Q R S T U V W X Y Z A B C D E F
h    H I J K L M N O P Q R S T U V W X Y Z A B C D E F G
i    I J K L M N O P Q R S T U V W X Y Z A B C D E F G H
j    J K L M N O P Q R S T U V W X Y Z A B C D E F G H I
k    K L M N O P Q R S T U V W X Y Z A B C D E F G H I J
l    L M N O P Q R S T U V W X Y Z A B C D E F G H I J K
m    M N O P Q R S T U V W X Y Z A B C D E F G H I J K L
n    N O P Q R S T U V W X Y Z A B C D E F G H I J K L M
o    O P Q R S T U V W X Y Z A B C D E F G H I J K L M N
p    P Q R S T U V W X Y Z A B C D E F G H I J K L M N O
q    Q R S T U V W X Y Z A B C D E F G H I J K L M N O P
r    R S T U V W X Y Z A B C D E F G H I J K L M N O P Q
s    S T U V W X Y Z A B C D E F G H I J K L M N O P Q R
t    T U V W X Y Z A B C D E F G H I J K L M N O P Q R S
u    U V W X Y Z A B C D E F G H I J K L M N O P Q R S T
v    V W X Y Z A B C D E F G H I J K L M N O P Q R S T U
w    W X Y Z A B C D E F G H I J K L M N O P Q R S T U V
x    X Y Z A B C D E F G H I J K L M N O P Q R S T U V W
y    Y Z A B C D E F G H I J K L M N O P Q R S T U V W X
z    Z A B C D E F G H I J K L M N O P Q R S T U V W X Y
```

of a stream cipher as an attempt to replace the true random key of the one-time pad by a pseudo-random key.

• If one uses several periods of the key stream sequence of the stream cipher, attacks such as the one used against the Vigenère cipher become possible. Therefore the period of any stream cipher must be high ($> 2^{64}$) and we should use only short messages ($< 2^{32}$ bits).

• The distinction between block ciphers and stream ciphers is merely a matter of taste. One can say the Vigenère cipher is a polyalphabetic cipher which change the

---

biased random sequence. In this case there is a real chance to do some analysis and there are examples of successful attacks against such pseudo one-time pads.

The one-time pad is a pure cipher, it does not secure the message against active attacks. So if you want to use a one-time pad, you should consider using it in addition to a perfect MAC to guarantee the authentication of the message.

Finally, even the best cipher does not help if you use a malfunction protocol. There are a lot of examples where a system has been broken by ignoring the cipher completely and just using a protocol failure. Section 9.2.3 contains an interesting example of this kind.

In a nutshell, security proofs are not worthless, but you must carefully check what the proof exactly says. It is often not what you really want (see also the discussion in Chap. 11).

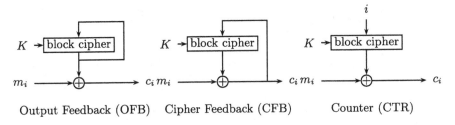

Output Feedback (OFB)   Cipher Feedback (CFB)   Counter (CTR)

Fig. 1.2 Stream-oriented block cipher modes

```
SECRETMANYYEARSAGOICONTRACTEDANINTIMACYWITHAMRWILLIAML
MANYYEARSAGOICONTRACTEDANINTIMACYWITHAMRWILLIAMLEGRAND
EEPPCXMRFYESITGNZFIEHRWRNKGXLMNKLPQFHCKNEBSLURITPRZAZV
```

Fig. 1.3 Encrypting a text with an auto key cipher

encryption function for every letter, but one could also say that it is a polygraphic cipher which work on blocks of the length of the keyword.

In modern cryptography block ciphers are normally used in a stream-oriented method. The naive idea of using a block cipher by applying it successively to the message blocks ($c_i = E(m_i, k)$) is called the electronic code book (ECB) mode. The disadvantage of this mode is that the same plaintext is always encrypted into the same ciphertext block, which leaks information. Figure 1.2 show three popular operation modes for block ciphers. In all these modes the block cipher is used as a source of pseudo-random numbers. For further reference, see [90].

When the important idea is the changing internal state we use the term 'stream cipher', and when it is the division of the plaintext into blocks we use the term 'block cipher'.

Several variants of the Vigenère cipher have been introduced to deal with the problem of the short period in the cipher. An interesting idea is the *auto key cipher*.

In the Vigenère cipher the keyword is repeated until it has the same length as the message. In an auto key cipher the message itself is used as part of the key

Example 1.2 We encrypt the same text as in Example 1.1 with an auto key cipher. The keyword is again "SECRET". In Fig. 1.3 you see the encryption. Note how the message is used as part of the key.

The Vigenère and auto key ciphers exhibit an important difference, which is used to classify stream ciphers. Either the cipher generates its key stream independently from the message or the message becomes a part of the feedback function. In the first case one speaks of a *synchronous stream cipher* and in the second case one speaks of a *self-synchronizing* or *asynchronous stream cipher*.

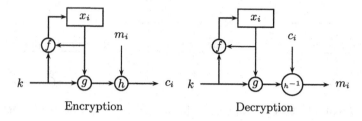

Fig. 1.4 A synchronous stream cipher

In general a synchronous stream cipher has the form

$$x_{i+1} = f(x_i, k),$$

$$z_i = g(x_i, k),$$

$$c_i = h(z_i, m_i),$$

where k denotes the key, x_i is the internal state at time i and m_i and c_i are the ith bit (letter) in the message and the ciphertext, respectively (see Fig. 1.4). f is the feedback function of the cipher, g is the key stream extractor and h combines the key stream $(z_i)_{i \in \mathbb{N}}$ with the message stream $(m_i)_{i \in \mathbb{N}}$. x_0 is called the initial state and may depend on the key.

In most applications we will take the 'exclusive or' operation as the combiner ($c_i = z_i \oplus m_i$) and the feedback and extraction function do not depend on the key ($x_{i+1} = f(x_i)$, $z_i = g(x_i)$), i.e. the key is only used to choose the initial internal state x_0. In this special case we speak of a *binary additive stream cipher*.

An important feature of synchronous stream ciphers is that they assure only the confidentially of the data, but not its integrity. An active attacker can simply flip the bits of the ciphertext, which flips the corresponding plaintext bits. To prevent active attacks one needs in addition a message authentication code (MAC). It is remarkable how many applications fail to observe this simple fact (see for example the GSM-protocol Chap. 8 or WEP Sect. 9.2.1).

To prevent active attacks and to transmit data over a noisy channel one must use Algorithm 1.1. The important part is that the error-correcting code must be applied last (see the lesson from the Enigma code (Sect. 1.4)).

Algorithm 1.1 Submitting data over a noisy channel using a synchronous stream cipher

---

1. Compute the hash h of the message m under a cryptographic hash function.
2. Encrypt m using the synchronous stream cipher. Append h to the ciphertext c.
3. Apply an error-correcting code to $c \| h$ and transmit the result.

A *self-synchronizing stream cipher* generates the keystream as a function of the key and a fixed number of preceding ciphertext digits (or, what is equivalent, a fixed number of preceding plaintext digits).

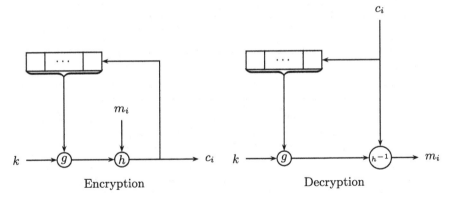

Fig. 1.5 A self-synchronizing stream cipher

The encryption has the form

$$x_i = (c_{i-1}, \ldots, c_{i-t}),$$
$$z_i = g(x_i, k),$$
$$c_i = h(z_i, m_i),$$

where the initial state $x_0 = (c_{-1}, \ldots, c_{-t})$ may depend on the key (see Fig. 1.5). The CFB-mode of block ciphers is an example of a self-synchronizing cipher.

Self-synchronizing stream ciphers have advantages over synchronous stream ciphers.

- A deletion or insertion of a bit in the ciphertext will cause only a finite number of plaintext bits to change, i.e. the cipher establishes proper decryption automatically after a loss of synchronization (the self-synchronizing property).
- If an attacker flips some bits in the ciphertext, the errors will propagate and several other bits in the plaintext will flip. Most likely this results in a nonsense text, i.e. we detect the active attack. So, in contrast to synchronous stream ciphers, one needs no extra hash function to secure the message against active attacks.
- Since every bit of the plaintext influences all subsequent bits of the ciphertext, the statistical properties of the plaintext are dispersed through the ciphertext. Hence self-synchronizing stream ciphers may be more resistant against attacks based on redundancy in the plaintext. The reader can try an experiment and attempt to break a Vigenère cipher and an auto key cipher (see Exercises 17.3 and 17.4). Most people find the first task easier.

However, self-synchronizing stream ciphers also have disadvantages. The separation of key stream generation and encryption in synchronous stream ciphers makes the implementation easier. It also makes the analysis of the cipher easier and helps in security proofs. So self-synchronizing stream ciphers may be more secure, but there is always a risk of large undetected security holes, since we understand them less. Most modern stream ciphers are synchronous stream ciphers.

1.3 History II: The Golden Age of Stream Ciphers

At the beginning of the 20th century cryptography took the step from simple cryptosystems which can be applied manually to complex systems which need machines to implement. Since this time, the question of whether a given cipher can fit on the available hardware has always been important for the success of the cipher.

The challenge was to implement the new cryptosystems on, for example, mechanical typewriters and telegraphs. People first began to experiment with electric typewriters where the keys connected to the output in some random fashion (Hebren 1915). However, such ciphers are only monoalphabetic. The next step was to put the wires on a rotor that change its position after each letter. Hebren advertised such a cipher in 1921 as "unbreakable", but it was still very weak. Combining several rotors of different speed finally gave a satisfactory system.

In the 1920s rotor machines were independently invented several times in different countries (Hugo Alexander, Netherlands; Arvid Gerhard Damm, Sweden; Arthur Scheribus, Germany) and quickly became a standard for cryptography. The fact that rotor machines work so well with telegraphy and typewriters, together with the high level of security that can be achieved by these machines, left almost no room for other types of cryptosystem.

The most famous rotor machine of all time is the German Enigma machine invented by the engineer Arthur Scheribus, who founded his *Chiffriermaschinen Aktiengeselschaft* in 1923. Despite all advertisements, the Enigma was not a commercial success at first. This changed in 1934 when Hitler started to rearm Germany and the Wehrmacht decided that the Enigma should become the new cryptography machine for the German army.

The main difference between the Enigma and other rotor machines of that time is that it reflects the signal at the end and sends it through the rotors a second time. This effectively doubles the number of rotors, but has the consequence that the cipher becomes involutionary, i.e. if X is sent to A then A must be sent to X. Being involutionary must not generally be regarded as a disadvantage for a cipher. The fact that decrypting and encrypting can be done by the same machine can be considered as positive. In fact many modern ciphers (including all binary additive stream-ciphers) have this property. However, in the case of the Enigma, it was a serious flaw which, together with other flaws, made it possible for the allies in the second world war to break the cipher. The cryptographic success of the allies had a significant impact on the course of the war.

1.4 Lessons from the Enigma

The Enigma had several flaws that could be used in cryptanalysis, but the operators also made several protocol failures. One interesting aspect is the following (see also [16]).

For each message the operator selects a message key consisting of three letters $\alpha\beta\gamma$. Then the message key is repeated ($\alpha\beta\gamma\alpha\beta\gamma$) and these six letters are encrypted with the current day's key and transmitted to another station. The repetition should help to detect transmission errors. This protocol violates the advice given in Algorithm 1.1 that the error detecting code should always be applied last and in this case the mistake led to the following attack developed by Polish cryptographers under Rejewski.

The technique of sending the signal through the rotors twice ensures that the Enigma applies a permutation of the 26 letters of the alphabet consisting of 13 disjoint cycles to the plaintext. This is itself already a weakness, since it is impossible to map a letter to itself. Thus the ciphertext leaks information about the plaintext, but we will not discuss this weakness.

Call a permutation of the 26 letters of the alphabet that consists of 13 disjoint cycles an *Enigma permutation*. The attack is based on the following lemma.

Lemma 1.1 *Let π and π' be two Enigma permutations. Then for every $l \in \mathbb{N}$ the permutation $\pi\pi'$ contains an even number of cycles of length l.*

If ($\alpha\beta$) is a transposition in π or π' then α and β lie in different cycles of $\pi\pi'$ of the same length.

Proof The enigma permutations partition the set of letters into parts of the form $\{p_1, \ldots, p_{2k}\}$ where π contains the involutions $(p_1 p_2), (p_3 p_4), \ldots (p_{2k-1} p_{2k})$ while π' contains the involutions $(p_{2k} p_1), (p_2 p_3), \ldots, (p_{2k-2} p_{2k-1})$.

On the set $\{p_1, \ldots, p_{2k}\}$ we find that $\pi\pi'$ is $(p_1 p_3 \ldots p_{2k-1})(p_2 p_4 \ldots p_{2k})$. Thus cycles of length l come in pairs. Furthermore an involution $(p_i p_{i+1})$ of π or π' has one letter in each cycle. □

Lets assume that the following 6-tuples are encrypted session keys from one day.

HKI CED	HTN CYA	HGI CCD	DPN BUA	WDB XAU
SHZ SHV	QGU QCN	UQT DBG	DEF BGH	EJN GOA
ZFN WLA	RDC OAY	GPR IUO	MSO EDR	YWW MWT
KNA LQM	SGK SCE	VFY ULC	BAM NZL	BAJ NZI
NIT PFG	JMH VTB	XPH AUB	TTT JYG	KWS LWW
ERV GXX	JTT VYG	PNJ TQI	ILM KNL	DSP BDJ
AIF ZFH	EAY GZC	ZAM WZL	ZFP WLJ	IOS KPW
UBC DRY	CZK RSE	LWM FWL	BFO NLR	VNV UQX
TWM JWL	JPF VUH	LST FDG	KST LDG	VNV UQX
JPF VUH	PND TQS	YNJ MQI	GYJ IJI	PSB TDU

What can we derive from this observation? Denote by $\pi_1, \ldots, \pi_6$ the six unknown permutations the Enigma performs with the day's key. Let us try to reconstruct $\pi_1\pi_4$. In the collection of the 50 session keys 24 different letters occur in the first position. We know from the observed message AIF ZFH that $\pi_1\pi_4$ maps A to Z and so on. Only the images of F and O are missing, but Lemma 1.1 gives us enough extra information to fill the gap. One obtains

$$\pi_1\pi_4 = (AZWX)(CROH)(BNPTJVUD)(EGIKLFYM)(Q)(S)$$

This is a lot of information. We certainly know that π_1 and π_4 interchange Q and S, so we have already determined some part of the session keys. For the other parts our uncertainly has decreased dramatically. For example, we know that if we see an A in the first position the session key must start with either C, R, O or H.

In addition the knowledge of $\pi_1\pi_4$ can help to determine the wires on the first rotor of the Enigma. (This was important before the allies were able to capture an Enigma machine.) Operation failures such as choosing weak keys like AAA help the cryptanalyst further. This short sketch is of course not a full cryptanalysis of the Enigma, but it shows how Polish cryptanalysts and later the English team at Bletchley Park could attack the Enigma. It also shows that choosing the wrong order of encryption and error detecting code is a serious mistake that helps the attacker. More about the cryptanalysis of the Enigma can be found in [220, 268].

At the time this attack was of course a military secret, but now the second world war is long over, the enemies have become friends and the military secret has become a textbook exercise. So why do people continue to repeat the errors from the Enigma and implement the same protocol failure in our modern mobile phones (Chap. 8) and computers (Sect. 9.2.1)?

1.5 History III: Towards Modern Cryptography

In the days of the rotor machines stream ciphers dominated cryptography. There was almost nothing else. Being at the very top is not always a good position, you can only lose. For stream ciphers modern cryptography is the story of decline. However, it is better to say that modern cryptography is a story of normalization. Block ciphers were underestimated for many centuries.

The change to electronic devices was no problem for stream ciphers. (Linear) feedback shift registers are perfectly suited to the new hardware and give satisfactory results.

The first setback for stream ciphers was the data encryption standard (DES) cipher in 1973, a block cipher. This was the first time that a cryptosystem had become a public standard. Naturally it drew much research interest.

In 1977, with the RSA cryptosystem, the first example of asymmetric cryptography was published. Asymmetric cryptography is today an important part of many protocols.

With the success of modern computers stream ciphers encountered more problems. A processor loads a word (or a block) into its registers, manipulates it and then writes it back. This fits perfectly with the idea of block ciphers, but less to the idea of a stream cipher. Many modern block ciphers (IDEA, AES, . . .) are perfectly adapted to software implementation.

In the 1990s stream ciphers had a renaissance in mobile devices (telephones, wireless LAN, bluetooth). The first generation of mobile devices had no general purpose processor and energy efficiency had top priority. Stream ciphers were perfect for this job. However, this renaissance was not without troubles.

One point was that often a design criterion for the new ciphers was: "Do not make it too safe." This was especially true for the ciphers A5/1 and A5/2 used in GSM mobile phones (see Chap. 8), and the first WLAN standard WEP (see Chap. 9) also contains some needless weaknesses. This gave many people the wrong impression that stream ciphers must be insecure.

The second point is that mobile devices rapidly become more powerful. A modern smart phone is just a very small general purpose computer. Together with the more powerful embedded processors, block ciphers become a more and more attractive solution for these devices.

1.6   When to Use Stream Ciphers?

Block ciphers are better understood than stream ciphers. The main reason is that for block ciphers it is easy to modularize the problem. One can study the operation mode without looking at the underlying cipher or one can look at a single round of a DES-like cipher and begin to study APN-functions. For stream ciphers there is almost no such modularity and often the key scheduling and the keystream generation interact in a complicated way. The result is that block ciphers are easier to use.

So if one has no special requirement, my advice is always: use a standard block cipher (AES is perfect), but use it in a stream cipher mode. (The ECB mode is worse, but the CBC mode also loses against most stream cipher modes with respect to information leakage and parallelism. This is especially true if you compare the CBC mode with the modern counter mode CTR, see Exercise 17.1.)

However, there are applications where this is not practicable (otherwise I would not have written this book). In embedded devices the goal is to save gates and energy. A shift register-based stream cipher needs fewer gates by several magnitudes than even a simple embedded CPU.

Stream ciphers can reach a higher speed than block ciphers. On a standard computer a factor of 3 is not unusual if one compares an implementation of AES with a stream cipher designed for software implementation. If one is willing to use specialized hardware, even more is possible, for example the cipher Trivium (Sect. 10.1) can generate 64 bits per clock cycle. This is far higher than anything which is possible with block ciphers.

At the time of writing, hard disk space is growing faster than CPU speed. This may be an indication that in future we will have a greater need for high speed ciphers. Another area in which stream ciphers may find an important application are RFID devices. Here low energy consumption is important and stream ciphers beat block ciphers in this aspect.

1.7   Outline of the Book

The book is divided into five parts. Part I covers the theory of shift register-based stream ciphers. Shift registers are perfect for specialized hardware and, despite all attempts to design good software stream ciphers, shift register-based stream ciphers are still the most important class of stream cipher.

Chapter 2 is a survey of linear feedback shift registers. It develops the theory of generating functions, describes the famous Berlekamp-Massey algorithm and covers implementation aspects.

Pure linear functions are weak as cryptographic functions. So one must use non-linear combinations of linear feedback shift registers to obtain good stream ciphers. Chapter 3 contains the basic concepts of non-linear combinations of linear feedback shift registers. It gives an overview of different attack classes and introduces basic concepts such as algebraic complexity and correlation immune functions.

With Chap. 4 we begin the cryptanalysis of stream ciphers. The first attack class we study are correlation attacks. These attacks try to use statistical abnormalities to recover a part of the internal state. This attack principle is old, but in recent years many improvements have been made.

Chapter 5 covers a relative new type of attack. Binary Decision Diagram-based attacks were introduced in 2002 by M. Krause. The idea behind these attacks is remarkably simple. The set of internal states that is consistent with the observed output sequence describes a Boolean function. BDDs are a tool to efficiently handle the Boolean functions. The attack successively computes BDDs that describe the internal state with increasing accuracy until it finally yields a unique solution. This is more efficient that the complete key search, but requires a lot of memory.

Chapter 6 covers algebraic attacks. The idea of these attacks is to express the stream cipher as a system of non-linear equations. The chapter has a short introduction to the branch of computer algebra which is used to solve such equations (especially Gröbner bases). Examples of algebraic attacks against real world ciphers complete the chapter.

Chapter 7 will introduce stream ciphers with irregular clock control. Irregular clock control is an attractive way to create strong ciphers and some of the simplest examples of stream ciphers with irregular clock control are still unbroken. The drawback is that it is very hard to prove any property of the cipher, which makes undetected weaknesses more likely. Ciphers with irregular clock control are also especially susceptible to side channel attacks. This may be the reason why most real world ciphers have regular clock control.

Part II contains the description and cryptanalysis of some special ciphers.

The ciphers A5/1 and A5/2 which are used in GSM security are presented in Chap. 8. They are shift register-based and we use them as real world examples for the attacks described in the first part.

Chapter 9 is about the cipher RC4. This cipher was optimized for use on 8-bit processors and is not based on shift registers. It is especially famous since it is used in the wireless LAN standard. When used correctly, RC4 is unbroken, but the key scheduling of RC4 is weak and its careless use allows related key attacks.

The ECRYPT Stream Cipher Project [85] was run from 2004 to 2008 to identify a portfolio of promising new stream ciphers. Chapter 10 describes some of the ciphers from this project as examples of modern stream cipher design.

Chapter 11 covers some ciphers which are provable as secure as some (hopefully) hard number theoretic problem. These ciphers are very secure, but unfortunately slow in comparison to other ciphers in this book. They are mostly used as part of a key generation protocol.

I assume that the reader is familiar with basic mathematics (number theory, algebra, combinatorics and statistics), but sometimes I have had to use more exotic concepts. The chapters in Part III collect some background material.

Exercises with solutions can be found in Part IV.

Implementation for most algorithms covered by this book can found at http:// cage.ugent.be/~klein/streamcipher. Chapter 19 gives an overview of the programs.

To document the programs I wrote a new literate programming tool. It is freely available and, perhaps after reading this book, the reader may want to use it for their own projects. Chapter 20 contains the user manual for this tool.

Part I
Shift Register-Based Stream Ciphers

Chapter 2
Linear Feedback Shift Registers

2.1 Basic Definitions

In a hardware realization of a finite state machine it is attractive to use flip-flops to store the internal state. With n flip-flops we can realize a machine with up to 2^n states. The update function is a Boolean function from $\{0, 1\}^n$ to $\{0, 1\}^n$. We can simplify both the implementation and the description if we restrict ourselves to feedback shift registers.

In a feedback shift register (see Fig. 2.1) we number the flip-flops $F_0, \ldots, F_{n-1}$. In each time step F_i takes the value of F_{i-1} for $i > 0$ and F_0 is updated according to the feedback function $f : \{0, 1\}^n \to \{0, 1\}$. We will always assume that the value of F_{n-1} is the output of the shift register.

Feedback shift registers are useful tools in coding theory, in the generation of pseudo-random numbers and in cryptography. In this chapter we will summarize all results on linear feedback shift registers relevant to our study of stream ciphers. For other applications of feedback shift registers I recommend the classical book of Solomon W. Golomb [115].

Mathematically the sequence $(a_i)_{i \in \mathbb{N}}$ generated by a shift register is just a sequence satisfying the n-term recursion

$$a_{i+n} = f(a_i, \ldots, a_{i+n-1}). \tag{2.1}$$

This definition is, of course, not restricted to binary sequences and most of our results will hold for shift register sequences defined over any (finite) field or sometimes even for sequences defined over rings.

We will call a shift register linear if the feedback function is linear. Thus:

Definition 2.1 A *linear feedback shift register* (LFSR) sequence is a sequence $(a_i)_{i \in \mathbb{N}}$ satisfying the recursion

$$a_{i+n} = \sum_{j=0}^{n-1} c_j a_{i+j}. \tag{2.2}$$

A. Klein, *Stream Ciphers*, DOI 10.1007/978-1-4471-5079-4_2,
© Springer-Verlag London 2013

Fig. 2.1 A feedback shift
register

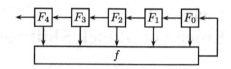

Since the next value depends only on the preceding n values, the sequence must become periodic. The state $(a_i, \ldots, a_{i+n-1}) = (0, \ldots, 0)$ leads to the constant sequence 0, thus the period of an LFSR sequence over $\mathbb{F}_q$ can be at most $q^n - 1$. If in addition $c_0 \neq 0$, we can extend the sequence backwards in time via

$$a_i = c_0^{-1}\left(a_{i+n} - \sum_{j=1}^{n-1} c_j a_{j+n}\right)$$

which proves that it is ultimately periodic.

As we have already seen in the introduction, a necessary condition for the security of a system is that the generated pseudo-random sequence has a large period. Thus the sequences of maximal period are of special interest.

Definition 2.2 An LFSR sequence over $\mathbb{F}_q$ with period $q^n - 1$ is called an *m-sequence* (maximal sequence).

2.2 Algebraic Description of LFSR Sequences

In this section we develop an algebraic description of LFSR sequences. We especially want to find a closed formula for an LFSR sequence. One way to reach this goal is to study the *companion matrix* of the LFSR sequence. We have

$$\begin{pmatrix} a_{k+1} \\ \vdots \\ a_{k+n-1} \\ a_{k+n} \end{pmatrix} = \begin{pmatrix} 0 & 1 & & 0 \\ \vdots & & \ddots & \\ 0 & 0 & & 1 \\ c_0 & c_1 & \cdots & c_{n-1} \end{pmatrix} \begin{pmatrix} a_k \\ \vdots \\ a_{k+n-2} \\ a_{k+n-1} \end{pmatrix} \tag{2.3}$$

and thus

$$\begin{pmatrix} a_k \\ \vdots \\ a_{k+n-2} \\ a_{k+n-1} \end{pmatrix} = \begin{pmatrix} 0 & 1 & & 0 \\ \vdots & & \ddots & \\ 0 & 0 & & 1 \\ c_0 & c_1 & \cdots & c_{n-1} \end{pmatrix}^k \begin{pmatrix} a_0 \\ \vdots \\ a_{n-2} \\ a_{n-1} \end{pmatrix}. \tag{2.4}$$

Transforming the companion matrix to Jordan normal form makes it easy to compute the k-th power and transforming it back gives a closed formula for the LFSR sequence.

In the next section we will take another approach that is based on generating functions.

2.2.1 Generating Functions

This section contains the part of the theory of generating functions that we need, but for those who want to learn more about generating functions, I recommend [119].

Definition 2.3 The *generating function* $A(z)$ associated to a sequence $(a_i)_{i \in \mathbb{N}}$ is the formal power series $A(z) = \sum_{i=0}^{\infty} a_i z^i$.

A generating function is useful because it describes an entire sequence with a single algebraic object.

By the recursion (2.2) we find:

$$A(z) - \sum_{j=0}^{n-1} c_j A(z) z^{n-j} = g(z)$$

$$\Longleftrightarrow \quad A(z)\left(1 - \sum_{j=0}^{n-1} c_j z^{n-j}\right) = g(z) \tag{2.5}$$

for some polynomial $g(z)$ of degree at most $n - 1$.

The polynomial $1 - \sum_{j=0}^{n-1} c_j z^{n-j}$ is important enough to deserve a name.

Definition 2.4 For an LFSR sequence with recursion formula (2.2) we call

$$f(z) = z^n - \sum_{j=0}^{n-1} c_j z^j \tag{2.6}$$

the *feedback polynomial* of the LFSR. The *reciprocal polynomial*[1] is denoted by

$$f^*(z) = z^n f\left(\frac{1}{z}\right) = 1 - \sum_{j=0}^{n-1} c_j z^{n-j}. \tag{2.7}$$

From Eq. (2.5) we derive a closed formula for the generation function of an LFSR sequence:

$$A(z) = \frac{g(z)}{f^*(z)}. \tag{2.8}$$

For the derivation of the closed form of a_i it is useful to begin with the case where the feedback polynomial $f(z)$ has no multiple roots.

---

[1] $f^*(z)$ is sometimes called the feedback polynomial. As the literature has not adopted a unique notation, it is important to check which notation is being used.

2.2.2 Feedback Polynomials Without Multiple Roots

Let $f(z)$ be a feedback polynomial without multiple roots and let $\xi_1, \ldots, \xi_n$ be the different zeros of $f(z)$. Then $f^*(z) = \prod_{j=1}^{n}(1 - z\xi_j)$ and thus we get the partial fraction decomposition

$$A(z) = \frac{g(z)}{f^*(z)} = \sum_{j=1}^{n} \frac{b_j}{1 - z\xi_j}. \tag{2.9}$$

All we need to obtain a closed formula from the partial fraction decomposition is the geometric sum

$$\sum_{i=0}^{\infty} z^i = \frac{1}{1-z}$$

and thus

$$A(z) = \sum_{j=1}^{n} \frac{b_j}{1 - z\xi_j}$$

$$= \sum_{j=1}^{n} b_j \sum_{i=0}^{\infty} (\zeta_j z)^i$$

$$= \sum_{i=0}^{\infty} \left(\sum_{j=0}^{n} b_j \zeta_j^i \right) z^i. \tag{2.10}$$

This gives us the closed formula

$$a_i = \sum_{j=0}^{n} b_j \zeta_j^i \tag{2.11}$$

for the LFSR sequence.

Formula (2.11) holds if the feedback polynomial has no multiple roots. For separable irreducible feedback polynomials we can transform (2.11) to the following theorem. Note that over finite fields and fields with characteristic 0 every polynomial is separable. We will not deal with other fields in this book.

Theorem 2.1 *Let $(a_i)_{i \in \mathbb{N}}$ be an LFSR sequence over $\mathbb{F}_q$ and let ξ be a zero of the irreducible feedback polynomial. Then*

$$a_i = \mathrm{Tr}_{\mathbb{F}_{q^n}/\mathbb{F}_q}(\alpha \xi^i) \tag{2.12}$$

for some $\alpha \in \mathbb{F}_{q^n}$.

Proof We have already proved that the sequence $(a_i)_{i\in\mathbb{N}}$ has a unique closed form (2.11). Since the feedback polynomial is irreducible, its zeros have the form ξ^θ where θ is an automorphism of $\mathbb{F}_{q^n}/\mathbb{F}_q$. But $a_i^\theta = a_i$ for all i. Thus Equation (2.11) is invariant under θ. Therefore the coefficients b_j are conjugated, i.e.

$$a_i = \sum_{\theta\in\mathrm{Aut}(\mathbb{F}_{q^n}/\mathbb{F}_q)} \alpha^\theta \left(\xi^\theta\right)^i = \mathrm{Tr}_{\mathbb{F}_{q^n}/\mathbb{F}_q}\left(\alpha\xi^i\right).$$

$\square$

Corollary 2.1 *Under the conditions of Theorem 2.1 the period of the sequence is the multiplicative order $o(\xi)$ of ξ.*

As already mentioned in the previous section, the period $q^n - 1$ is of special interest. Thus the following feedback polynomials are special.

Definition 2.5 An irreducible polynomial of degree n over $\mathbb{F}_q$ is *primitive* if the order of its zeros is $q^n - 1$.

2.2.3 Feedback Polynomials with Multiple Roots

Now we want to determine all possible periods of LFSR sequences.

First we consider the easy case where the feedback polynomial is reducible, but has no multiple roots. In this case we can factor the feedback polynomial f and write the generating function (see Eq. (2.8)) of $(a_i)_{i\in\mathbb{N}}$ as

$$A(z) = \frac{g(z)}{f^*(z)} = \sum_{j=1}^{k} \frac{g_j(z)}{f_j^*(z)}$$

where the polynomials f_j are the different irreducible factors of the feedback polynomial f.

Thus the sequence $(a_i)_{i\in\mathbb{N}}$ can be represented as a sum of k LFSR sequences $(a_i^{(j)})_{i\in\mathbb{N}}$ with irreducible feedback polynomial. By Corollary 2.1 the period of $(a_i^{(j)})_{i\in\mathbb{N}}$ divides $q^{n_j} - 1$ where $n_j = \deg f_j$ and hence the sequence $(a_i)_{i\in\mathbb{N}} = \sum_{j=1}^{k}(a_i^{(j)})_{i\in\mathbb{N}}$ has period

$$p = \mathrm{lcm}(\pi_1, \ldots, \pi_k)$$

where π_j is the period of $(a_i^{(j)})_{i\in\mathbb{N}}$.

To analyze the case of multiple roots we need an additional tool. In this case the partial fraction decomposition of the generation function yields:

$$A(z) = \frac{g(z)}{f^*(z)} = \sum_{j=1}^{n_1} \frac{b_{j,1}}{1 - z\xi_j} + \sum_{j=1}^{n_2} \frac{b_{j,2}}{(1 - z\xi_j)^2} + \cdots + \sum_{j=1}^{n_r} \frac{b_{j,r}}{(1 - z\xi_j)^r}$$

with $n_1 \geq n_2 \geq \cdots \geq n_r \geq n_{r+1} = 0$ where $\xi_{n_k+1}, \ldots, \xi_{n_k}$ are roots of f of multiplicity k. So to get a closed formula we need in addition the power series of $\frac{1}{(1-z)^k}$.

We can find the power series either by computing the $(k-1)$th derivative of $\frac{1}{1-z} = \sum_{i=0}^{\infty} z^i$ or we use the binomial theorem

$$(1+x)^r = \sum_{i=0}^{\infty} \binom{r}{i} x^i.$$

For a negative integer we get

$$\frac{1}{(1-z)^k} = \sum_{i=0}^{\infty} \binom{-k}{i} (-1)^i z^i$$

$$= \sum_{i=0}^{\infty} \binom{k+i-1}{i} z^i$$

$$= \sum_{i=0}^{\infty} \binom{k+i-1}{k-1} z^i.$$

This leads to the closed formula

$$a_i = \sum_{j=0}^{n_1} b_{j,1} \zeta_j^i + \sum_{j=0}^{n_2} b_{j,2} \binom{i+1}{1} \zeta_j^i + \cdots + \sum_{j=0}^{n_k} b_{j,k} \binom{i+k-1}{k-1} \zeta_j^i$$

$$= \sum_{j=0}^{n_1} b'_{j,1} \zeta_j^i + \sum_{j=0}^{n_2} b'_{j,2} i \zeta_j^i + \cdots + \sum_{j=0}^{n_k} b'_{j,k} i^{k-1} \zeta_j^i \tag{2.13}$$

where the last transformation uses the fact that $\binom{k-1+i}{k-1}$, $k = 1, \ldots, n$, is a basis for the polynomials of degree less than k. Note that the converse is also true. Given a sequence in the form of Eq. (2.13) we can reverse all previous steps and find the linear recurrence satisfied by that sequence.

From Eq. (2.13) we can immediately see the period of the sequence $(a_i)_{i \in \mathbb{N}}$. The power series $(\zeta_j^i)_{i \in \mathbb{N}}$ has a period π_i where $\pi_i | q^{n_j} - 1$ and n_j is the degree of the minimal polynomial of ζ_j. And since we are working in $\mathbb{F}_q$, the period of a polynomial series $(i^k)_{i \in \mathbb{N}}$ is the characteristic p of $\mathbb{F}_q$. Thus

$$\pi = p \operatorname{lcm}(\pi_1, \ldots, \pi_k)$$

where $\pi_1, \ldots, \pi_k$ are the different orders of $\zeta_1, \ldots, \zeta_{n_1}$.

We summarize the results of this section in the following theorem.

Theorem 2.2 Let $(a_i)_{i \in \mathbb{N}}$ be an LFSR sequence over $\mathbb{F}_q$, $q = p^e$. Then the period π of $(a_i)_{i \in \mathbb{N}}$ is either

$$\pi = \operatorname{lcm}(\pi_1, \ldots, \pi_k) \tag{2.14}$$

where $\pi_j | q^{n_j} - 1$ and $\sum_{j=1}^{k} n_j \leq n$ or

$$\pi = p \operatorname{lcm}(\pi_1, \ldots, \pi_k) \tag{2.15}$$

where $\pi_j | q^{n_j} - 1$ and $n_1 + \sum_{j=1}^{k} n_j \leq n$.

Proof We have already proved that the period π must have either the form (2.14) or (2.14). Now we prove the converse that for each such π there is an LFSR sequence with period πa.

Let π be of the form (2.14). Choose $\zeta_j \in \mathbb{F}_{q^{n_j}}$ such that ζ_j has order π_j. Without loss of generality we may assume that $\mathbb{F}_q(\zeta_j) = \mathbb{F}_{q^{n_j}}$, if not just replace n_j by a smaller n'_j. The sequence

$$x_i = \sum_{j=1}^{k} \operatorname{Tr}_{\mathbb{F}_{q^{n_j}}/\mathbb{F}_q}\left(\zeta_j^i\right)$$

is a linear shift register sequence with feedback polynomial

$$f(z) = \prod_{j=1}^{k} \prod_{l=0}^{n_j-1} \left(1 - z\zeta_j^{q^l}\right).$$

The sequence ζ_j has period π since the "subsequences" ζ_j^i and hence $\operatorname{Tr}_{\mathbb{F}_{q^{n_j}}/\mathbb{F}_q}(\zeta_j^i)$ have period π_j $(1 \leq j \leq k)$.

If π is of the form (2.15), we find that the sequence

$$x_i = i \operatorname{Tr}_{\mathbb{F}_{q^{n_j}}/\mathbb{F}_q}\left(\zeta_1^i\right) + \sum_{j=2}^{k} \operatorname{Tr}_{\mathbb{F}_{q^{n_j}}/\mathbb{F}_q}\left(\zeta_j^i\right)$$

is a linear shift register sequence with feedback polynomial

$$f(z) = \left(\prod_{l=0}^{n_j-1} \left(1 - z\zeta_1^{q^l}\right)^2\right)\left(\prod_{j=1}^{k} \prod_{l=0}^{n_j-1} \left(1 - z\zeta_j^{q^l}\right)\right)$$

and period $\pi = p \operatorname{lcm}(\pi_1, \ldots, \pi_k)$. The additional factor p is for the period of the polynomial i in $\mathbb{F}_q$. $\qquad\square$

2.2.4 LFSR Sequences as Cyclic Linear Codes

Another description of LFSR sequences is based on coding theory.

The LFSR defines a linear mapping from its initial state $(a_0, \ldots, a_{n-1})$ to its output sequence $(a_i)_{i \in \mathbb{N}}$. For fixed N we may interpret the mapping

$$C : (a_0, \ldots, a_{n-1}) \mapsto (a_0, \ldots, a_{N-1})$$

as a linear code of length N and dimension n.

A parity check matrix of the code is

$$H = \begin{pmatrix} c_0 & \cdots & c_{n-1} & -1 & 0 & \cdots & & & 0 \\ 0 & c_0 & \cdots & c_{n-1} & -1 & 0 & \cdots & & 0 \\ & & & & \ddots & & \ddots & \ddots & \\ 0 & & \cdots & & 0 & c_0 & \cdots & c_{n-1} & -1 \end{pmatrix}. \tag{2.16}$$

If we look at a full period of the LFSR, i.e. if we choose $N = p$, then the resulting linear code is cyclic and $f^*(z)$ is its parity check polynomial.

The code C also has a unique *systematic generator matrix*

$$G = \begin{pmatrix} 1 & & 0 & c_{n,0} & \cdots & c_{N-1,0} \\ & \ddots & & \vdots & & \vdots \\ 0 & & 1 & c_{n,n-1} & \cdots & c_{N-1,n-1} \end{pmatrix}. \tag{2.17}$$

We have $(a_0, \ldots, a_{N-1}) = (a_0, \ldots, a_{n-1})G$, i.e.

$$a_k = \sum_{i=0}^{n-1} c_{k,i} a_i. \tag{2.18}$$

We will use this linear representation of the element a_k in terms of the initial state in several attacks.

2.3 Properties of m-Sequences

2.3.1 Golomb's Axioms

Linear shift register sequences of maximal length (m-sequences) have many desirable statistical properties.

The best known of these properties is that they satisfy Golomb's axioms for pseudo-random sequences [115].

We study a periodic binary sequence $(a_i)_{i \in \mathbb{N}}$ with period length p. Then the three axioms for $(a_i)_{i \in \mathbb{N}}$ to be a pseudo-random sequence are:

(G1)   In every period the number of ones is nearly equal to the number of zeros, more precisely the difference between the two numbers is at most 1:

$$\left| \sum_{i=1}^{p} (-1)^{a_i} \right| \leq 1.$$

(G2)   For any k-tuple b, let $N(b)$ denote the number of occurrences of the k-tuple b in one period.
Then for any k with $1 \leq k \leq \log_2 p$ we have

$$\left| N(b) - N(b') \right| \leq 1$$

for any k-tuples b and b'.

(G2')   A sequence of consecutive ones is called a *block* and a sequence of consecutive zeros is called a *gap*. A *run* is either a block or a gap.
In every period, one half of the runs has length 1, one quarter of the runs has length 2, and so on, as long as the number of runs indicated by these fractions is greater than 1.
Moreover, for each of these lengths the number of blocks is equal to the number of gaps.

(G3)   The auto-correlation function

$$C(\tau) = \sum_{i=0}^{p-1} (-1)^{a_i} (-1)^{a_{i+\tau}}$$

is two-valued.

Axiom (G1) is called the *distribution test*, Axiom (G2) is the *serial test* and Axiom (G3) is the *auto-correlation test*. In [115] Golomb uses (G2') instead of (G2). Axiom (G2) was introduced in [169] and is in some respects more useful than the original axiom.

The distribution test (G1) is a special case of the serial test (G2). However, (G1) is retained for historical reasons, and sequences which satisfy (G1) and (G3), but not (G2), are also important.

Theorem 2.3 (Golomb [115]) *Every m-sequence satisfies* (G1)–(G3).

Proof An m-sequence is characterized by the fact that the internal state of the linear feedback shift register runs through all elements of $\mathbb{F}_2^n \setminus \{(0, \ldots, 0)\}$. Since at any time the next n output bits form the current internal state, this means that $(a_t, \ldots, a_{t+n-1})$ runs over all elements of $\mathbb{F}_2^n \setminus \{(0, \ldots, 0)\}$ where t runs from 0 to $2^n - 1$. This proves

$$N(a_1, \ldots, a_k) = \begin{cases} 2^{n-k} - 1 & \text{for } a_1 = \cdots = a_k = 0, \\ 2^{n-k} & \text{otherwise.} \end{cases}$$

Thus an m-sequence passes the serial test for blocks of length up to n and hence it satisfies (G2) and (G1).

A run of length k is just a subsequence of the form $1, 0, 0, \ldots, 0, 1$ with k zeros and a block of length k is a subsequence of the form $0, 1, 1, \ldots, 1, 0$. We have already proved that an m-sequence contains exactly 2^{n-k-2} subsequences of type $k \le n - 2$. This is the statement of (G2$'$).

We find $C(0) = 2^n - 1$ as one value of the auto-correlation function. We now prove $C(\tau) = -1$ for $0 < \tau < 2^n - 1$. By Theorem 2.1 we have $a_i = \mathrm{Tr}_{\mathbb{F}_{2^n}/\mathbb{F}_2}(\alpha \xi^i)$ for a primitive element ξ of $\mathbb{F}_{2^n}$ and $a_{i+\tau} = \mathrm{Tr}_{\mathbb{F}_{2^n}/\mathbb{F}_2}(\alpha' \xi^i)$. Note that we have the same ξ in both equations, since $(a_i)_{i \in \mathbb{N}}$ and $(a_{i+\tau})_{i \in \mathbb{N}}$ satisfy the same recurrence. Thus $a_i + a_{i+\tau} = \mathrm{Tr}_{\mathbb{F}_{2^n}/\mathbb{F}_2}((\alpha + \alpha')\xi^i)$ and hence $(a_i + a_{i+\tau})_{i \in \mathbb{N}}$ is also an m-sequence. By (G1) we have

$$C(\tau) = \sum_{i=0}^{p-1} (-1)^{a_i + a_{i+\tau}} = -1.$$

Thus the auto-correlation function takes just the two values $2^n - 1$ and -1. □

Besides the Golomb axioms, m-sequences also satisfy other interesting equations:

Theorem 2.4 *Every m-sequence satisfies*:

(a) *For every $0 < k < 2^n - 1$ there exists a δ for which*

$$a_i + a_{i+k} = a_{i+\delta}$$

for all $i \in \mathbb{N}$. This is called the shift-and-add property.
(b) *There exists a τ such that*

$$a_{i2^j + \tau} = a_{i+\tau}$$

for all $i, j \in \mathbb{N}$. This is called the constancy on cyclotomic cosets.

Proof We have already used and proved the shift-and-add property when we demonstrated that an m-sequence satisfies the auto-correlation test.

By Theorem 2.1 we know that $a_i = \mathrm{Tr}_{\mathbb{F}_{2^n}/\mathbb{F}_2}(\alpha \xi^i)$ for some $\alpha \in \mathbb{F}_{2^n}$ and a primitive $\xi \in \mathbb{F}_{2^n}$. We choose τ such that $\xi^\tau = \alpha^{-1}$.
Then

$$a_{i+\tau} = \mathrm{Tr}_{\mathbb{F}_{2^n}/\mathbb{F}_2}\left(\alpha \xi^{i+\tau}\right)$$

$$= \mathrm{Tr}_{\mathbb{F}_{2^n}/\mathbb{F}_2}\left(\xi^i\right)$$

$$= \mathrm{Tr}_{\mathbb{F}_{2^n}/\mathbb{F}_2}\left(\xi^{i2^j}\right) \quad \text{since } x \mapsto x^{2^j} \text{ is an automorphism of } \mathbb{F}_{2^n}/\mathbb{F}_2$$

$$= \mathrm{Tr}_{\mathbb{F}_{2^n}/\mathbb{F}_2}\left(\alpha \xi^{i2^j + \tau}\right)$$

$$= a_{i2^j + \tau}.$$

  □

The shift-and-add property is of special interest since it characterizes the m-sequences uniquely.

Theorem 2.5 *Every sequence which satisfies the shift-and-add property is an m-sequence.*

Proof Let $A = (a_i)_{i \in \mathbb{N}}$ be a sequence of period p which has the shift-and-add property. Then the p shifts of the sequence, together with the zero sequence, form an elementary Abelian group. It follows that $p + 1 = 2^n$ for some $n \in \mathbb{N}$. Let A_k denote the sequence $(a_{i+k})_{i \in \mathbb{N}}$. Any n successive shifts of the sequence A form a basis of the elementary Abelian group, thus we can write A_n as a linear combination of $A_0, \ldots, A_{n-1}$, i.e.

$$A_n = \sum_{k=0}^{n-1} c_k A_k.$$

Reading the last equation element-wise gives

$$a_{i+n} = \sum_{k=0}^{n-1} c_k a_{i+k},$$

i.e. the sequence A satisfies a linear recurrence. Since the period of A is $p = 2^n - 1$, it is an m-sequence.   □

2.3.2 Sequences with Two Level Auto-Correlation

It is a natural question whether the converse of Theorem 2.3 holds. Golomb conjectured that it does and indicated in a passage of his book (Sect. 4.7 in [115]) that he had a proof, but the actual answer turns out to be negative (see also [114]).

To put this answer in a bigger context we will study sequences which satisfy Axiom (G3), which have a strong connection to design theory. We make the following definition.

Definition 2.6 Let G be an additive group of order v and let D be a k-subset of G.

D is called a (v, k, λ)-*difference set* of G, if for every element $h \neq 0$ in G the equation

$$h = d - d'$$

has exactly λ solutions with $d, d' \in D$. If $G = \mathbb{Z}/v\mathbb{Z}$ is a cyclic group we speak of a cyclic (v, k, λ)-*difference set*.

The connection between sequences satisfying (G3) and difference sets is given by the following theorem.

Theorem 2.6 *The following statements are equivalent.*

(1) *There exists a periodic sequence of period length v over $\mathbb{F}_2$ with two level auto-correlation and k ones in its period.*
(2) *There exists a cyclic (v, k, λ)-difference set.*

Proof Let $a_0, \ldots, a_{v-1}$ be the period of a sequence with two level auto-correlation. This means the auto-correlation function satisfies $C(0) = v$ and $C(\tau) = x < v$ for $1 \leq \tau \leq v - 1$.

For $1 \leq \tau \leq v - 1$ let $\lambda_\tau = |\{i \mid a_i = a_{i+\tau} = 1, 0 \leq i \leq v - 1\}|$. Then

$$\left|\{i \mid a_i = 1, a_{i+\tau} = 0, 0 \leq i \leq v - 1\}\right| = k - \lambda_\tau,$$

$$\left|\{i \mid a_i = 0, a_{i+\tau} = 1, 0 \leq i \leq v - 1\}\right| = k - \lambda_\tau,$$

$$\left|\{i \mid a_i = a_{i+\tau} = 0, 0 \leq i \leq v - 1\}\right| = v + \lambda_\tau - 2k.$$

Thus $x = \lambda_\tau + (v + \lambda_\tau - 2k) - 2(k - \lambda_\tau) = v - 4(k - \lambda_\tau)$, i.e. $\lambda_\tau = \lambda$ independent of τ.

Let $D = \{i \mid a_i = 1\}$. Then $h = d - d'$ has exactly $\lambda = \lambda_h$ solutions with $d, d' \in D$.

For the converse, let D be a cyclic (v, k, λ)-difference set and define the sequence (a_i) by $a_i = 1$ if $i \in D$ and $a_i = 0$ for $i \notin D$. The definition of a difference set says that there are λ indices with $a_i = a_{i+\tau} = 1$ and, as above, we obtain $C(\tau) = \lambda_\tau + (v + \lambda_\tau - 2k) - 2(k - \lambda_\tau) = v - 4(k - \lambda_\tau)$ for $1 \leq t \leq v - 1$, i.e. the auto-correlation function is two leveled.   □

If we apply Theorem 2.6 to an m-sequence we obtain a difference set with parameters $v = 2^n - 1$, $k = 2^{n-1}$ and $\lambda = 2^{n-3}$. This is an example of a Hadamard difference set.

Definition 2.7 A $(4n - 1, 2n - 1, n - 1)$-difference set is called a *Hadamard difference set.*

Hadamard difference sets are of course strongly connected with the well-known Hadamard matrices.

Definition 2.8 A *Hadamard matrix* of order n is an $n \times n$ matrix H with entries ± 1 which satisfies $HH^t = nI$.

The connection is that we can construct a $(4n) \times (4n)$ Hadamard matrix from a $(4n - 1, 2n - 1, n - 1)$-difference set. Let D be a $(4n - 1, 2n - 1, n - 1)$-difference set over the group $G = \{g_1, \ldots, g_{4n-1}\}$ and define $H = (h_{i,j})_{i,i=0,\ldots,4n-1}$ by

$$h_{ij} = \begin{cases} 1 & \text{if } i = 0 \text{ or } j = 0, \\ 1 & \text{if } i, j \geq 1 \text{ and } g_i + g_j \in D, \\ 0 & \text{if } i, j \geq 1 \text{ and } g_i + g_j \notin D. \end{cases}$$

Then a short calculation shows that H is a Hadamard matrix.

The opposite implication is false since the Hadamard difference set needs the group G acting on it, whereas a Hadamard matrix need not have any symmetries. Nevertheless, we see that there are many links between pseudo-random sequences and other interesting combinatorial objects. (Besides those we have mentioned here, there are also strong links to designs and coding theory.)

This is not the place to go deeper into the theory of Hadamard Matrices, but to answer the question we posed at the beginning of this section we mention that Cheng and Golomb [51] use the Hadamard $(127, 63, 31)$-difference set of type E (given by Baumert [17]) to construct the sequence:

11111011110011111110010010111010101110001100000100110111100110 00

1101101110100100011010000101010011010010100011101100001010000 00

which satisfies (G1), (G2) and (G3) but is not an m-sequence.

2.3.3 Cross-Correlation of m-Sequences

Sequences with low correlation are intensively studied in the literature (see [126] for an overview). Since m-sequences have ideal auto-correlation properties, it is interesting to study the cross-correlation function for pairs of m-sequences. Many families of low correlation functions have been constructed in this way.

We use Eq. (2.12) to represent the m-sequence. We can shift it so that we get the form

$$a_i = \text{Tr}(\xi^i).$$

The second m-sequence can be assumed, without loss of generality, to be

$$b_i = \text{Tr}(\xi^{di})$$

for some d with $\gcd(d, q^n - 1) = 1$. The cross-correlation function depends only on d. This motivates the following definition:

Definition 2.9 Let ξ be a primitive element of $\mathbb{F}_{q^n}$ and let $\omega \in \mathbb{C}$ be a q-th root of unity. For each d with $\gcd(d, q^n - 1) = 1$ we define the *cross-correlation function*

$$\theta_{1,d}(\tau) = \sum_{x \in \mathbb{F}_{q^n}^*} \omega^{\text{Tr}(\xi^\tau x - x^d)}.$$

We will not need $\theta_{1,d}$ in the following, so we keep this section short and just sketch a link to bent functions, which are important in cryptography. We state the following without proof.

Theorem 2.7 (Gold [106], Kasami [144]) *Let $q = 2$ and $\omega = -1$. For $1 \leq k \leq n$ let $e = \gcd(n, k)$. If n/e is odd and $d = 2^k + 1$ or $d = 2^{2k} - 2^k + 1$, then $\theta_{1,d}$ takes the following three values:*

- $-1 + 2^{(n+e)/2}$ *is taken* $2^{n-e-1} + 2^{(n-e-2)/2}$ *times.*
- -1 *is taken* $2^n - 2^{n-e} - 1$ *times.*
- $-1 - 2^{(n+e)/2}$ *is taken* $2^{n-e-1} - 2^{(n-e-2)/2}$ *times.*

A pair of m-sequences whose cross-correlation function takes only the three values -1, $-1 + 2^{\lfloor (n+2)/2 \rfloor}$ and $-1 - 2^{\lfloor (n+2)/2 \rfloor}$ is called a *preferred pair*. Theorem 2.7 allows the construction of preferred pairs for n not divisible by 4. For $n \equiv 0 \bmod 4$ preferred pairs do not exist (see [185]).

What makes this interesting for cryptographic purposes is the following. To avoid attacks such as differential cryptanalysis [26] or linear cryptanalysis [179], the S-box of a block cipher must be as far from a linear mapping as possible. The appropriate measure of distance between two functions is provided by the Walsh transform.

Definition 2.10 Let $f : \mathbb{F}_{2^n} \to \mathbb{F}_2$. The *Walsh transform* of f is

$$f^W(a) = \sum_{x \in \mathbb{F}_{2^n}} (-1)^{f(x) + \mathrm{Tr}(ax)}.$$

The image of f^W is the *Walsh spectrum* of f.

A linear function will contain $\pm 2^n$ in its Walsh spectrum. A function provides the strongest resistance against a linear cryptanalysis if its Walsh spectrum contains only values of small absolute value. One can prove that the Walsh spectrum of f must contain at least a value of magnitude $2^{\lceil n/2 \rceil}$ (see, for example, [48]).

Definition 2.11 For even n, a function $f : \mathbb{F}_{2^n} \to \mathbb{F}_2$ is *bent* if $f^W(x) \leq 2^{n/2}$ for all $x \in \mathbb{F}_{2^n}$.

For odd n, we call a function *almost bent* if $f^W(x) \leq 2^{(n+1)/2}$ for all $x \in \mathbb{F}_{2^n}$.

From two m-sequences $a_i = \mathrm{Tr}(\xi^i)$ and $b_i = \mathrm{Tr}(\xi^{di})$ we construct a Boolean function $f : \mathbb{F}_2^n \to \mathbb{F}_2^n$ by defining $f(\xi^i) = \xi^{di}$ and $f(0) = 0$. Applying this construction to a preferred pair with odd n results in an almost bent function.

2.4 Linear Complexity

2.4.1 Definition and Basic Properties

We have seen in the previous section that m-sequences have very desirable statistical proprieties. However, as linear functions, LFSR sequences are unusable as cryptographic pseudo-random generators. First note that the first n output bits of an LFSR form its initial state, thus a LFSR fails against a known plaintext attack. Even if the feedback polynomial of the LFSR is unknown to the cryptanalyst, the system is still insecure. The first n output bits give us the initial state and the next n output bits give us n equations of the form

$$\begin{pmatrix} a_{n-1} & a_{n-2} & \cdots & a_0 \\ a_n & a_{n-1} & \ddots & \\ \vdots & & \ddots & \\ a_{2n-2} & a_{2n-3} & \cdots & a_{n-1} \end{pmatrix} \begin{pmatrix} c_{n-1} \\ c_{n-2} \\ \vdots \\ c_0 \end{pmatrix} = \begin{pmatrix} a_n \\ a_{n+1} \\ \vdots \\ a_{2n-1} \end{pmatrix}. \tag{2.19}$$

The determination of the unknown feedback coefficients c_i therefore only requires the solution of a system of n linear equations. A matrix of the form given in Eq. (2.19) is called a *Toeplitz matrix*. Toeplitz matrices appear in many different contexts. A lot is known about the solution of Toeplitz systems (see, for example, [142]).

In Sect. 2.4.2 we will learn a quadratic algorithm that computes not only the solution of the system (2.19), but gives us a lot of extra information. So an LFSR is not secure, even if its feedback polynomial is secret.

On the other hand, it is clear that every periodic sequence can be generated by a linear feedback shift register—simply take an LFSR of the same size as the period. It is therefore natural to use the length of the shortest LFSR that generates the sequence as a measure of its cryptographic security.

Definition 2.12 The *linear complexity* $\mathcal{L}((a_i)_{i=0,\ldots,n-1})$ of a finite sequence $a_0, \ldots, a_{n-1}$ is the length of the shortest linear feedback shift register that produces that sequence.

The following theorem summarizes some very basic properties of linear complexity which follows directly from the definition.

Theorem 2.8

(a) *A sequence of length n has linear complexity at most n:*

$$\mathcal{L}\big((a_i)_{i=0,\ldots,n-1}\big) \leq n.$$

(b) *The linear complexity of the subsequence $(a_i)_{i=0,\ldots,k-1}$ of $(a_i)_{i=0,\ldots,n-1}$ satisfies*

$$\mathcal{L}\big((a_i)_{i=0,\ldots,k-1}\big) \leq \mathcal{L}\big((a_i)_{i=0,\ldots,n-1}\big).$$

Proof

(a) The first n output bits of a shift register of length n is simply its initial state. Thus every shift register of length n can produce any finite sequence of length n as output.

Fig. 2.2 The sum of two
LFSRs

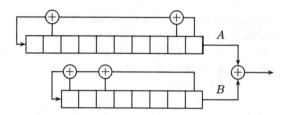

(b) Any shift register that produces the sequence $(a_i)_{i=0,\dots,n-1}$ also produces the
subsequence $(a_i)_{i=0,\dots,k-1}$.

$\square$

It is noteworthy that the bound of Theorem 2.8 (a) is sharp, as the following
example shows.

Lemma 2.1 *The sequence* $0,\dots,0,1$ *($n-1$ zeros) has linear complexity* n.

Proof Assume that the sequence is generated by a shift register of length $k < n$.
Since the first k symbols are 0 the shift register must be initialized with 0. But any
LFSR initialized with 0 generates the constant sequence 0. $\square$

Lemma 2.1 demonstrates that a high linear complexity is just a necessary but not
sufficient condition for a good pseudo-random sequence.

Next we study the sum of two LFSR sequences, which is generated by the device
shown in Fig. 2.2.

Theorem 2.9 *For two sequences* (a_i) *and* (b_i) *we have*

$$\mathcal{L}(a_i + b_i) \le \mathcal{L}(a_i) + \mathcal{L}(b_i).$$

Proof Consider the generating functions

$$A(z) = \frac{g_A(z)}{f_A^*(z)} \quad \text{and} \quad B(z) = \frac{g_B(z)}{f_B^*(z)}$$

of the two LFSR sequences. Then the sum $(a_i + b_i)$ has the generating function
$S(z) = A(z) + B(z)$.
    Thus

$$S(z) = \frac{g_a(z)f_B^*(z) + g_B(z)f_A^*(z)}{f_A^*(z)f_B^*(z)}$$

which implies by Sect. 2.2.1 that $(a_i + b_i)$ can be generated by an LFSR with feedback polynomial $f_A(z)f_B(z)$, i.e.

$$\mathcal{L}(a_i + b_i) \le \mathcal{L}(a_i) + \mathcal{L}(b_i).$$

(Note that $\mathrm{lcm}(f_A(z), f_B(z))$ is the feedback polynomial of the minimal LFSR that generates $(a_i + b_i)$.)   □

Theorem 2.9 shows that a linear combination of several linear feedback shift registers does not result in a cryptographic improvement. In fact there is a much more general theorem: Any circuit of n flip-flops which contains only XOR-gates can be simulated by an LFSR of size at most n (see [46]).

Theorem 2.9 has a corollary, which will be crucial for the next section.

Corollary 2.2 *If*

$$\mathcal{L}\big((a_i)_{i=0,\dots,n-2}\big) < \mathcal{L}\big((a_i)_{i=0,\dots,n-1}\big)$$

then

$$\mathcal{L}\big((a_i)_{i=0,\dots,n-1}\big) \geq n - \mathcal{L}\big((a_i)_{i=0,\dots,n-2}\big). \tag{2.20}$$

Proof Let (b_i) be the sequence $0, \dots, 0, 1$ ($n-1$ zeros) and let $a'_i = a_i + b_i$.

Since $\mathcal{L}((a_i)_{i=0,\dots,n-2}) < \mathcal{L}((a_i)_{i=0,\dots,n-1})$, the minimal LFSR that produces the sequence $(a_i)_{i=0,\dots,n-2}$ will produce $a_{n-1} + 1$ as nth output, i.e.

$$\mathcal{L}\big((a_i)_{i=0,\dots,n-2}\big) = \mathcal{L}\big((a'_i)_{i=0,\dots,n-1}\big).$$

Applying Theorem 2.9 to the sequences (a_i) and (a'_i) we obtain

$$\mathcal{L}\big((a_i)_{i=0,\dots,n-1}\big) + \mathcal{L}\big((a'_i)_{i=0,\dots,n-1}\big) \leq \mathcal{L}\big((b_i)_{i=0,\dots,n-1}\big) = n,$$

where the last equality is due to Lemma 2.1.   □

In the next section we will prove that we even have equality in (2.20).

2.4.2 The Berlekamp-Massey Algorithm

In 1968 E.R. Berlekamp [20] presented an efficient algorithm for decoding BCH-codes (an important class of cyclic error-correcting codes). One year later, Massey [178] noticed that the decoding problem is in its essential parts equivalent to the determination of the shortest LFSR that generates a given sequence.

We present the algorithm, which is now known as Berlekamp-Massey algorithm, in the form which computes the linear complexity of a binary sequence.

In cryptography we are interested only in binary sequences, in contrast to coding theory where the algorithm is normally used over larger finite fields. To simplify the notation we specialize Algorithm 2.1 to the binary case.

Theorem 2.10 *In the notation of Algorithm 2.1, L_i is the linear complexity of the sequence $x_0, \dots, x_{i-1}$ and f_i is the feedback polynomial of the minimal LFSR that generates $x_0, \dots, x_{i-1}$.*

Algorithm 2.1 The Berlekamp-Massey algorithm

1: {initialization}
2: $f_0 \leftarrow 1, L_0 \leftarrow 0$
3: $f_{-1} \leftarrow 1, L_{-1} \leftarrow 0$
4: {Compute linear complexity}
5: **for** i **from** 0 **to** $n-1$ **do**
6:    $L_i = \deg f_i$
7:    $d_i \leftarrow \sum_{j=0}^{L_i} \text{coeff}(f_i, L_i - j)x_{i-j}$
8:    **if** $d_i = 0$ **then**
9:       $f_{i+1} \leftarrow f_i$
10:    **else**
11:       $m \leftarrow \begin{cases} \max\{j \mid L_j < L_{j+1}\} & \text{if } \{j \mid L_j < L_{j+1}\} \neq \emptyset \\ -1 & \text{if } \{j \mid L_j < L_{j+1}\} = \emptyset \end{cases}$
12:       **if** $m - L_m \geq i - L_i$ **then**
13:          $f_{i+1} \leftarrow f_i + X^{(m-L_m)-(i-L_i)} f_m$
14:       **else**
15:          $f_{i+1} \leftarrow X^{(i-L_i)-(m-L_m)} f_i + f_m$
16:       **end if**
17:    **end if**
18: **end for**

Proof As a first step we will prove by induction on i that f_i is a feedback polynomial of an LFSR that produces $x_0, \ldots, x_{i-1}$. In the second step we prove the minimality.

We start the induction with $i = 0$. The empty sequence has, by definition, linear complexity 0 and the "generating LFSR" has feedback polynomial 1.

Now suppose that f_i is the feedback polynomial of an LFSR which generates $x_0, \ldots, x_{i-1}$. We prove that f_{i+1} is the feedback polynomial of an LFSR which generates $x_0, \ldots, x_i$. In line 7, Algorithm 2.1 tests if the sequence $x_0, \ldots, x_i$ is also generated by the LFSR with feedback polynomial f_i. If this is the case we can keep the LFSR.

Now consider the case that the LFSR with feedback polynomial f_i fails to generate $x_0, \ldots, x_i$. In this case we need to modify the LFSR. To this we use in addition to the LFSR with feedback polynomial $f_i(x) = \sum_{i=1}^{L_i} a_i x^i$ the latest time step m in which the linear complexity of the sequence was increased. Let $f_m(x) = \sum_{j=0}^{L_m} b_j x^j$ be the feedback polynomial for that time step. For the first steps in which no such time step is present, we use the conventional values $m = -1$, $f_{-1} = 1$, $L_{-1} = 0$. The reader can easily check that the following argument works with this definition.

With these two shift registers we construct the automaton described in Fig. 2.3.

In the lower part we see the feedback shift register with feedback polynomial f_i. It generates the sequence $x_0, \ldots, x_i, x_{i+1} + 1$. So at time step $i - L_i$ it computes the wrong feedback $x_{i+1} + 1$. We correct this error with the feedback register shown in the upper part. We use the feedback polynomial f_m to test the sequence $x_0, \ldots$. In the first time steps this test will output 0. At time step $m - L_m$ we will use at first

Fig. 2.3 Construction for the
Berlekamp-Massey algorithm

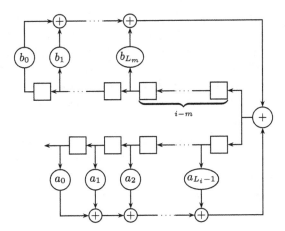

the value x_{m+1} and the test will output 1. We will feed back this 1 in such a way that it cancels with the 1 in the wrong feedback $x_{i+1} + 1$ of the first LFSR. To achieve this goal we have to delay the upper part by $i - m$.

We can combine the two registers into one LFSR by reflecting the upper part to the lower part. Here we must distinguish between the cases $m - L_m \geq i - L_i$ and $m - L_m \leq i - L_i$. (For $m - L_m = i - L_i$ both cases give the same result.)

If $L_m + (i - m) \geq L_i$, then the upper part contains more flip-flops. In this case the combination of both LFSRs looks like Fig. 2.4 (a).

If $L_m + (i - m) \geq L_i$, then the lower part contains more flip-flops. In this case the combination of both LFSRs looks like Fig. 2.4 (b).

In both cases the reader can check that the diagrams in Fig. 2.4 are described algebraically by the formulas given in lines 13 and 15 of the algorithm.

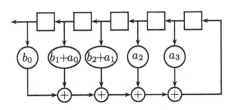

(a) The case $L_m + (i - m) \geq L_i$ (drawing with $L_i = 4$, $L_m = 2$, $i - m = 3$)

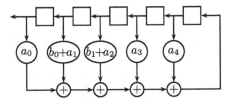

(b) The case $L_m + (i - m) \geq L_i$ (drawing with $L_i = 5$, $L_m = 1$, $i - m = 3$)

Fig. 2.4 Combination of the two LFSRs of Fig. 2.3

At this point we have proved that f_i is a feedback polynomial of an LFSR that generates $x_0, \ldots, x_i$. Now we prove the minimality.

At time $t = 0$ we have the empty sequence and the empty shift register, which is clearly minimal. Now look at the step $i \to i+1$. If $\deg f_{i+1} = \deg f_i$ then f_{i+1} is clearly minimal (part (b) of Theorem 2.8).

The only interesting case is when $\deg f_{i+1} > \deg f_i$. This happens if and only if $d_i \neq 0$ and $m - L_m < i - L_i$.

In this case we may apply Corollary 2.2 to conclude

$$\mathcal{L}(x_0, \ldots, x_i) \geq i + 1 - L_i.$$

To prove $\deg f_{i+1} = L_{i+1} = \mathcal{L}(x_0, \ldots, x_i)$ it suffices to show $L_{i+1} = i + 1 - L_i$.
    But

$$L_{i+1} = (i - L_i) + (m - L_m) + \deg f_i$$
$$= i - m + L_m$$
$$= (i + 1) - (m + 1 - L_m).$$

Since we have chosen m in such a way that $\deg f_m < \deg f_{m+1}$, we may use the induction hypothesis on that step and conclude $L_{m+1} = m + 1 - L_i$. But again, by choice of m, we have $\deg f_{m+1} = \deg f_{m+2} = \cdots = \deg f_i$. So we get

$$L_{i+1} = (i + 1) - (m + 1 - L_m)$$
$$= (i + 1) - L_i. \tag{2.21}$$

By Corollary 2.2 this proves $L_{i+1} = \mathcal{L}(x_0, \ldots, x_i)$.   □

As part of the proof we have improved the statement of Corollary 2.2.

Corollary 2.3 *If*

$$\mathcal{L}\big((a_i)_{i=0,\ldots,n-2}\big) < \mathcal{L}\big((a_i)_{i=0,\ldots,n-1}\big)$$

then

$$\mathcal{L}\big((a_i)_{i=0,\ldots,n-1}\big) = n - \mathcal{L}\big((a_i)_{i=0,\ldots,n-2}\big).$$

Proof The only case in which $\mathcal{L}((a_i)_{i=0,\ldots,n-2}) < \mathcal{L}((a_i)_{i=0,\ldots,n-1})$ is treated in line 15 of Algorithm 2.1. In the proof of Theorem 2.10 we have shown (Eq. (2.21)) that in this case $L_n = n - L_{n-1}$.   □

In each loop of the Berlekamp-Massey algorithm we need $O(L_i) = O(i)$ time steps, thus the computation of the linear complexity of a sequence of length n takes just $O(n^2)$ time steps. In comparison, the solution of a system of n linear equations, which we used in the beginning of this section, needs $O(n^3)$ time steps. In addition it has the advantage that it computes the linear complexity of all prefixes of the sequence. This is helpful if we want to study the linear complexity profile of the sequence (see Sect. 2.5).

2.4.3 Asymptotic Fast Computation of Linear Complexity

The Berlekamp-Massey algorithm is already very fast, so it may come as a surprise that we can compute the linear complexity even more rapidly.

One can lower the complexity of the computation of $\mathcal{L}(a_0, \ldots, a_{n-1})$ from $O(n^2)$ to $O(n \log n (\log \log(n))^2)$ (see, for example, [142]). The following algorithm is my own development and has the advantage that it has a small constant in the O-term and it computes, in addition to the feedback polynomial f_n, the whole sequence $\mathcal{L}(a_0, \ldots, a_{i-1})$, $i \leq n$.

The idea is a divide and conquer variant of the original Berlekamp-Massey algorithm. The main part is the function Massey(i, i') which allows us to go from time step i directly to time step i'.

We use the following notation: For any time step i, let f_i be the minimal feedback polynomial of the sequence $a_0, \ldots, a_{i-1}$. The degree of f_i is L_i and by $m(i)$ we denote the largest $m < i$ with $L_m < L_i$. If such an m does not exist we use the conventional values $m(i) = -1$ and $f_{-1} = 1$.

By $A(z) = \sum_{j=0}^{N} a_j z^j$ we denote the generating function of the sequence A_i and by $A^*(z) = z^N A(1/z)$ we denote the reciprocal polynomial.

To simplify the presentation of Algorithm 2.2 we take the values $m(i)$, L_i and so on as known. We will see later how to compute these values quickly.

Lines 3 and 4 need some further explanation. As we can see in the final recursion step (line 8), we access only the coefficient $N - i + L_i$ of D_i. So there is no need to compute the full polynomial $D_i = A^* f_i$. We will prove in Theorem 2.13 that only the coefficients from $z^{N-i'+L_m+2}$ to $z^{\max\{L_i-i, L_{m(i)}-m(i)\}}$ (inclusive) are needed when we call Massey(i, i'). Since for a sequence with typical complexity profile $m \approx i$ and $L_i \approx L_m$, this means that we need only about $i' - i$ coefficients. For the complexity of the algorithm it is crucial to implement this optimization.

Before we prove the correctness of the algorithm we look at its running time. Let $T(d)$ be the running time of Massey($i, i+d$).

For $d > 1$ we have two recursive calls of the function Massey and the computations in lines 3, 4 and 6. This leads to the recursion

$$T(d) = 2T(d/2) + 4M(d/2, d/4) + 8M(d/4, d/4).$$

We neglect the time for the additions and memory access in this analysis. Asymptotically it does not matter anyway, and computer experiments show that even for small d it has very little impact on the constants.

$M(k, l)$ denotes the complexity of multiplying a polynomial of degree k by a polynomial of degree l.

Even with "school multiplication" $M(k, l) = kl$ we get $T(n) = 2n^2$ which is not too bad in comparison with the original Berlekamp-Massey algorithm, which needs $\approx n^2/4$ operations.

With multiplication based on the fast Fourier transform we get $T(n) = O(n \log^2 n \log \log n)$. However, the constant is not so good in this case.

Algorithm 2.2 Massey(i, i')

Require: The check polynomials $D_i = A^* f_i$ and $D_{m(i)} = A^* f_{m(i)}$, $L_i = \deg f_i$, $L_{m(i)} = \deg f_{m(i)}$

Ensure: Polynomials g_{00}, g_{01}, g_{10} and g_{11} with $f_{m(i')} = f_{m(i)} g_{00} + f_i g_{01}$ and $f_{i'} = f_{m(i)} g_{10} + f_i g_{11}$.

1: **if** $i' - i > 1$ **then**
2:     Call Massey$(i, \frac{i+i'}{2})$ to get polynomials $g'_{00}, g'_{01}, g'_{10}$ and g'_{11}
3:     $D_{m(\frac{i+i'}{2})} := D_i g'_{00} + D_{m(i)} g'_{01}$ {Compute just the coefficients needed later!}
4:     $D_{\frac{i+i'}{2}} := D_i g'_{10} + D_{m(i)} g'_{11}$ {Compute just the coefficients needed later!}
5:     Call Massey$(\frac{i+i'}{2}, i')$ to get polynomials $g''_{00}, g''_{01}, g''_{10}$ and g''_{11}
6:     Compute $g_{kj} = g'_{0j} g''_{k0} + g'_{1j} g''_{k1}$ for $j, k \in \{0, 1\}$
7: **else**
8:     **if** $\operatorname{coeff}(D_i, N - i + L_i) = 1$ **then**
9:         $g_{00} = g_{11} = 1, g_{01} = g_{10} = 0$
10:    **else**
11:        **if** $m(i) - L_{m(i)} > i - L_i$ **then**
12:            $g_{00} = 1, g_{01} = 0, g_{10} = x^{(m(i) - L_{m(i)}) - (i - L_i)}, g_{11} = 1$
13:        **else**
14:            $g_{00} = 0, g_{01} = 1, g_{10} = 1, g_{11} = x^{(i - L_i) - (m(i) - L_{m(i)})},$
15:        **end if**
16:    **end if**
17: **end if**

With different multiplication algorithms such as, for example, Karatsuba (see also Sect. 11.3), we can get $T(n) = O(n^{1.59})$ with a very good constant. Fortunately there are plenty of good libraries which implement fast polynomial arithmetic, and the designers of the libraries have performed timings to choose the best multiplication algorithm for each range of n. So we can just implement Algorithm 2.2 and be sure that the libraries will guarantee an asymptotic running time of $O(n \log^2 n \log \log n)$ and select special algorithms with good constants for small n. For the timings at the end of this section we choose NTL [244] as the underling library for polynomial arithmetic.

We prove by induction that the polynomials g_{00}, g_{01}, g_{10} and g_{11} have the desired property.

Theorem 2.11 *Algorithm 2.2 computes polynomials* g_{00}, g_{01}, g_{10} *and* g_{11} *with* $f_{m(i')} = f_{m(i)} g_{00} + f_i g_{01}$ *and* $f_{i'} = f_{m(i)} g_{10} + f_i g_{11}$.

Proof We prove the theorem by induction on $d = i' - i$.

For $d = 1$ lines 8–16 are just a reformulation of lines 8–17 of Algorithm 2.1. Note that we flipped the sequence A_i and thus we have to investigate the bit at position $N - i + L_i$ instead of the bit at position i.

Now we have to check the case $d > 1$. By induction the call of Massey$(i, \frac{i+i'}{2})$ gives us polynomials g'_{00}, g'_{01}, g'_{10} and g'_{11} with $f_{m(\frac{i+i'}{2})} = f_{m(i)}g'_{00} + f_i g'_{01}$ and $f_{\frac{i+i'}{2}} = f_{m(i)}g'_{10} + f_i g'_{11}$.

To prove that the call of Massey$(\frac{i+i'}{2}, i')$ in line 5 gives the correct result, we have to show that the values $D_{m(\frac{i+i'}{2})}$ and $D_{\frac{i+i'}{2}}$ computed in lines 3 and 4 satisfy the requirements of the algorithm.

In line 3 we compute $D_{m(\frac{i+i'}{2})}$ as

$$D_{m(\frac{i+i'}{2})} = D_i g_{10} + D_{m(i)} g_{11}$$

$$= A^* f_i g_{01} + A^* f_{m(i)} g_{01} \quad \text{(required input form)}$$

$$= A^* f_{m(\frac{i+i'}{2})} \quad \text{(induction)}$$

and similarly $D_{\frac{i+i'}{2}} = A^* f_{\frac{i+i'}{2}}$. Thus we meet the requirement for Massey$(\frac{i+i'}{2}, i')$ and by induction we obtain polynomials g''_{00}, g''_{01}, g''_{10} and g''_{11} with $f_{m(i')} = f_{m(\frac{i+i'}{2})}g''_{00} + f_{\frac{i+i'}{2}} g''_{01}$ and $f_{i'} = f_{m(\frac{i+i'}{2})}g''_{10} + f_{\frac{i+i'}{2}} g''_{11}$.

Thus

$$f_{m(i')} = f_{m(\frac{i+i'}{2})}g''_{00} + f_{\frac{i+i'}{2}} g''_{01}$$

$$= \left(f_{m(i)}g'_{00} + f_i g'_{01}\right)g''_{00} + \left(f_{m(i)}g'_{10} + f_i g'_{11}\right)g''_{01}.$$

This proves $g_{00} = g'_{00}g''_{00} + g'_{10}g''_{01}$ and so on, i.e. the polynomials computed in line 6 satisfy the statement of the theorem.   □

In order to do the test in line 11 of Algorithm 2.2 we need to know $m(i)$, L_i and $L_{m(i)}$. We assume that the algorithm receives these numbers as input and we must prove that we can rapidly compute $m(i')$, $L_{i'}$ and $L_{m(i')}$.

In Algorithm 2.2 we have at any time $f_{i'} = f_{m(i)}g_{10} + f_i g_{11}$. We must compute $L_{i'} = \deg f_i$. The problem is that we cannot compute $f_{i'}$ for all i', since it will take $O(n^2)$ steps just to write the results. However, we don't have to compute $f_{i'}$ to determine $\deg f_{i'}$, as the following theorem shows.

Theorem 2.12 *Using the notation of Algorithm* 2.2

$$\deg f_{i'} = \deg f_i + \deg g_{11}$$

and if $m(i') > m(i)$ then

$$\deg f_{m(i')} = \deg f_i + \deg g_{01}.$$

Proof It is enough to prove $\deg f_{i'} = \deg f_i + \deg_{g_{11}}$, since the second part follows from the first simply by changing i to $m(i)$.

We will prove $\deg(f_{m(i)}g_{10}) < \deg(f_i g_{11})$, which implies the theorem. The proof is by induction on $d = i' - i$. This time we do not go directly from d to $2d$, but instead we will go from d to $d + 1$.

For $d = 1$ the only critical part is line 12. In all other cases we have $\deg f_i > \deg f_{m(i)}$ and $\deg g_{11} > \deg g_{10}$.

However, if we use line 12, then $\deg f_{i+1} = \deg f_i$, since $m(i) - L_{m(i)} > i - L_i$ and hence $\deg f_{i+1} = \deg f_i + \deg g_{11}$.

Now look at the step d to $d + 1$ as described in the algorithm. Let g'_{00}, g'_{01}, g'_{10} and g''_{11} be the polynomials with $f_{m(i+d)} = f_{m(i)}g'_{00} + f_i g'_{01}$ and $f_{i+d} = f_{m(i)}g'_{10} + f_i g'_{11}$. By induction, $\deg f_{i+d} = \deg(f_i g'_1) > f_{m(i)}g'_0$. Now observe how f_{i+d+1} is related to f_i. If we use line 9 in Algorithm 2.1 then $f_{i+d+1} = f_{i+d}$ and there is nothing left to prove.

If we use line 13 in Algorithm 2.1, then $\deg f_{i+d+1} = \deg f_{i+d}$ (since $m - L_m > i - L_i$) and hence $g'_1 = 1$ and $\deg f_{i+d+1} = \deg(f_{i+d}g''_1) = \deg(f_i g'_1 g''_1) = \deg f_i + \deg g_{11}$.

If we are in the case of line 15 in Algorithm 2.1 then $\deg g''_1 > \deg g''_0 = 0$ and hence $\deg f_i g''_1 > \deg f_{m(i)}g''_0$ and thus $\deg f_{i+d+1} = \deg(f_{i+d}g''_1) = \deg f_i + \deg g_{11}$.

This proves the theorem. $\square$

Theorem 2.12 allows us to compute $L_{i'} = \deg f_{i'}$ and $L_{m(i')} = \deg f_{m(i')}$ with just four additions from L_i and $L_{m(i)}$. Since $\deg f_i \leq i \leq n$ the involved numbers have at most $\log(n)$ bits, i.e. we need just $O(\log n)$ extra time steps per call of Massey(i, i').

The last thing we have to explain is how the algorithm computes the function $m(i)$. At the beginning $m(0) = -1$ is known. If $i' - i = d = 1$ we obtain $m(i + 1)$ as follows: If we use line 9 or line 12 then $m(i + 1) = m(i)$, and if we use line 14 then $m(i + 1) = i$.

If $d > 1$ then we obtain the value $m(\frac{i+i'}{2})$ needed for the call of Massey$(\frac{i+i'}{2}, i')$ resulting from the call of Massey$(i, \frac{i+i'}{2})$.

Similarly the algorithm can recursively compute $L_i = \deg f_i$ and $L_{m(i)} = \deg f_{m(i)}$, which is even faster than using Theorem 2.12.

Finally, we show that we can trim the Polynomials D_i and $D_{m(i)}$. We used these to get the sub-quadratic bound for the running time.

Theorem 2.13 *The function* Massey(i, i') *needs only the coefficients between* $z^{N-i'+L_{m(i)}+2}$ *and* $z^{\max\{L_i - i, L_{m(i)} - m(i)\}}$ *(inclusive) from the polynomials* D_i *and* $D_{m(i)}$.

Proof In the final recursion steps the algorithm will have to access coeff$(D_j, N - j + L_j)$ for each j in $\{i, \ldots, i' - 1\}$. So we have to determine which parts of the polynomials D_i and $D_{m(i)}$ are needed to compute the relevant coefficients.

Let $g_0^{(j)}$ and $g_1^{(j)}$ be the polynomials with $D_j = D_{m(i)}g_0^{(j)} + D_i g_1^{(j)}$.

By Theorem 2.12 we know that $\deg f_i + \deg g_1^{(j)} = \deg_{f_j}$ and $\deg f_m + \deg g_0^{(j)} < \deg_{f_j}$.

Thus $\max\{\deg g_0^{(j)}, \deg g_1^{(j)}\} \le L_j - L_m - 1$. Therefore we need only the coefficients from $z^{(N-j+L_j)-(L_j-L_m-1)} = z^{N-j+L_m+1}$ to z^{N-j+L_j} of D_i and D_m to compute $\mathrm{coeff}(D_j, N - j + L_j)$. (Note that in the algorithm we compute D_j not directly, but in intermediate steps so as not to change the fact that $\mathrm{coeff}(D_j, N - j + L_j)$ is affected only by this part of the input.)

Since $j < i'$ we see that we need no coefficient below $z^{N-i'+L_m+2}$.

To get an upper bound we have to bound $L_j - j$ in terms of L_i, L_m and i. By induction on j we prove

$$\max\{L_j - j, L_{j(m)} - m(j)\} \le \max\{L_i - i, L_{m(i)} - m(i)\}.$$

For $j = i$ this is trivial. Now we prove

$$\max\{L_{j+1} - (j+1), L_{m(j+1)} - m(j+1)\} \le \max\{L_j - j, L_{m(j)} - m(j)\}.$$

To this end, we study what happens in lines 9, 13 and 15 of Algorithm 2.1.

In the first two cases we have $L_{j+1} = L_j$, $m(j+1) = m(j)$ and the inequality is trivial. In the last case $L_{j+1} = L_j + (j - L_j) - (m(j) - L_{m(j)})$ and $m(j+1) = j$ thus

$$\max\{L_{j+1} - (j+1), L_{m(j+1)} - m(j+1)\} = \max\{L_{m(j)} - m(j) - 1, L_j - j\}.$$

This proves $L_j - j < \max\{L_i - i, L_{m(i)} - m(i)\}$, which gives the desired upper bound $N - j + L_j$. $\square$

We have seen that the algorithm keeps track of the values $m(i)$, L_i and $L_{m(i)}$. So the only thing we have to do to get the full sequence $L_1, \ldots, L_n$ is to output $L_{\frac{i+i'}{2}}$ when the algorithm reaches line 5. This costs no extra time.

If we are interested in the feedback polynomials we have to do more work. We need an array R of size n. Each time the algorithm reaches its end (line 17) we store the values $(i, g_{00}, g_{01}, g_{10}, g_{11})$ at $R[i']$.

When the computation of the linear complexity is finished the array R is completely filled. Now we can compute the feedback polynomials by the following recursive function (Algorithm 2.3).

Finally, we can also use the algorithm in an iterative way. If we have observed N bits, we can call Massey$(0, N)$ to compute the linear complexity of the sequence $a_0, \ldots, a_{N-1}$. We will remember $m(N)$ (computed by the algorithm), $f_N = g_{10} + g_{11}$ and $f_{m(N)} = g_{00} + g_{01}$. If we later observe N' extra bits of the sequence, we can call Massey$(N, N + N')$ to get the linear complexity of $a_0, \ldots, a_{N+N'-1}$.

In the extreme case we can always stop after one extra bit. In this case the algorithm will of course need quadratic time, since it must compute all intermediate feedback polynomials. Computer experiments show that the new algorithm beats

---

Algorithm 2.3 feedback(i')

1: Get the values $i, g_{00}, g_{01}, g_{10}, g_{11}$ from $R[i']$.
2: **if** $i = 1$ **then**
3:    $f_{i'} = g_{10} + g_{11}$, $f_{m(i')} = g_{00} + g_{01}$
4: **else**
5:    Obtain f_i and $f_{m(i')}$ by calling feedback(i).
6:    $f_{i'} = f_{m(i)}g_{10} + f_i g_{11}$, $f_{m(i')} = f_{m(i)}g_{00} + f_i g_{01}$
7: **end if**

Table 2.1 Tests for the algorithms

	n					
	100	500	1000	5000	10000	100000
Algorithm 2.1	0.00002	0.00065	0.0036	0.098	0.39	36.11
Algorithm 2.2	0.001	0.00157	0.0042	0.049	0.094	0.94

the original Berlekamp-Massey algorithm if it receives at least 5000 bits at once. The full speed is reached only if we receive the input in one step.

Table 2.1 shows that the asymptotic fast Berlekamp-Massey algorithm beats the classical variant even for very small n.

2.4.4 Linear Complexity of Random Sequences

Since we want to use linear complexity as a measure of the randomness of a sequence, it is natural to ask what the expected linear complexity of a random sequence is.

Theorem 2.14 (Rueppel [228]) *Let* $1 \leq L \leq n$. *The number* $N(n, L)$ *of binary sequences of length* n *having linear complexity exactly* L *is* $N(n, L) = 2^{\min\{2n-2L, 2L-1\}}$.

Proof We are going to find a recursion for $N(n, L)$. For $n = 1$ we have $N(1, 0) = 1$ (the sequence 0) and $N(1, 1) = 1$ (the sequence 1).

Now consider a sequence $a_0, \ldots, a_{n-1}$ of length n and linear complexity L. Let $f(z)$ be a feedback polynomial of a minimal LFSR generating $a_0, \ldots, a_{n-1}$.

We have one way to extend the sequence $a_0, \ldots, a_{n-1}$ by an a_n without changing the feedback polynomial $f(z)$. (This already proves $N(n + 1, L) \geq N(n, L)$.)

Now consider the second possible extension $a_0, \ldots, a_{n-1}, \overline{a_n}$. By Corollary 2.3 we have either

$$\mathcal{L}(a_0, \ldots, a_{n-1}, \overline{a_n}) = \mathcal{L}(a_0, \ldots, a_{n-1}) = L$$

or

$$\mathcal{L}(a_0, \ldots, a_{n-1}, \overline{a_n}) = n + 1 - L > L.$$

If $L \geq \frac{n+1}{2}$, the second case is impossible, i.e. we have $N(n + 1, L) \geq 2N(n, L)$ for $L \geq \frac{n+1}{2}$.

Now let $L < \frac{n+1}{2}$ and let m be the largest index with $L_m = \mathcal{L}(a_0, \ldots, a_{m-1}) < L$. Then by Corollary 2.3 we find

$$L_m = m - \mathcal{L}(a_0, \ldots, a_{m-2}) > \mathcal{L}(a_0, \ldots, a_{m-2})$$

and hence $L_m > m/2$. Therefore $n - L > \frac{n+1}{2} > m/2 > m - L_m$, which means by Algorithm 2.1 that

$$\mathcal{L}(a_0, \ldots, a_{n-1}, \overline{a_n}) > \mathcal{L}(a_0, \ldots, a_{n-1}) = L.$$

This proves the recursion

$$N(n + 1, L) = \begin{cases} 2N(n, L) + N(n, n + 1 - L) & \text{if } 2L > n + 1, \\ 2N(n, L) & \text{if } 2L = n + 1, \\ N(n, L) & \text{if } 2L < n + 1. \end{cases}$$

With this recursion it is just a simple induction to prove

$$N(n, L) = 2^{\min\{2n - 2L, 2L - 1\}}. \qquad \square$$

With Theorem 2.14 we need only elementary calculations to obtain the expected linear complexity of a finite binary sequence.

Theorem 2.15 (Rueppel [228]) *The expected linear complexity of a binary sequence of length n is*

$$\frac{n}{2} + \frac{4 + (n \,\&\, 1)}{18} - 2^{-n}\left(\frac{n}{3} + \frac{2}{9}\right).$$

Proof For even n we get

$$\sum_{i=1}^{n} i N(n, l) = \sum_{i=1}^{n/2} i \cdot 2^{2i-1} + \sum_{i=0}^{n/2-1} (n - i)2^{2i}$$

$$= \left[n\frac{2^n}{3} + \frac{2^{n+1}}{9} + \frac{2}{9} \right] + \left[n\frac{2^n}{6} + \frac{2^{n+2}}{9} - \frac{n}{3} - \frac{4}{9} \right]$$

$$= n2^{n-1} + \frac{2^{n+1}}{9} - \frac{n}{3} - \frac{2}{9}$$

and similarly for odd n we get

$$\sum_{i=1}^{n} iN(n,l) = \sum_{i=1}^{(n-1)/2} i \cdot 2^{2i-1} + \sum_{i=0}^{(n-1)/2} (n-i)2^{2i}$$

$$= n2^{n-1} + \frac{5}{18}2^n - \frac{n}{3} - \frac{2}{9}.$$

Multiplying by the probability 2^{-n} for a sequence of length n we get the expected value. ☐

It is also possible to determine the expected linear complexity of a periodic sequence of period n with random elements. However, since in cryptography a cipher stream is broken if we are able to observe more than one period (see the Vigenère cipher in Sect. 1.1), this kind of result is of less interest. We state the following without proof.

Theorem 2.16 *A random periodic sequence with period n has expected linear complexity*:

(a) $n - 1 + 2^{-n}$ *if n is power of* 2;
(b) $(n-1)(1 - \frac{1}{2^{o(2,n)}}) + \frac{1}{2}$ *if n is an odd prime and* $o(2,n)$ *is the order of* 2 *in* $\mathbb{F}_n^{\times}$.

Proof

(a) See Proposition 4.6 in [228].
(b) See Theorem 3.2 in [186].

☐

2.5 The Linear Complexity Profile of Pseudo-random Sequences

2.5.1 Basic Properties

We have introduced linear complexity as a measure of the cryptographic strength of a pseudo-random sequence. However, a high linear complexity is only a necessary but not sufficient condition for cryptographic strength. Take for example the sequence

$$1010111100010011010111100$$

which is generated by an LFSR with feedback polynomial $z^4 + z + 1$. It is a weak key stream and its linear complexity is 4. By changing just the last bit of the sequence the linear complexity rises to 22 (see Corollary 2.3), but changing just one bit does not make a keystream secure.

One way to improve linear complexity as a measure of the randomness of a sequence is to look at the linear complexity profile.

Fig. 2.5 The linear
complexity profile of
1010111100010011010111101

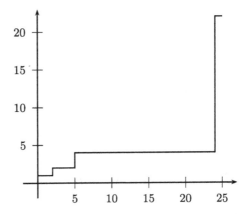

Fig. 2.6 A typical linear
complexity profile

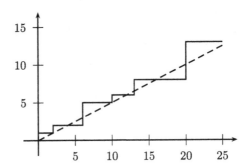

Definition 2.13 The linear *complexity profile* of a binary sequence $(a_n)_{n \in \mathbb{N}}$ is the
function $LP : \mathbb{N}^+ \to \mathbb{N}$ with $n \mapsto \mathcal{L}(a_0, \ldots, a_{n-1})$.

If we draw the linear complexity profile for the sequence

$$1010111100010011010111101$$

we see (Fig. 2.5) that the linear complexity jumps with the last bit to the high value
22. Prior to this we have the low value 4, which means that the sequence is a weak
key stream.

By Theorem 2.15 the expected linear complexity of a sequence of length n is
about $\frac{n}{2}$, i.e. the linear complexity profile of a good pseudo-random sequence should
lie around the line $n \mapsto n/2$ as shown in Fig. 2.6.

In the remaining part of this section we will study sequences with a linear complexity profile which is "as good as possible".

Definition 2.14 A sequence $(a_n)_{n \in \mathbb{N}}$ has a *perfect linear complexity profile* if

$$\mathcal{L}(a_0, \ldots, a_{n-1}) = \left\lfloor \frac{n+1}{2} \right\rfloor.$$

The linear complexity profile is *good* if

$$\left| \mathcal{L}(a_0, \ldots, a_{n-1}) - \frac{n}{2} \right| = O\big(\log(n)\big).$$

H. Niederreiter [198] classified all sequences with a good linear complexity profile by means of continued fractions. We will follow his proof in the remaining part of the section.

2.5.2 Continued Fractions

In this section we classify all sequences with a good linear complexity profile. To that end we establish a connection between the continued fraction expansion of the generation function and the complexity profile.

Consider the field $F((z^{-1})) = \{\sum_{i \geq n} a_i z^{-i} \mid n \in \mathbb{Z}, a_i \in F\}$ of formal Laurent series in z^{-1}. For $S = \sum_{i \geq n} a_i z^{-i} \in F((z^{-1}))$ we denote by $[S] = \sum_{0 \geq i \geq n} a_i z^{-i}$ the polynomial part of S.

A continued fraction is an expression of the form

$$b_0 + \cfrac{a_1}{b_1 + \cfrac{a_2}{b_2 + \cfrac{a_3}{b_3 + \ddots}}}$$

For continued fractions we use the compact notation of Pringsheim:

$$b_0 + \frac{a_1|}{|b_1} + \frac{a_2|}{|b_2} + \frac{a_3|}{|b_3} + \cdots$$

For $S \in F((z^{-1}))$ recursively define

$$A_i = \left[R_{i-1}^{-1} \right], \qquad R_i = R_{i-1}^{-1} - A_i \quad \text{for } i \geq 0 \tag{2.22}$$

with $R_{-1} = S_0$. This gives the continued fraction expansion

$$S = A_0 + \frac{1|}{|A_1} + \frac{1|}{|A_2} + \cdots \tag{2.23}$$

of S.

The term

$$A_0 + \frac{1|}{|A_1} + \cdots + \frac{1|}{|A_k} = \frac{P_k}{Q_k}$$

with $P_k, Q_k \in F[z]$ is called the kth *convergent fraction* of f.

Let us recall some basic properties of continued fractions (see, for example, [211]).

The polynomials P_k, Q_k satisfy the recursion

$$6P_{-1} = 1, \quad P_0 = A_0, \quad P_k = A_k P_{k-1} + P_{k-2}, \tag{2.24}$$

$$Q_{-1} = 0, \quad Q_0 = 1, \quad Q_k = A_k Q_{k-1} + Q_{k-2}. \tag{2.25}$$

In addition we have the identities

$$P_{k-1} Q_k - P_k Q_{k-1} = (-1)^k, \tag{2.26}$$

$$\gcd(P_k, Q_k) = 1, \tag{2.27}$$

$$S = \frac{P_k + R_k P_{k-1}}{Q_k + R_k Q_{k-1}}. \tag{2.28}$$

The above identities hold for every continued fraction. The next identities use the degree function and hold only for continued fractions defined over $F((z^{-1}))$. Using the recursion (2.25) we get $\deg Q_i = \sum_{j=1}^{i} \deg A_i$ for $j \geq 1$.

Lemma 2.2 *For all $j \in \mathbb{N}$ we have*

$$\deg(Q_j S - P_j) = -\deg(Q_{j+1}).$$

Proof We prove this by induction on j. For $j = 0$ this follows immediately from Eq. (2.28) with

$$-\deg R_0 = \deg R_0^{-1} = \deg\left[R_0^{-1}\right] = \deg A_1 = \deg Q_1.$$

Now let $j \geq 1$. By Eq. (2.28) we have

$$SQ_j - P_j = B_j(SQ_{j-1} - P_{j-1}).$$

By induction $\deg SQ_{j-1} - P_{j-1} = \deg Q_j$ and since $\deg B_j = -\deg A_{j+1}$ and $\deg Q_{j+1} = \deg A_{j+1} + \deg Q_j$ we get

$$\deg(SQ_j - P_j) = -\deg Q_{j+1}. \qquad \square$$

The connection of linear complexity and the Berlekamp-Massey algorithm with continued fractions and the Euclidean algorithm has been observed several times. The formulation of the following theorem is from [197].

Theorem 2.17 *Let $(a_n)_{n \in \mathbb{N}}$ be a sequence over the field F and let $S(z) = \sum_{j=0}^{-\infty} a_j z^{-j-1}$. Let $\frac{P_k}{Q_k}$ be the kth convergent fraction of the continued fraction expansion of S.*

Then for every $n \in \mathbb{N}^+$ the linear complexity $L_n = \mathcal{L}(a_0, \ldots, a_{n-1})$ is given by $L_n = 0$ for $n < \deg Q_0$ and $L_n = \deg Q_j$ where $j \in \mathbb{N}$ is determined by

$$\deg Q_{j-1} + \deg Q_j \leq n < \deg Q_j + \deg Q_{j+1}.$$

Proof By Lemma 2.2 we have

$$\deg\left(S - \frac{P_j}{Q_j}\right) = -\deg Q_j - \deg Q_{j+1}.$$

This means that the first $\deg Q_j + \deg Q_{j+1}$ elements of the sequence with the rational generating function $\frac{P_j}{Q_j}$ coincide with $(a_n)_{n=0,\ldots,\deg Q_j + \deg Q_{j+1}-1}$.

But the rational generating function $\frac{P_j}{Q_j}$ belongs to an LFSR with feedback polynomial Q_j^*, which proves that

$$L_n \leq \deg Q_j \quad \text{for } n < \deg Q_j + \deg Q_{j+1}. \tag{2.29}$$

This already establishes one part of the theorem. Now we prove the equality. That $L_n = 0$ if $n < \deg Q_0$ is just a reformulation of the fact that $\deg Q_0$ denotes the number of leading zeros in the sequence $(a_n)_{n\in\mathbb{N}}$.

By induction we know now that $L_n = \deg Q_j$ for $\deg Q_{j-1} + \deg Q_j \leq n < \deg Q_j + \deg Q_{j+1}$.

If k is the smallest integer with $L_k > \deg Q_j$ then by Corollary 2.3 we have $L_k = k - \deg Q_j$. The only possible value of k for which L_k satisfies Eq. (2.29) is $k = \deg Q_j + \deg Q_{j+1}$. Thus $k = \deg Q_j + \deg Q_{j+1}$ and $L_k = \deg Q_{j+1}$. By Eq. (2.29) we get $L_n = \deg Q_{j+1}$ for $\deg Q_j + \deg Q_{j+1} \leq n < \deg Q_{j+1} + \deg Q_{j+2}$, which finishes the induction. □

2.5.3 Classification of Sequences with a Perfect Linear Complexity Profile

By Theorem 2.17 it is easy to characterize sequences with a good linear complexity profile in terms of continued fractions (see [197, 198]). As a representative of all results of this kind, we present Theorem 2.18 which treats the case of a perfect linear complexity profile.

Theorem 2.18 (see [197]) *A sequence* $(a_n)_{n\in\mathbb{N}}$ *has a perfect linear complexity profile if and only if the generating function* $S(z) = \sum_{j=0}^{-\infty} a_j z^{-j-1}$ *is irrational and has a continued fraction expansion*

$$S = \frac{1\,|}{|A_1} + \frac{1\,|}{|A_2} + \frac{1\,|}{|A_3} + \cdots$$

with $\deg A_i = 1$ *for all* $i \geq 1$.

Proof A perfect linear complexity profile requires that the linear complexity grows at most by 1 at each step. By Theorem 2.17 this means that the sequence $\deg Q_i$ contains all positive integers, i.e. $\deg A_i = 1$ for all continued fraction denominators A_i.

On the other hand, $\deg A_i = 1$ for all i implies $\deg Q_i = i$ and by Theorem 2.17 we get $L_i = \lfloor \frac{i+1}{2} \rfloor$.   □

By Theorem 2.18 we can deduce a nice characterization of sequences with a perfect linear complexity profile.

Theorem 2.19 (see [274]) *The binary sequence $(a_i)_{i \in \mathbb{N}}$ has a perfect linear complexity profile if and only if it satisfies $a_0 = 1$ and $a_{2i} = a_{2i-1} + a_i$ for all $i \geq 1$.*

Proof Define the operation $D : \mathbb{F}_2((z^{-1})) \to \mathbb{F}_2((z^{-1}))$ by

$$D : T \mapsto z^{-1} T^2 + (1 + z^{-1}) T + z^{-1}.$$

A short calculation reveals the following identities:

$$D(T + U + V) = D(T) + D(U) + D(V) \quad \text{for } T, U, V \in \mathbb{F}_2((z^{-1})),$$

$$D(T^{-1}) = D(T) T^{-2} \quad \text{for } T \in \mathbb{F}_2((z^{-1})),$$

$$D(z) + D(c) = c + 1 \quad \text{for } c \in \mathbb{F}_2.$$

Now assume that the sequence $(a_i)_{i \in \mathbb{N}}$ has a perfect linear complexity profile and let

$$S(z) = \sum_{j=0}^{\infty} a_j z^{-j-1} = \frac{1 \mid}{\mid A_1} + \frac{1 \mid}{\mid A_2} + \frac{1 \mid}{\mid A_3} + \cdots$$

be the corresponding generating function with its continued fraction expansion.

By Theorem 2.18 we have $A_i = z + a_i$ with $a_i \in \mathbb{F}_2$. By Eq. (2.22) we have $R_i^{-1} = R_{i-1} - A_j$ and hence

$$D(R_{i-1}) R_{i-1}^{-2} = D(R_{i-1}^{-1}) = D(R_i + z + a_i) = D(R_i) + 1 + a_i.$$

By definition, $S = R_{-1}$, and by induction on i we have

$$D(S) = \sum_{j=0}^{i-1} (a_j + 1) \prod_{k=-1}^{j-1} R_k^2 + \prod_{k=-1}^{i-1} R_k^2 D(R_i).$$

We can turn $\mathbb{F}_2((z^{-1}))$ into a metric space by defining $d(Q, R) = 2^{-\deg(Q-R)}$. Since $\deg R_i < 0$ for all i we get

$$\lim_{i \to \infty} \prod_{k=-1}^{i-1} R_k^2 D(R_i) = 0$$

and hence

$$D(S) = \sum_{j=0}^{\infty} (a_j + 1) \prod_{k=-1}^{j-1} R_k^2.$$

Since all summands lie in $\mathbb{F}_2((z^{-2}))$, we get $D(S) = U^2$ for some $U \in \mathbb{F}_2((z^{-1}))$ or equivalently

$$S^2 + (z+1)S + 1 = zU^2. \tag{2.30}$$

Comparing the coefficients of z^0 we get $a_0 = 1$, and comparing the coefficients of z^{2i} ($i \in \mathbb{N}^+$) we get $a_i + a_{2i-1} + a_{2i} = 0$.

For the opposite direction note that the recursion $a_0 = 1$ and $a_i + a_{2i-1} + a_{2i} = 0$ imply that Eq. (2.30) is satisfied for some suitable $U \in \mathbb{F}_2((z^{-1}))$.

Assume that the linear complexity profile of the sequence is not perfect.

Then we find an index j with $\deg A_j > 1$ and by Lemma 2.2 we have

$$\deg(SQ_j - P_j) = -\deg Q_{j+1} < -\deg Q_j - 1.$$

It follows that

$$\begin{aligned}
\deg&\left(P_j^2 + (x+1)P_j Q_j + Q_j^2 + xU^2\right) \\
&= \deg\left(Q^2 S^2 - P_j^2 + (x+1)Q_j(SQ_j - P_j)\right) \\
&\leq \max\left\{\deg\left(Q^2 S^2 - P_j^2\right) \deg\left((x+1)Q_j(SQ_j - P_j)\right)\right\} \\
&< 0. \tag{2.31}
\end{aligned}$$

In particular the constant term $P_j(0)^2 + P_j(0)Q_j(0) + Q_0^2$ is 0, but this implies $P_j(0) = Q_j(0) = 0$ and hence $\gcd(P_j, Q_j) \neq 1$, contradicting Eq. (2.27).

This proves that the sequence that satisfies the recurrence given in Theorem 2.19 has a perfect linear complexity profile. □

We remark that even a sequence with a perfect linear complexity profile can be highly regular. For example, consider the sequence $(a_i)_{i \in \mathbb{N}}$ with $a_0 = 1$, $a_{2j} = 1$ for $j \in \mathbb{N}$ and $a_j = 0$ for $j \in \mathbb{N} \setminus \{1, 2^k \mid k \in \mathbb{N}\}$ given by Dai in [71]. This sequence is obviously a very weak key stream, but as one can check by Theorem 2.19, it has an optimal linear complexity profile.

2.6 Implementation of LFSRs

This book is about the mathematics of stream ciphers, but in cryptography mathematics is not everything. We have mentioned in the introduction that LFSRs are popular because they can be implemented very efficiently. This section should justify this claim.

Fig. 2.7 The Fibonacci
implementation of an LFSR

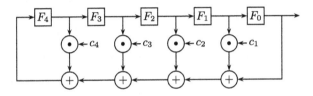

Fig. 2.8 The Galois
implementation of an LFSR

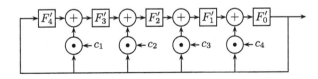

2.6.1 Hardware Realization of LFSRs

For implementing LFSRs in hardware there are two basic strategies. Either we convert Fig. 2.1 directly into the hardware, which then looks like Fig. 2.7. If the feedback polynomial is fixed we can save the AND-gates (multiplication).

This way to implement an LFSR is called the *Fibonacci implementation* or sometimes the *simple shift register generator* (SSRG).

The alternative implementation (see Fig. 2.8) is called the *Galois implementation*
(alternative names: *multiple-return shift register generator* (MRSRG) or *modular
shift register generator* (MSRG)).

The advantage of the Galois implementation is that every signal must pass
through at most one XOR-gate. By contrast, with a dense feedback polynomial the
feedback signal in the Fibonacci implementation must pass through approximately
$n/2$ XOR-gates.

As indicated in the figure, one has to reverse the feedback coefficients in the
Galois implementation. The internal states of the Fibonacci implementation and the
Galois implementation are connected by Theorem 2.20.

Theorem 2.20 *The Galois implementation generates the same sequence as the Fibonacci implementation if it is initialized with $F_i' = \sum_{j=0}^{i} F_{i-j} c_{n-j}, (0 \leq i \leq n)$
where $F_0, \ldots, F_{n-1}$ is the initial state of the Fibonacci implementation of the LFSR.*

Proof We have $F_0' = F_0$, so the next output bit is the same in both implementations.

We must prove that the next state $\hat{F}_{n-1}, \ldots, \hat{F}_0$ of the Fibonacci implementation
and the next state $\hat{F}_{n-1}', \ldots, \hat{F}_0'$ of the Galois implementation again satisfy $\hat{F}_i' =
\sum_{j=0}^{i} \hat{F}_{i-j} c_{n-j}, (0 \leq i \leq n)$.

For $i \leq n - 2$ we get

$$\hat{F}'_i = F'_{i+1} + c_{n-1-i} F'_0$$

$$= \sum_{j=0}^{i+1} F_{i+1-j} c_{n-j} + c_{n-1-i} F'_0$$

$$= \sum_{j=0}^{i} F_{i+1-j} c_{n-j}$$

$$= \sum_{j=0}^{i} \hat{F}_{i-j} c_{n-j}.$$

For $i = n - 1$ we have

$$\hat{F}'_{n-1} = F'_0 = F_0$$

$$= \sum_{i=0}^{n-1} c_i F_i + \sum_{i=1}^{n-1} c_i F_i$$

$$= \hat{F}_{n-1} + \sum_{i=1}^{n-1} c_i \hat{F}_{i-1}.$$

So both implementations give the same result.   □

2.6.2 Software Realization of LFSRs

Now we look at software implementation of LFSRs. All modern processors have instructions that help to achieve fast implementations. The main problem with software implementations is that we lose the advantage of low power consumption that we have with specialized hardware. Also, block ciphers with implementations based on table look-up become a good alternative.

2.6.2.1 Bit-Oriented Implementation

We first describe a bit-oriented implementation. The advantage is that we have a direct simulation of the LFSR in the software. For optimal performance it would be better to use a byte-oriented implementation.

We use an array of words $w_0, \ldots, w_{\hat{n}}$ which represent the internal state of the shift register. We can easily implement the shift operation on this bit field by calling the shift and rotation instructions of our processor (see Algorithm 2.4

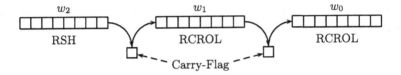

Fig. 2.9 Right shift over several words

and Fig. 2.9). Unfortunately, even a language like C does not provide direct access to the rotation operations. So we have to use hand-written assembler code. A portable C-implementation of the shift operation is a bit slower. Good code can be found in [255], which contains many tips about implementing cryptographic functions.

Algorithm 2.4 Right shift over several words

1: RSH $w_{\hat{n}-1}$ {Right shift by 1}
2: **for** $k \leftarrow \hat{n} - 1$ **to** $\hat{n} - 1$ **do**
3:    RCROL w_k {Right roll by 1, use the carry flag}
4: **end for**

If the feedback polynomial has only a few non-zero coefficients we can compute the feedback value by $f = x[f_1] + \cdots + x[f_k]$. However, if the feedback polynomial has many non-zero coefficients, this method is too slow. A better technique is to store the feedback polynomial in the bit field f. We compute $x \& f$ and count the number of non-zero bits in $x \& f$ modulo 2. The operation of counting the numbers of set bits in a bit field is known as *sideway addition* or *population count* (see Sect. 12.1.2). If our computer supports this operation we should use it, otherwise we should use Algorithm 2.5 which directly computes the number of non-zero bits in $x \& f$ modulo 2 (see also Sect. 5.2 in [276]).

Algorithm 2.5 Sideway addition mod 2 (32 bit version)

Ensure: $y = x_0 + \cdots + x_{31} \mod 2$
1: $y \leftarrow x \oplus (x \gg 1)$
2: $y \leftarrow y \oplus (y \gg 2)$
3: $y \leftarrow a(y \& a) \mod 2^{32}$ {with $a = (11111111)_{16}$}
4: $y \leftarrow (y \gg 28) \& 1$

Theorem 2.21 *Given the input* $x = (x_0, \ldots, x_{n-1})_2$, *Algorithm 2.5 computes* $x_0 + x_1 + \cdots + x_n \mod 2$.

Proof After the first line we have $y_0 + y_2 + \cdots + y_{62} = x_0 + \cdots + x_{63}$ and after the second line we have

$$y_0 + y_4 + \cdots + y_{60} = x_0 + \cdots + x_{63}.$$

Since $y \& a$ has at most 8 non-zero bits, we can be sure that the multiplication does not overflow, i.e.

$$a(y \& a) = \left(\sum_{i=0}^{7} y_{4j}, \sum_{i=0}^{6} y_{4j}, \ldots, y_0 + y_4, y_0 \right)_4 .$$

In the final step we extract the bit $\sum_{i=0}^{7} y_{4j} \bmod 2$.   □

If we work with 64-bit words, we must add an extra shift $y \leftarrow y \oplus (y \gg 4)$ after line two and use the mask $a = (11111111)_{256}$ instead of $a = (11111111)_{16}$.

Some processors (such as the IA32 family) have a slow multiplication routine (≈ 10 clock cycles, while the shift and XOR takes only 1 clock cycle). In this case Algorithm 2.6, which avoids multiplication, may be faster.

Algorithm 2.6 Sideway addition mod 2 (without multiplication)

Require: $x = (x_{n-1} \ldots x_0)$ is n-bit word
Ensure: $y = SADD(x) \mod 2$
  1: $y \leftarrow x$
  2: **for** $k \leftarrow 0$ **to** $\lfloor (\log_2(n-1) \rfloor$ **do**
  3:     $y \leftarrow y \oplus (y \gg 2^k)$
  4: **end for**
  5: $y \leftarrow y \& 1 \ \{ y \leftarrow y \mod 2 \}$

2.6.2.2 Word-Oriented Implementation

The bitwise generation of an LFSR sequence is attractive for simulating hardware realizations. However, on most computers it will be faster to generate the sequence word-wise. Let s be the word size of our computer, i.e. $s = 8$ on a small embedded processor (such as Intel's MCS-51 series) or $s = 32$ or $s = 64$ on a Desktop machine. We assume that s is a power of 2 and that $s = 2^d$.

For simplicity we assume that the length n of the feedback shift register is divisible by the word size s. Let $n = s\hat{n}$.

Let $c_{j,k}$ be the coefficients of the generator matrix associated with the LFSR (see Eq. (2.17)). Define

$$f_i \left(a_0 + \cdots + 2^7 a_7 \right) = \bigoplus_{k=0}^{7} \left(\sum_{j=0}^{s-1} 2^j a_k c_{8i+k,n+j} \right). \tag{2.32}$$

Let $x = (x_{n-1}, \ldots, x_0)_2$ be the internal state of the LFSR. Then the next word $x' = (x_{n+s}, \ldots, x_n)_2$ is

$$x' = \left(\bigoplus_{i=0}^{n-1} x_i c_{i,n+s}, \ldots, \bigoplus_{i=0}^{n-1} x_i c_{i,n} \right)_2$$

$$= \bigoplus_{i=0}^{n-1} \sum_{j=0}^{s-1} 2^j x_i c_{i,j}. \tag{2.33}$$

(This is just the definition of the generator matrix.)
Now write $x = (\hat{x}_{\hat{n}-1}, \ldots, \hat{x})$ and regroup the sum in Eq. (2.33), yielding:

$$x' = \bigoplus_{i=0}^{\hat{n}-1} f_i(\hat{x}_i). \tag{2.34}$$

Equation (2.34) gives us a table look-up method to compute the word of the LFSR sequence. We just have to pre-compute the functions f_i and evaluate Eq. (2.33). Algorithm 2.7 shows this in pseudo-code.

Algorithm 2.7 LFSR byte-oriented implementation (table look-ups)

1: **output** w_0
2: $w \leftarrow f_0(w_0)$
3: **for** $k \leftarrow 1$ **to** $\hat{n} - 1$ **do**
4:    $w \leftarrow w \oplus f_i$
5:    $w_{i-1} \leftarrow w_i$
6: **end for**
7: $w_{\hat{n}-1} \leftarrow w$

Algorithm 2.7 uses huge look-up tables. This may be a problem in embedded devices. In this case we can use the following algorithm that is based on the idea that we can use byte operations to compute several sideway additions simultaneously and which needs no look-up table.

The core of our program is Algorithm 2.8, which takes 2^k words $w_0, \ldots, w_{2^k-1}$ of 2^k bits each and computes the word $y = (y_{2^k-1} \cdots y_0)$ with $y_i = \text{SADD}(w_i)$ mod 2. (SADD denotes the sideway addition.)

Theorem 2.22 *The result $y = PSADD(d, 0)$ of Algorithm 2.8 satisfies $y_i = \text{SADD}(w_i)$ mod 2.*

Proof We prove by induction on d' that $y^{(d',k)} = PSADD(d', k)$ satisfies

$$\bigoplus_{j=0}^{2^{d-d'}} y^{(d',k)}_{i+j2^{d'}} = \text{SADD}(w_{k+i}) \mod 2 \tag{2.35}$$

Algorithm 2.8 Parallel sideway addition mod 2

Ensure: $PSADD(d, 0)$ returns the word
$\quad (SADD(w_{2^s-1}) \bmod 2, \ldots, SADD(w_0) \bmod 2)_2$
1: **if** $d' = 0$ **then**
2: $\quad$ **return** w_i
3: **else**
4: $\quad y \leftarrow PSADD(d' - 1, k)$
5: $\quad y \leftarrow (y \gg 2^{d'-1} \oplus y)$
6: $\quad y' \leftarrow PSADD(d' - 1, k + 2^{d-1})$
7: $\quad y' \leftarrow (y' \ll 2^{d'-1} \oplus y')$
8: $\quad$ **return** $(y \mathbin{\&} \mu_{d'-1}) \mid (y' \mathbin{\&} \overline{\mu_{d'-1}})$
9: **end if**

---

for $i \in \{0, \ldots, 2^{d'} - 1\}$.

For $d' = 0$ we have simply $y^{(0,k)} = w_k$ and Eq. (2.35) is trivial.

If $d' > 0$ we call $PSADD(d' - 1, k)$ in line 4 and the return value satisfies Eq. (2.35). The shift and XOR operation in step 4 give us a word $y = y^{(d'-1,k)}$ which satisfies

$$\bigoplus_{j=0}^{2^{d-d'+1}} y_{i+j2^{d'-1}}^{(d'-1,k)} = SADD(w_{k+i}) \quad \bmod 2$$

for $i \in \{0, \ldots, 2^{d'-1} - 1\}$. The shift and XOR operation in line 5 computes the sums $y_{i+(2j)2^{d'-1}}^{(d'-1,k)} \oplus y_{i+(2j+1)2^{d'-1}}^{(d'-1,k)}$. Thus after line 5 the word y satisfies

$$\bigoplus_{j=0}^{2^{d-d'}} y_{i+j2^{d'}} = SADD(w_{k+i}) \quad \bmod 2$$

for $i \in \{0, \ldots, 2^{d'-1} - 1\}$.

Similarly we process the word y' in the lines 6 and 7 and we have

$$\bigoplus_{j=0}^{2^{d-d'}} y'_{i+2^{d'-1}+j2^{d'}} = SADD(w_{(k+2^{d-1})+i}) \quad \bmod 2$$

for $i \in \{0, \ldots, 2^{d'-1} - 1\}$. (That is, we use a left shift in line 7 instead of a right shift, resulting in the $+2^{d-1}$ term in the index of y'.)

Finally we use in line 8 the mask μ_{d-1} (see Sect. 12.1.1) to select the right bits from words w and w'. $\qquad\square$

A problem is how to find the right input words w_i for Algorithm 2.8. One possibility is to pre-compute feedback polynomials $f_0, \ldots, f_{s-1}$ in Algorithm 2.7. This

---

Algorithm 2.9 LFSR update with parallel sideway addition mod 2

Require: x is the internal state of the LFSR
Ensure: y is the next s bits in the LFSR sequence
1: **for** $i \leftarrow 0$ **to** $n-1$ **do**
2:    $w_i \leftarrow$ byte-wise XOR of $(x \gg i) \& f$
3: **end for**
4: Compute y with $y_i = \text{SADD}(w_i) \mod 2$
5: $s \leftarrow 0, z \leftarrow y$
6: **for** $i \leftarrow 1$ **to** $n-1$ **do**
7:    $s \leftarrow s \oplus f[n-i+1] = 1$
8:    $y \leftarrow y \oplus (z \gg i)$
9: **end for**

Table 2.2 Speed of different LFSR implementations (128 bit LFSR)

Bitwise generation	74.7 Mbits/sec
bytewise generation (table look-up)	666 Mbits/sec
bytewise generation (PSADD)	84.4 Mbits/sec

Table 2.3 Speed of an LFSR with feedback polynomial $z^{127} + z + 1$

Generic bitwise generation	74.7 Mbits/sec
trinomial bitwise generation	132 Mbits/sec
Algorithm 2.10	15300 Mbits/sec

needs ns bits in a look-up table, but with a few extra operations we can avoid storing the extra polynomials $f_1, \ldots, f_{s-1}$.

We can use Algorithm 2.8 to compute the next 2^k bits in the LFSR sequence as follows. The internal state of our LFSR is stored in the bit field x.

The idea of Algorithm 2.9 is that $SADD((X \gg i) \& f) \mod 2$ is almost x_{n+i}. The only thing that is missing is $x_n f_{n-i+1} \oplus \cdots \oplus x_{n+i-1} f_{n-1}$. The loop in lines 6–9 computes this correction term.

All of the above implementations were designed for arbitrary feedback polynomials $f(z) = z^n - \sum_{j=0}^{n} c_j z^j$. However, if we choose a feedback polynomial with few coefficients, with the additional property that $f(z) - z^n$ has a low degree, we can obtain a very fast implementation. This is especially true if we use a trinomial $f(z) = z^n + z^k + 1$ as a feedback polynomial.

Algorithm 2.10 describes how we can compute the next $n - k$ bits of the LFSR sequence. The internal state of the LFSR is denoted by x.

Such a special algorithm is of course much faster than the generic algorithms. However, feedback polynomials of low weight do not only help to speed up the implementation of an LFSR, there are also special attacks against stream ciphers based on these LFSRs (see Sect. 4.3). One should keep this in mind when designing an LFSR-based stream cipher. Most often, the extra speed up of an LFSR with a sparse feedback polynomial is not worth the risk of a special attack.

Algorithm 2.10 Generating an LFSR sequence with the feedback polynomial $z^n + z^k + 1$

1: $y \leftarrow ((x \gg k) \oplus x) \& (2^k - 1)$
2: **output** y
3: $x \leftarrow (x \gg k) | (y \ll k)$

We close this section with some timings for the implementation of our algorithms. All programs run on a single core of a 32-bit Intel Centrino Duo with 1.66 Gz.

Chapter 3
Non-linear Combinations of LFSRs

3.1 De Bruijn Sequences

We have seen in the previous chapter that linear feedback shift register sequences have some very desirable statistical properties, but provide no security against cryptographic attacks.

One approach to this problem is to allow non-linear feedback functions in the shift registers. We can describe a general shift register sequence by a graph of 2^n vertices labeled by the words of $\{0, 1\}^n$. The edges show possible transitions of a shift register, i.e. we have an edge from the vertex $(a_0, \ldots, a_{n-1})$ to $(a_1, \ldots, a_{n-1}, 0)$ and to $(a_1, \ldots, a_{n-1}, 1)$. These graphs are now named after de Bruijn for his work in [73], but they also appeared at the same time in the work of Good [117]. Figure 3.1 shows the smallest de Bruijn graphs.

We obtain the state diagram of a non-linear feedback shift register out of the corresponding de Bruijn graph by selecting for every vertex one of the two outgoing edges.

As in the case of linear shift register sequences, the sequences of maximal period are of special interest and furthermore we can prove that a sequence of maximal period will satisfy the second Golomb axiom (see Theorem 2.3). Since a non-linear shift register can use the state $0, \ldots, 0$ we can even prove that every n-tuple has to occur as a subsequence. For the following we take this property as a definition.

Definition 3.1 A *de Bruijn sequence* of order n is a periodic sequence with period length 2^n which contains every n-tuple as a subsequence.

The following lemma connects the de Bruijn sequences with the de Bruijn graphs.

Lemma 3.1 *A de Bruijn sequence of order n corresponds to a Hamiltonian cycle (a cycle which visits every vertex exactly once) in the de Bruijn graph D_n and to an Eulerian cycle (a cycle which visits every edge exactly once) in the de Bruijn graph D_{n-1}.*

A. Klein, *Stream Ciphers*, DOI 10.1007/978-1-4471-5079-4_3,
© Springer-Verlag London 2013

Fig. 3.1 The smallest de
Bruijn graphs

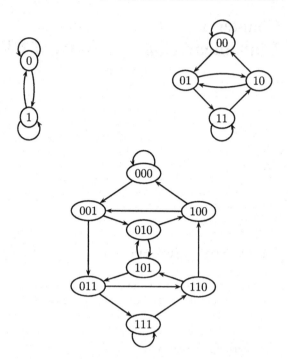

Proof We can map any sequence to a path in a de Bruijn Graph and conversely a
path in a de Bruijn Graph defines a binary sequence.

By definition, a period of a de Bruijn sequence contains every n-tuple exactly
once as a subsequence. So if we map it to the de Bruijn graph D_n we visit every
vertex $(a_1, \ldots, a_n)$ exactly once, i.e. we have a Hamiltonian cycle. And in the opposite direction, a Hamiltonian cycle defines a periodic sequence which contains every
n-tuple exactly once as a subsequence, i.e. a de Bruijn sequence.

If we map the de Bruijn sequence to D_{n-1} we visit every edge

$$(a_1, \ldots, a_{n-1}) \to (a_2, \ldots, a_n)$$

exactly once, i.e. we have an Eulerian cycle.   □

It is easy to prove the existence of de Bruijn sequences. We already know that
an m-sequence runs through all vertices of the de Bruijn graph except for the vertex $0 \ldots 0$. The only thing we have to do to obtain a de Bruijn sequence from
an m-sequence is to replace the subsequence of $n - 1$ zeros by a subsequence
of n zeros, i.e. we replace the transition $10 \ldots 0 \to 0 \ldots 01$ with the transitions
$10 \ldots 0 \to 0 \ldots 0 \to 0 \ldots 01$.

Now we want to count the number of de Bruijn sequences. This was done by
de Bruijn in 1946 (see [73]), however the problem of enumerating the de Bruijn
sequences had already been solved in 1894 by Fly Sainte-Marie [231] as a solution
to a problem proposed by de Rivière [223] (see [74]).

Theorem 3.2 *The number of de Bruijn sequences of order n is $2^{2^{n-1}}$.*

By Lemma 3.1 we must count the number of Eulerian cycles in the de Bruijn graph D_{n-1}. We will use the following theorem to do this.

Let us first recall some basic notions from graph theory. A *directed graph* consists of a *vertex set* V and an *edge set* $E \subset V \times V$. The edge (a, b) is said to start at a and end at b. The graph is *balanced* if for every vertex v the number of edges starting at v is equal to the number of edges ending at v. The graph is *connected* if for every two vertices $v \neq v'$, there exists a path $v_0 = v, v_1, \ldots, v_{n-1}, v_n = v'$ such that (v_{i-1}, v_i) is an edge for $i = 1, \ldots, n$. A *tree* is an unorientated connected graph which contains no cycle. In a orientated tree in addition a vertex is distinguished as root an all edges points towards the root. A vertex u' of a tree with root lies above a vertex u if the path from u' to the root contains u. A *spanning subtree* of G is a subgraph of G which is a tree and contains all vertices of G.

Theorem 3.3 (Aardenne-Ehrenfest, de Bruijn [1]) *Let G be a connected, balanced, directed graph with vertex set V. Let $e = (v, w)$ be an edge and let $\tau(G, v)$ denote the number of oriented spanning subtrees of G with root v. Then the number $\epsilon(G, e)$ of Eulerian tours of G starting at the edge e is*

$$\epsilon(G, e) = \tau(G, v) \prod_{u \in V} (\text{outdeg}(u) - 1)!$$

Proof Let $E = e_1, \ldots, e_q$ be an Eulerian tour in G. For each vertex $u \neq v$ we denote by $e(u)$ the last exit from u, i.e. $e(u) = e_i$ with $\text{init}(e_i) = u$ and $\text{init}(e_j) \neq u$ for $j > i$.

We claim that the edges $e(u)$, $u \in V \setminus \{v\}$, form an oriented spanning subtree T of G with root v. To prove this, it suffices to note the following.

1. For every vertex u of $V \setminus \{v\}$ we selected exactly one leaving edge.
2. T contains no cycle, since if we have a path from u to u' in T, the last exit from u' must occur after the last exit from u.

Both observations together imply that T must be an oriented tree with root v.

Thus we can associate with every Eulerian tour starting at e a directed tree with vertex v.

In the converse direction, we start with a directed tree T with vertex v and construct an Eulerian tour in which the edges of T are the last exits. The construction is simple. In the first step we walk along the edge e. In each of the following steps we choose any possible edge to continue the tour, the only restriction being that we never choose an edge in T unless there is no other possible alternative. Since G is a balanced graph our tour can only stop at v, at which point all edges of v will have been used.

We claim that no matter which choices are made, this tour must be Eulerian, i.e. we must use all edges of G. Assume otherwise. Then there are unused edges. If $u \neq v$ is a vertex and if we haven't used all outgoing edges from u, then we haven't used the edge $e(u) \in T$, because such exit edges are saved for last. Let $e(u) = (u, u')$. We haven't used all incoming edges of u' in our tour, so we haven't used all outgoing edges of u'. This proves that every vertex of T above u which follows u have not yet been used. In particular we haven't used all outgoing edges of v. But then the tour is not finished, a contradiction.

Thus we have associated with each Eulerian tour the tree of last exits. Conversely we have shown how to obtain all Eulerian tours from a given last exit tree. For each tree and each vertex u of G there are $(\text{outdeg}(u) - 1)!$ choices to be made, i.e. there are $\prod_{u \in V} (\text{outdeg}(u) - 1)!$ Eulerian tours for each possible last exit tree. □

To use Theorem 3.3 we need a formula for $\tau(G, v_k)$. Theorem 3.4 connects $\tau(G, v_k)$ with the Laplacian matrix of G.

For a graph G with vertex set $V = \{v_1, \ldots, v_m\}$ the *Laplacian matrix* L is the $n \times n$ matrix with

$$L_{i,j} = \begin{cases} -m_{i,j} & \text{if } i \neq j \text{ and there are } m_{i,j} \text{ edges from } v_i \text{ to } v_j, \\ \text{outdeg}(v_i) - m_{i,i} & \text{if } i = j \text{ and there are } m_{i,i} \text{ loops from } v_i \text{ to } v_i. \end{cases}$$

If all vertices have the same out-degree d, then $L = dI_n - A$, where $A = (m_{i,j})_{1 \leq i, j \leq n}$ is the *adjacency matrix* of G.

Theorem 3.4 (see [267], Theorem 3.6) *Let G be a directed graph with vertex set $V = \{v_1, \ldots, v_n\}$. Let L be the Laplacian matrix of G and let L_0 be the matrix L with the k-th row and k-th column deleted.*

Then

$$\tau(G, v_k) = \det L_0.$$

If, furthermore, G is balanced and $\mu_1, \ldots, \mu_{n-1}, \mu_n = 0$ are the eigenvalues of L, then

$$\tau(G, v_k) = \frac{1}{n} \mu_1 \cdots \mu_{n-1}.$$

Proof We prove $\tau(G, v_k) = \det L_0$ by induction on the number of edges of G.

First note that if G is disconnected, $\tau(G, v_k) = 0$, G_1 is the component containing v_k and G_2 is the rest of the graph, then we have $\det L_0 = \det L_0(G_1) \det L(G_2) = 0$, since the Laplacian matrix $L(G_2)$ has eigenvalue 0. So we turn our attention to connected graphs.

For connected graphs we prove the theorem by induction on the number of edges of G. The smallest number of edges of G is $n - 1$, since G is connected and in this case the undirected graph corresponding to G is a tree. If G is not an oriented tree with root v_k, then there exists a vertex v_j with out-degree 0. In this case, L_0 contains a zero row, i.e. $\det L_0 = 0 = \tau(G, v_k)$.

If, on the other hand, G is an oriented tree with root v_k (and hence $\tau(G, v_k) = 1$), then there is an ordering of the vertex set $V \setminus \{v_k\}$ such that L_0 is an upper triangular matrix with ones on the main diagonal, i.e. $\det L_0 = 1 = \tau(G, v_k)$.

Now suppose that G has $m > n - 1$ edges and that the theorem holds for all graphs with at most $m - 1$ edges. We can further assume that G has no edge starting at v_k, since such an edge neither contributes to $\tau(G, v_k)$ nor to L_0. Since G has at least n edges, there must be a vertex $v_j \neq v_k$ with out-degree at least 2. Let e be one of these edges and G_1 be the graph with e removed. Let G_2 be the graph where all the edges starting from v_j, except e, have been removed.

By induction, $\det L_0(G_1) = \tau(G_1, v_k)$ and $\det L_0(G_2) = \tau(G_2, v_k)$.

Since an oriented tree with root v_k contains exactly one edge starting at v_j, we have $\tau(G, v_k) = \tau(G_1, v_k) + \tau(G_1, v_k)$. On the other hand, the multi-linearity of the determinant implies $\det L_0(G) = \det L_0(G_1) + \det L_0(G_2)$. So by induction $\det L_0 = \tau(G, v_{,k})$.

Now we prove $\det L_0 = \frac{1}{n} \mu_1 \cdots \mu_{n-1}$ for a balanced digraph G. Note first that for a balanced digraph the Laplacian matrix L has the property that every row sum and every column sum is 0.

We compute the characteristic polynomial $\chi(L) = \det(L - x I_n)$. Add all rows except the last row to the last row. This operation does not change the determinant and the last row is now $(-x, \ldots, -x)$, since the column sums of L are 0. We can put the $-x$ before the determinant. Now add all columns except the last column to the last column. The determinant is not changed and, since the row sums of L are 0, we have the transformed the last column to $(-x, \ldots, -x, n)^t$. Developing the determinant by the last column shows that the coefficient of x in $\chi(L) = \det(L - x I_n)$ is $-n \det L_0$.

On the other hand $\chi(L) = \prod_{j=1}^{n} (\mu_j - x)$ and hence the coefficient of x is $-\sum_{j=1}^{n} \prod_{k \neq j} \mu_k = -\prod_{j=1}^{n-1} \mu_j$, which proves the second part of the theorem.   □

Now we are ready to count the de Bruijn sequences.

Proof of Theorem 3.2 Let A be the adjacency matrix of the de Bruijn graph D_{n-1}, i.e. let $A = (a_{ij})$ with

$$a_{ij} = \begin{cases} 1 & \text{if } (V_i, v_j) \text{ is an edge,} \\ 0 & \text{otherwise.} \end{cases}$$

We will determine the eigenvalues of A.

Note that two vertices $(a_1, \ldots, a_{n-1})$ and $(b_1, \ldots, b_{n-1})$ of the de Bruijn graph D_{n-1} are connected by exactly one path of length $n - 1$, namely the path $(a_1, \ldots, a_{n-1})$, $(a_2, \ldots, a_{n-2} b_1)$, $\ldots$, $(a_{n-1}, b_1, \ldots, b_{n-2})$, $(b_1, \ldots, b_{n-1})$. Thus $A^{n-1} = J$, where J is the $2^{n-1} \times 2^{n-1}$ matrix with all entries 1. As J has eigenvalue 0 with multiplicity $2^{n-1} - 1$ and the eigenvalue 2^{n-1} once, the eigenvalues of A must be 0 ($2^{n-1} - 1$ times) and 2λ (once) where λ is a 2^{n-1}th root of unity.

Since D_{n-1} has two loops, the trace of A is 2, and hence $\lambda = 1$.

Thus $L_0 = 2I - A$ has the eigenvalues 2 ($2^{n-1} - 1$ times) and 0 (once).

Fig. 3.2 The Geffe generator

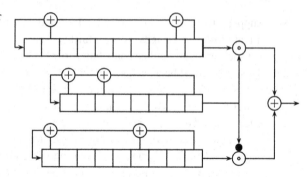

By Theorem 3.4

$$\tau(D_{n-1}, v_k) = \frac{1}{2^{n-1}} \prod_{j=1}^{2^{n-1}-1} 2 = 2^{2^{n-1}-n}$$

which is, by Theorem 3.3, the number of Eulerian cycles in the de Bruijn Graph D_{n-1}. Each Eulerian cycle leads to 2^n possible de Bruijn sequences (with different starting points).   □

3.2 A Simple Example of a Non-linear Combination of LFSRs

In the previous section we have learned about de Bruijn sequences which have some desirable statistical properties, but we don't have a fast algorithm to generate such sequences (see also Exercise 17.9).

An approach that retains the best parts of LFSRs (statistical properties, fast generation, simple design) but removes their cryptographic weakness is to use a non-linear combination of linear feedback shift registers. This technique yields very good results and many stream ciphers are based on this idea. For the remaining part of the chapter we will only consider stream ciphers of this type.

A very simple generator of this kind was described by Geffe [104] (see Fig. 3.2).

A *Geffe generator* consists of three linear feedback shift registers. The output of one of the three registers is used to decide which of the other two LFSRs is to be used. In Fig. 3.2 we see that the output of the first LFSR will be the output of the Geffe generator if the second LFSR produces a 1. If the second LFSR produces a 0 we will see the output of the third LFSR as the output of the Geffe generator.

The Geffe generator exhibits many interesting ideas. Its output sequence is a mixture of two LFSR sequences and it will inherit most of their statistical properties. The non-linear elements will guarantee a high linear complexity of the resulting sequence. (In the case of Fig. 3.2, the linear complexity is 161, as we will prove later in this chapter.)

However, the Geffe generator fails to produce a secure keystream, as we will see in the next section. In this chapter we will first learn the basic attack strategies against stream ciphers based on non-linear combinations of LFSRs and then we will learn how to select effective non-linear functions for the combination which can avoid these attacks.

3.3 Different Attack Classes

3.3.1 Time-Memory Trade-off Attacks

The simplest possible attack against any cipher is a complete search over all possible keys. If stream ciphers had no specialities at this point, this fact would not have been worth mentioning.

First we can search the internal state instead of the key. This limits the effective key size by the internal state. Further, there exists a time-memory trade-off, apparently first mentioned by Golic [109].

Assume that the cipher has 2^s internal states. We generate a list of 2^a random states and the next few bits generated by the cipher form these states. Once we have the list, we can attack the cipher by searching in the output sequence for a subsequence that corresponds to one of the states in our list. If we have an output sequence of length 2^b and $a + b > s$, it is very probable that we can find a good subsequence.

Thus the internal state of the cipher must be significantly larger than the key size. The cipher A5/1 (see Sect. 8.3) has only 2^{64} internal states, which makes it vulnerable against this attack.

3.3.2 Algebraic Attacks

Let $(x_i)_{i \in \mathbb{N}}$ be the sequence generated by the Geffe generator and let $(a_i)_{i \in \mathbb{N}}$, $(b_i)_{i \in \mathbb{N}}$ and $(c_i)_{i \in \mathbb{N}}$ be the corresponding LFSR sequences. Then we can express the plugging diagram of Fig. 3.2 by

$$x_i = a_i b_i + c_i (b_i + 1)$$

where all operations are performed over $\mathbb{F}_2$.

With the representation of an LFSR sequence as a linear code (see Sect. 2.2.4, Eq. (2.18)) we can write

$$a_i = \sum_{j=0}^{L_a - 1} \alpha_{j,i} a_i, \qquad b_i = \sum_{j=0}^{L_b - 1} \beta_{j,i} b_i, \qquad c_i = \sum_{j=0}^{L_c - 1} \gamma_{j,i} c_i$$

with known coefficients $\alpha_{j,i}$, $\beta_{j,i}$ and $\gamma_{j,i}$. Thus we can express x_i as a non-linear function of the unknown initial values $a_0, \ldots, a_{L_a}, b_0, \ldots, b_{L_b}, c_0, \ldots, c_{L_c}$.

This lead us to the so-called algebraic attacks. The attacker tries to solve a system of non-linear equations.

The basic approach to solve such systems is to replace all products of variables by a new artificial variable and then solve the resulting system of linear equations.

A stream cipher is strong against algebraic attacks if the non-linear function used in the combination has a high degree. To be precise, the smallest degree of an annihilator of the combination function must be large (see Sect. 6.2).

We will deal with algebraic attacks in detail in Chap. 6.

3.3.3 Correlation Attacks

Correlation attacks follow a divide and conquer principle.

In the case of the Geffe Generator we observe the following: The output of the Geffe generator is always equal to the output of the first shift register if the second LFSR outputs 1, but if the second LFSR outputs 0, we still have a 1/2 chance that the third LFSR, and therefore the Geffe generator, will produce the same output as the first LFSR. Thus

$$P(x_i = a_i) \approx \frac{3}{4}.$$

An attacker can use this correlation in the following way: Instead of searching over all $2^{L_A+L_B+L_C}$ possible internal states he enumerates only all 2^{L_A} internal states of the first LFSR and checks if the corresponding sequence (a_i) is correlated to the output of the Geffe generator. Once he knows a part of the internal state, recovering the remaining parts is easy.

We will treat such attacks in Chap. 4. Later in this chapter we will define the notion of a k-correlation immune function as a measure of resistance against these attacks.

3.4 Non-linear Combinations of Several LFSR Sequences

In this section we study the combination of LFSR sequences in general. Thus let $(a_i^{(j)})_{i \in \mathbb{N}}$, $(1 \leq j \leq k)$, be different LFSR sequences and let $C : \{0, 1\}^k \to \{0, 1\}$ be a Boolean function. We define a new sequence $(b_i)_{i \in \mathbb{N}}$ by

$$b_i = C(a_i^{(1)}, \ldots, a_i^{(k)})$$

for all $i \in \mathbb{N}$.

We have to ask ourselves: Can we give necessary and sufficient criteria for the sequence $(b_i)_{i \in \mathbb{N}}$ to be good for cryptographic purposes?

We start by determining the linear complexity of the sequence $(b_i)_{i \in \mathbb{N}}$. Let L_j be the linear complexity of $(a_i^{(j)})_{i \in \mathbb{N}}$ and let f_i denote the corresponding feedback polynomial.

For the case $k = 2$ and $C(a^{(1)}, a^{(2)}) = a^{(1)} + a^{(2)}$ we solved this problem in Theorem 2.9. The next basic function we have to consider is multiplication.

3.4.1 The Product of Two LFSRs

For this section, let $k = 2$ and $C(a^{(1)}, a^{(2)}) = a^{(1)}a^{(2)}$. Note that this combiner is it-self not suitable for cryptographic purposes. If $a^{(1)}$ and $a^{(2)}$ are independent random values with $P(a^{(1)} = 1) = P(a^{(2)} = 1) = \frac{1}{2}$, the output $C(a^{(1)}, a^{(2)})$ will take the value 1 only with probability $\frac{1}{4}$. This is clearly unacceptable for a key stream. We study multiplication only because it is needed as basic operation in more complex combiners.

We are especially interested in the cases with the highest possible linear complexity. The following theorem characterizes these cases.

Theorem 3.5 *Let* $(a_i^{(1)})_{i \in \mathbb{N}}$ *and* $(a_i^{(2)})_{i \in \mathbb{N}}$ *be two LFSR sequences. Let* f_1 *and* f_2 *be the feedback polynomials of the sequences and let* $L_i = \deg f_i$. *The linear complexity* L *of the sequence* $(b_i)_{i \in \mathbb{N}} = (a_i^{(1)} a_i^{(2)})_{i \in \mathbb{N}}$ *satisfies*

$$L \leq L_1 L_2 \tag{3.1}$$

with equality if and only if at most one of the polynomials f_1 *and* f_2 *has a multiple root and* $\zeta_1 \zeta_2 = \zeta_1' \zeta_2'$ *for zeros* ζ_1, ζ_1' *of* f_1 *and zeros* ζ_2, ζ_2' *of* f_2 *implies* $\zeta_1 = \zeta_1'$ *and* $\zeta_2 = \zeta_2'$.

Proof We use the closed formula (2.13) to express $a_i^{(1)}$ and $a_i^{(2)}$, i.e. we have

$$a_i^{(l)} = \sum_{j=0}^{n_1^{(l)}} b_{j,1}^{(l)} \left(\zeta_j^{(l)}\right)^i + \sum_{j=0}^{n_2^{(l)}} b_{j,2}^{(l)} i \left(\zeta_j^{(l)}\right)^i + \cdots + \sum_{j=0}^{n_{k_l}^{(l)}} b_{j,k}^{(l)} i^{k-1} \left(\zeta_j^{(l)}\right)^i \quad \text{for } l = 1, 2$$

with $n_1^{(l)} \geq n_2^{(l)} \geq \cdots \geq n_{k_i}^{(l)}$. The $\zeta_j^{(l)}$ are the roots of the polynomial $f^{(l)*}$, i.e.

$$f^{(l)} = \prod_{k'=1}^{k_l} \prod_{j=1}^{n_{k'}} \left(1 - z\zeta_j^{(l)}\right) = \prod_{j=1}^{n_1^{(i)}} \left(1 - z\zeta_j^{(l)}\right)^{e_j^{(l)}}$$

where $e_j^l = \max\{i \mid n_i^{(l)} \geq j\}$ is the multiplicity of the root $\zeta_j^{(l)}$ of $f^{(l)}$.

Multiplying the closed formulas for $a_i^{(1)}$ and $a_i^{(2)}$ we get

$$a_i^{(1)} a_i^{(2)} = \sum_{j=0}^{n_1^l} \sum_{k=0}^{n_1^2} P_{j,k}(i) \left(\zeta_j^{(1)} \zeta_k^{(2)}\right)^i$$

where $p_{j,k}$ is a polynomial of degree $(e_j^{(i)} - 1) + (e_k^{(2)} - 1)$. From the closed form we see that the feedback polynomial of an LFSR generating $(a_i^{(1)} a_i^{(2)})_{i \in \mathbb{N}}$ is

$$f(z) = \prod_{j=1}^{n_1} \left(1 - z\zeta_j^{(1)}\zeta_k^{(2)}\right)^{(e_j^{(1)} + e_k^{(2)} - 1)}.$$

If all products $\zeta_j^{(1)}\zeta_j^{(2)}$ are different then $f(z)$ is also the feedback polynomial of the minimal LFSR generating $(a_i^{(1)} a_i^{(2)})_{i \in \mathbb{N}}$.

Since $\sum_{j=1}^{n_1^{(1)}} \sum_{k=1}^{n_1^{(2)}} (e_j^{(1)} + e_k^2 - 1) \le \sum_{j=1}^{n_1^{(1)}} e_j^1 \sum_{k=1}^{n_1^{(2)}} e_k^2 = L_1 L_2$ with equality if and only if either all $e_j^{(1)} = 1$ or all $e_k^{(2)} = 1$, this is the desired bound for the linear complexity of $(a_i^{(1)} a_i^{(2)})_{i \in \mathbb{N}}$.   □

If the feedback polynomials f_1 and f_2 are irreducible or primitive we can simplify Theorem 3.5.

Corollary 3.6 Let $(a_i^{(1)})_{i \in \mathbb{N}} \in \mathbb{F}^{\mathbb{N}}$ and $(a_i^{(2)})_{i \in \mathbb{N}} \mathbb{F}^{\mathbb{N}}$ be two LFSR sequences of linear complexity L_1 and L_2, respectively. Let L be the linear complexity of the sequence $(b_i)_{i \in \mathbb{N}} = (a_i^{(1)} a_i^{(2)})_{i \in \mathbb{N}}$. Then $L = L_1 L_2$

(a) if $\gcd(L_1, L_2) = 1$ and the feedback polynomials f_1 and f_2 are irreducible, separable and do not have roots $\zeta \ne \zeta'$ with $\zeta \zeta'^{-1} \in \mathbb{F}$; or
(b) if $1 < L_1 < L_2$ and the feedback polynomials are primitive.

Proof

(a) The condition $\gcd(L_1, L_2) = 1$ implies that $\mathbb{F} = \mathbb{F}(f_1) \cap \mathbb{F}(f_2)$. Hence $\zeta_1 \zeta_2 = \zeta_1' \zeta_2'$ with roots ζ_1, ζ_1' of f_1 and roots ζ_2, ζ_2' of f_2 implies $\zeta_1 \zeta_1'^{-1} = \zeta_2' \zeta_2^{-1} \in \mathbb{F}(f_1) \cap \mathbb{F}(f_2) = \mathbb{F}$. By assumption this implies $\zeta_1 = \zeta_1'$ and $\zeta_2 = \zeta_2'$. Thus the conditions of Theorem 3.5 hold and $L = L_1 L_2$.
(b) Let $\mathbb{F} = \mathbb{F}_q$, $n = \text{lcm}(L_1, L_2)$ and let ζ be a primitive element of $\mathbb{F}_{q^n}$. Then $\zeta^{(q^n-1)/(q^{L_1}-1)}$ is a primitive element of $\mathbb{F}_{q^{L_1}}$ and $\zeta^{(q^n-1)/(q^{L_2}-1)}$ is a primitive element of $\mathbb{F}_{q^{L_2}}$.

The zeros of the primitive polynomial f_1 are of the form $\zeta^{(q^n-1)/(q^{L_1}-1)a_1 q^x}$ with $x \in \{0, \dots, L_1 - 1\}$ and a_1 is some constant relatively prime to $q^n - 1$. Similarly the zeros of f_2 are of the form $\zeta^{(q^n-1)/(q^{L_2}-1)a_2 q^y}$ with $y \in \{0, \dots, L_2 - 1\}$ and a_2 is some constant relatively prime to $q^n - 1$.

The existence of two different zeros ζ_1, ζ_1' of f_1 and two different zeros ζ_2, ζ_2' with $\zeta_1 \zeta_1'^{-1} = \zeta_2^{-1} \zeta_2'$ is equivalent to

$$\zeta^{(q^n-1)/(q^{L_1}-1)a_1 q^x(q^{x'}-1)} = \zeta^{\pm(q^n-1)/(q^{L_2}-1)a_2 q^y(q^{y'}-1)} \qquad (3.2)$$

with $x, x' < L_1$ and $y, y' < L_2$.

Since ζ is a generator of the cyclic group $\mathbb{F}_q^\times$, we get

$$\frac{q^n - 1}{q^{L_1} - 1} a_1 q^x \left(q^{x'} - 1\right) \equiv \frac{q^n - 1}{q^{L_2} - 1} a_2 q^y \left(q^{y'} - 1\right) \quad \mathrm{mod}\ q^n - 1.$$

Multiplication by $q^{L_1} - 1$ and the inverse element of $a_2 q^y$ yields

$$0 \equiv \frac{q^n - 1}{q^{L_2} - 1} \left(q^{L_1} - 1\right)\left(q^{y'} - 1\right) \quad \mathrm{mod}\ q^n - 1$$

and hence

$$q^{L_2} - 1 \mid \left(q^{L_1} - 1\right)\left(q^{y'} - 1\right).$$

But this is false since

$$r = \left(q^{L_1} - 1\right)\left(q^{y'} - 1\right) - q^{L_1 + y' - L_2}\left(q^{L_2} - 1\right) = q^{L_1 + y' - L_2} + 1 - q^{L_1} - q^{y'}$$

lies between $q^{L_2} - 1$ and $-(q^{L_2} - 1)$ and is non-zero.

Thus we have verified the conditions of Theorem 3.5, i.e. $L = L_1 L_2$. $\square$

Theorem 3.5 suffices if we are only interested in the linear complexity of the sequence, but sometimes we also want the feedback polynomial of the resulting sequence. The proof of Theorem 3.5 is constructive in the sense that we have given a formula for the feedback polynomial of the sequence $(b_i)_{i \in \mathbb{N}}$. The disadvantage of that formula is that we have to compute the roots of the polynomials f_1 and f_2. Hence we must leave $\mathbb{F}_q$ and work in a bigger field $\mathbb{F}_{q^n}$. This is computationally very expensive, so it is desirable to find a way to do the computation in the ground field $\mathbb{F}_q$.

The right tool for this is the resultant. Let us recall the basic properties of resultants.

Let $f(z) = a_n z^n + \cdots + a_0 = a_n(z - \alpha_1) \cdots (z - \alpha_n)$ and $g(z) = b_m z^m + \cdots + b_0 = b_m(z - \beta_1) \cdots (z - \beta_m)$ be two polynomials. Then the *resultant* of $f(z)$ and $g(z)$ is

$$\mathrm{Res}\big(f(z), g(z)\big) = \begin{vmatrix} a_n & a_{n-1} & \cdots & a_0 & & & \\ & a_n & a_{n-1} & \cdots & a_0 & & \\ & & \ddots & \ddots & & \ddots & \\ & & & a_n & a_{n-1} & \cdots & a_0 \\ b_m & b_{m-1} & \cdots & b_0 & & & \\ & b_m & b_{m-1} & \cdots & b_0 & & \\ & & \ddots & \ddots & & \ddots & \\ & & & b_n & b_{m-1} & \cdots & b_0 \end{vmatrix}$$

$$= a_n^m b_m^n \prod_{i=0}^{n} \prod_{j=0}^{m} (\alpha_i - \beta_j).$$

Theorem 3.7 describes how to use the resultant to compute the feedback polynomial of a product of two LFSRs.

Theorem 3.7 *Let $(a_i^{(1)})_{i \in \mathbb{N}}$ and $(a_i^{(2)})_{i \in \mathbb{N}}$ be two (binary) LFSR sequences. Let f_1 and f_2 be the feedback polynomials of the sequences. Let $(b_i)_{i \in \mathbb{N}} = (a_i^{(1)} a_i^{(2)})_{i \in \mathbb{N}}$.*
Then

$$f(z) = \text{Res}_x\left(f_1(x), f_2^*(zx)\right) \tag{3.3}$$

is the feedback polynomial of an LFSR generating $(b_i)_{i \in \mathbb{N}}$. (Here $f_2^*(x) = f_2(1/x)x^{d_2}$ denotes the reciprocal polynomial to $f_2(x)$—see also Definition 2.4.) Under the condition given in Theorem 3.5, this is the feedback polynomial of the minimal LFSR generating $(b_i)_{i \in \mathbb{N}}$.

Proof Let $f_1(z) = \prod_{j=1}^{L_1}(1 - z\zeta_j^{(1)})$ and $f_2(z) = \prod_{j=1}^{L_2}(1 - z\zeta_j^{(2)})$. In the proof of Theorem 3.5 we saw that $f(z) = \prod_{j=1}^{L_1}\prod_{k=1}^{L_2}(1 - z\zeta_j^{(1)}\zeta_k^{(2)})$ is the feedback polynomial of an LFSR generating $(b_i)_{i \in \mathbb{N}}$.

The polynomial $f_1(x)$ has zeros $\frac{1}{\zeta_j^{(1)}}$ and leading coefficient $\prod_{j=1}^{L_1}\zeta_j^{(1)}$. The polynomial $f_2^*(x)$ has leading coefficient 1 and zeros $\zeta_k^{(2)}$ and hence $f_2^*(zx)$ has zeros $\frac{\zeta_k^{(2)}}{z}$ and leading coefficient z^{L_2} (as a polynomial in x).

Thus the resultant of $f_1(x)$ and $f_2^{(*)}(zx)$ is

$$\text{Res}_x\left(f_1(x), f_2^*(zx)\right) = \left(\prod_{j=1}^{L_1}\zeta_j^{(1)}\right)^{L_2} z^{L_2 L_1} \prod_{j=1}^{L_1}\prod_{k=1}^{L_2}\left(\frac{1}{\zeta_j^{(1)}} - \frac{\zeta_k^{(2)}}{z}\right)$$

$$= \prod_{j=1}^{L_1}\prod_{k=1}^{L_2}(z - \zeta_j^{(1)}\zeta_k^{(2)}),$$

i.e. it is the polynomial determined in the proof of Theorem 3.5.   □

3.4.2 General Combinations

With Theorem 2.9 and Theorem 3.5 we have covered the two basic operations that allow us to determine the linear complexity of arbitrary non-linear combinations of linear feedback shift registers. We summarize the results in Theorem 3.8 (see also [230]).

Theorem 3.8 *For a Boolean function $C : \mathbb{F}_2^n \to \mathbb{F}_2$ in algebraic normal form, let $\hat{C} : \mathbb{Z} \to \mathbb{Z}$ be the function we obtain if we replace the operations in $\mathbb{F}_2$ by operations in $\mathbb{Z}$.*

Let $(x_i^{(k)})_{i \in \mathbb{N}}$ *be LFSRs with linear complexity* L_i. *Then the linear complexity of the sequence* $C(x_i^{(1)}, \ldots, x_i^{(n)})$ *is at most* $\hat{C}(L_1, \ldots, L_n)$.

If, in addition, the feedback polynomials f_i *of the LFSR* $(x_i^{(k)})_{i \in \mathbb{N}}$ *are all primitive and of pairwise different degree, then the linear complexity of the sequence* $C(x_i^{(1)}, \ldots, x_i^{(n)})$ *is exactly* $\hat{C}(L_1, \ldots, L_n)$.

Proof The upper bound follows directly from the upper bound given in Theorem 2.9 and Theorem 3.5.

The proof that the upper bound is sharp for primitive feedback polynomials of pairwise different degree is a variation of the proof for two polynomials (Corollary 3.6).

Let $L = \mathrm{lcm}(L_1, \ldots, L_n)$ and let ζ be a primitive element of $\mathbb{F}_{q^L}$. Then $\zeta^{(q^L - 1)/(q^{L_i} - 1)}$ is a primitive element of $\mathbb{F}_{q^{L_i}}$ and the zeros of f_i have the form $\zeta^{(q^L - 1)/(q^{L_i} - 1)} a_i q^x$ for some constant a_i relatively prime to $q^L - 1$.

Now let $(\zeta_1, \ldots, \zeta_n) \neq (\zeta_1', \ldots, \zeta_n')$ where ζ_i and ζ_i' are either roots of f_i or 1. We have to prove that

$$\zeta_1 \cdots \zeta_n \neq \zeta_1' \cdots \zeta_n'.$$

Assume conversely that $\zeta_1 \cdots \zeta_n = \zeta_1' \cdots \zeta_n'$. Without loss of generality we can assume that $L_n > \cdots > L_1$ and let n' be the largest index with $\zeta_{n'} \neq \zeta_{n'}'$.

This leads to the equation

$$\zeta^{(q^L - 1)/(q^{L_1 - 1})a_1 q^{x_1} b_1} \cdots \zeta^{(q^L - 1)/(q^{L_{n'-1} - 1})a_1 q^{x_{n'-1}} b_{n'-1}}$$
$$= \zeta^{(q^L - 1)/(q^{L_{n'} - 1})a_{n'} q^{x_{n'}} b_{n'}} \tag{3.4}$$

where the factor b_i is either $\pm(q^{x_i'} - 1)$ for some $x_i' \in \{1, \ldots, L_i - 1\}$ if ζ_i and ζ_i' are different roots of f_i. If either ζ_i or ζ_i' is 1 then $b_i = \pm 1$ and $b_i = 0$ if $\zeta_i = \zeta_i'$.

Compare Eq. (3.4) to Eq. (3.2) in the proof of Corollary 3.6.

We continue as in the proof of Corollary 3.6 and obtain

$$(q^L - 1)/(q^{L_1} - 1)a_1 q^{x_1} b_1 + \cdots + (q^L - 1)/(q^{L_{n'-1}} - 1)a_{n'-1} q^{x_{n-1}} b_1$$
$$\equiv (q^L - 1)/(q^{L_{n'}} - 1)a_{n'} q^{x_1} b_{n'} \quad \mathrm{mod}\ q^L - 1,$$

since ζ is a generator of the cyclic group $\mathbb{F}_{q^L}$.

Multiplying by $(q^{L_1} - 1) \cdots (q^{L_{n'-1}} - 1)$ we get

$$0 \equiv (q^L - 1)/(q^{L_{n'}} - 1)a_n q^{x_1} b_n (q^{L_1} - 1) \cdots (q^{L_{n'-1}} - 1) \quad \mathrm{mod}\ q^L - 1$$

and hence

$$q^{L_{n'}} - 1 \mid b_n (q^{L_1} - 1) \cdots (q^{L_{n'-1}} - 1). \tag{3.5}$$

By Zsigmondy's theorem (see Sect. 13.4) we know that $q^{L_{n'}} - 1$ contains a prime factor not in $\prod_{j=1}^{L_{n'}-1} (q^j - 1)$ except for the cases $q = 2$, $L_{n'} = 6$ or where $q + 1$ is a power of 2 and $L_{n'} = 2$.

Fig. 3.3 A simple non-linear
filter

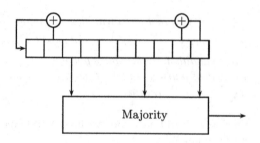

Equation (3.5) is obviously false if such a prime factor exists. In the case $L_{n'} = 2$, i.e. if only two factors exist, we have proved the result in Corollary 3.6.

So the only case left is $q = 2$, $L_{n'} = 6$, i.e. we have feedback polynomials of degree 2, 3 and 6. For this case we can check directly that the products of different roots are different.   □

3.5 Non-linear Filters

A special case of a non-linear combination of feedback shift registers is a non-linear filter. In this case we have just a single LFSR sequence $(a_i)_{i \in \mathbb{N}}$ and define a new sequence by $b_i = C(a_{i+t_0}, a_{i+t_1}, a_{i+t_{k-1}})$ for a non-linear function C and some constants $0 = t_0 < t_1 < \cdots < t_{k-1}$.

Figure 3.3 shows an example of a non-linear filter. For an example from a real world application, see Sect. 8.2.1.

The main advantage of non-linear filters is that against algebraic attacks they are almost as resistant as non-linear combinations of different LFSRs, but they use less internal memory. A disadvantage is that their statistical behavior is harder to analyze and so correlation attacks become more dangerous.

The upper bound of Theorem 3.8 of course still holds for non-linear filters, but we cannot expect it to be sharp. The following theorem improves the bound on the linear complexity of non-linear filters.

Theorem 3.9 *Let $(a_i)_{i \in \mathbb{N}} \in \mathbb{F}_2^{\mathbb{N}}$ be a binary LFSR sequence with primitive feedback polynomial of degree L over $\mathbb{F}_q$. Let C be a Boolean function of degree d in k variables.*

Then for $0 = t_0 < t_1 < \cdots < t_{k-1}$ the linear complexity of the sequence $b_i = C(a_{i+t_0}, a_{i+t_1}, a_{i+t_{k-1}})$ is bounded by

$$\mathcal{L}\big((b_i)_{i \in \mathbb{N}}\big) \le \sum_{j=1}^{d} \binom{L}{j}.$$

Proof Let $f(x) = \prod_{j=0}^{L-1}(x - \zeta^{2^j})$ be the feedback polynomial of the LFSR generating $(a_i)_{i \in \mathbb{N}}$.

By Theorem 3.5 the feedback polynomial of the product of d LFSR sequences with feedback polynomial f has zeros of the form $\zeta^{2^{t_1}+\cdots+2^{t_d}} = \zeta^a$, where a is a number which has at most d ones in its binary representation.

By Theorem 2.9, any linear combination of products of up to d LFSRs with feedback polynomial f can be generated by an LFSR with feedback polynomial

$$g(x) = \prod(x - \zeta^a)$$

where the product runs over all a which have at most d ones in its binary representation. The degree of g is $\sum_{j=1}^{d} \binom{L}{j}$. $\qquad\square$

For constructing strong ciphers, lower bounds are of greater interest than upper bounds. In the sequel we will derive such a lower bound.

We start with a lemma that allows us find roots of the minimal polynomial of the product of distinct phases of an LFSR sequence.

Lemma 3.10 *Let $(a_i)_{i\in\mathbb{N}}$ be a binary LFSR sequence generated by an LFSR of degree L with primitive feedback polynomial f and let ζ be a root of f.*

Let g be the minimal feedback polynomial of the sequence $(b_i)_{i\in\mathbb{N}}$ defined by

$$b_i = a_{i+t_0} \cdots a_{i+t_{k-1}}.$$

Let $0 < e < 2^L$ be a number whose binary representation needs k ones, i.e. $e = 2^{e_0} + \cdots + 2^{e_{k-1}}$ with $0 \le e_0 < e_1 < \cdots < e_{k-1} < L$.

Then ζ^e is a root of g if and only if the determinant

$$D_e = \begin{vmatrix} \zeta^{t_0 2^{e_0}} & \zeta^{t_1 2^{e_0}} & \cdots & \zeta^{t_{k-1} 2^{e_0}} \\ \zeta^{t_0 2^{e_1}} & \zeta^{t_1 2^{e_1}} & \cdots & \zeta^{t_{k-1} 2^{e_1}} \\ \vdots & \vdots & \ddots & \vdots \\ \zeta^{t_0 2^{e_{k-1}}} & \zeta^{t_1 2^{e_{k-1}}} & \cdots & \zeta^{t_{k-1} 2^{e_{k-1}}} \end{vmatrix}$$

is not equal to 0.

Proof Without loss of generality we may assume that

$$a_i = \mathrm{Tr}_{\mathbb{F}_{2^L}/\mathbb{F}_2}\left(\zeta^i\right)$$

(see Theorem 2.1).

Then

$$b_i = \prod_{j=0}^{k-1} a_{i+t_j}$$

$$= \prod_{j=0}^{k-1} \mathrm{Tr}_{\mathbb{F}_{2^L}/\mathbb{F}_2}\left(\zeta^{i+t_j}\right)$$

$$= \prod_{j=0}^{k-1} \left(\zeta^{i+t_j} + \zeta^{2(i+t_j)} + \cdots + \zeta^{2^{L-1}(i+t_j)}\right).$$

ζ^e is a root of the minimal feedback polynomial for the sequence $(b_i)_{i\in\mathbb{N}}$ if in the expansion $b_i = \sum_e A_e \zeta^{ei}$ the coefficient A_e of ζ^{ei} is non-zero.

Expanding the product and comparing the coefficients we get for $e = 2^{e_0} + \cdots + 2^{e_{k-1}}$, that

$$A_e = \sum_{\sigma \in S_n} \prod_{j=0}^{k-1} \zeta^{2^{e_j} t_{\sigma(j)}} = \mathrm{per}\begin{pmatrix} \zeta^{t_0 2^{e_0}} & \zeta^{t_1 2^{e_0}} & \cdots & \zeta^{t_{k-1} 2^{e_0}} \\ \zeta^{t_0 2^{e_1}} & \zeta^{t_1 2^{e_1}} & \cdots & \zeta^{t_{k-1} 2^{e_1}} \\ \vdots & \vdots & \ddots & \vdots \\ \zeta^{t_0 2^{e_{k-1}}} & \zeta^{t_1 2^{e_{k-1}}} & \cdots & \zeta^{t_{k-1} 2^{e_{k-1}}} \end{pmatrix}$$

where $\mathrm{per}(M)$ denotes the permanent of the matrix M.

For characteristic 2, addition and subtraction is the same, and hence the permanent and determinant of a matrix coincide, which completes the proof.   □

Corollary 3.11 *Let $(a_i)_{i\in\mathbb{N}}$ be a binary LFSR sequence generated by an LFSR of degree L and let*

$$b_i = a_i a_{i+j} \cdots a_{i+(k-1)j}.$$

Then $\mathcal{L}((b_i)_{i\in\mathbb{N}}) \geq \binom{L}{k}$.

Proof With the notation of Lemma 3.10 we must verify that

$$D_e = \begin{vmatrix} \zeta^{0\cdot j 2^{e_0}} & \zeta^{1\cdot j 2^{e_0}} & \cdots & \zeta^{(k-1)\cdot j t_{k-1} 2^{e_0}} \\ \zeta^{0\cdot j 2^{e_1}} & \zeta^{1\cdot j t_1 2^{e_1}} & \cdots & \zeta^{(k-1)\cdot j 2^{e_1}} \\ \vdots & \vdots & \ddots & \vdots \\ \zeta^{0\cdot j 2^{e_{k-1}}} & \zeta^{1\cdot j t_1 2^{e_{k-1}}} & \cdots & \zeta^{(k-1)\cdot j 2^{e_{k-1}}} \end{vmatrix}$$

is non-zero.

But this is a Vandermonde determinant, thus

$$D_e = \prod_{0 \leq t < t' < k} \left(\zeta^{j 2^{e_{t'}}} - \zeta^{j 2^{e_t}}\right) \neq 0.$$

Hence for every $e = 2^{e_0} + \cdots + 2^{e_{k-1}}$ with $0 \le e_0 < e_1 < \cdots < e_{k-1} < L$ the minimal polynomial of $(b_i)_{i \in \mathbb{N}}$ has ζ^e as a zero.   □

Corollary 3.12 (see [21]) *Let $(a_i)_{i \in \mathbb{N}} \in \mathbb{F}_2^{\mathbb{N}}$ be a binary LFSR sequence with primitive feedback polynomial of degree L.*

Let C be a Boolean function which contains the kth order term

$$a_i a_{i+j} \cdots a_{i+(k-1)j}$$

and an arbitrary number of terms of lower order.

Then the linear complexity of the sequence $b_i = C(a_i, a_{i+1}, a_{i+k})$ is bounded by

$$\mathcal{L}\big((b_i)_{i \in \mathbb{N}}\big) \ge \binom{L}{d}.$$

Proof Let ζ be a root of the feedback polynomial of the sequence $(a_i)_{i \in \mathbb{N}}$.

As we have seen in Corollary 3.11, ζ^e is a root of the feedback polynomial of the sequence $b_i' = a_i a_{i+j} \cdots a_{i+(k-1)j}$ if e has binary weight k.

Let $b_i'' = b_i - a_i a_{i+j} \cdots a_{i+(k-1)j} = C'(a_i)$ where C' is a function of degree $k' < k$. By Theorem 2.9 and Theorem 3.5 the feedback polynomial of $(b_i'')_{i \in \mathbb{N}}$ has only zeros of the form $\zeta^{e'}$ where e' has binary weight at most k'.

Thus, by Theorem 2.9, the feedback polynomial of the sum $b_i = b_i' + b_i''$ has at least the roots ζ^e where the binary weight of e is k, since these roots occur only in the summand b_i' and cannot cancel.   □

In [228] (page 84) this Corollary is generalized a little to allow certain combinations of kth order terms.

3.6 Correlation Immune Functions

3.6.1 Definition and Alternative Characterizations

We learned in Sect. 3.3.3 that correlations between the input and the output of a non-linear combiner can be explored by an attacker. This motivates the following definition:

Definition 3.2 Let $X_1, \ldots, X_n$ be uniformly independent and identically distributed (henceforth, iid.) random variables on $\mathbb{F}_q$.

A function $f : \mathbb{F}_q^n \to \mathbb{F}_q$ is *correlation immune* of order t if for every set $\{i_1, \ldots, i_t\}$ of size t the random variable

$$f(X_1, \ldots, X_n)$$

is stochastically independent from

$$(X_{i_1}, \ldots, X_{i_t}).$$

The function f is *balanced* if $P(f(X_{i_1}, \ldots, X_{i_n}) = \alpha) = \frac{1}{q}$.
A correlation immune function of order t that is balanced is called t-*resilient*.

There are several equivalent characterizations of correlation immune functions (see [41, 118]). The stochastic characterization chosen for Definition 3.2 emphasizes the idea that we want to avoid dependencies between the input and the output of the non-linear functions. The next lemma provides a purely combinatorial characterization in terms of orthogonal arrays.

Definition 3.3 An *orthogonal array* A of size m, n constraints, strength t and index λ over an alphabet $\mathcal{F}$ of size q is an $m \times n$ array of elements in $\mathcal{F}$ with the following properties:

1. No two rows of A are the same.
2. For any subset $\mathcal{F}^t$ of t columns of A, each of the q^t vectors of $\mathcal{F}^t$ appears exactly λ times as a row. (Clearly $m = \lambda q^t$.)

Lemma 3.13 *The function* $C : \mathbb{F}_q^n \to \mathbb{F}_q$ *is correlation immune of order* t *if and only if for every* $y \in \mathbb{F}_q$ *the set* $f^{-1}(y)$ *forms the rows of an orthogonal array of strength* t.

Proof Let $X_1, \ldots, X_n$ be uniformly iid. random variables on $\mathbb{F}_q$, then

$$P\big(f(X_1, \ldots, X_n) = y\big) = \frac{|f^{-1}(y)|}{q^n}.$$

For a set $i_1, \ldots, i_t$ of t columns, and for $\hat{x}_{i_1}, \ldots, \hat{x}_{i_t} \in \mathbb{F}_q$, let λ denote the number of solutions of $f(x_1, \ldots, x_n) = y$ with $x_{i_k} = \hat{x}_{i_k}$ for $1 \leq k \leq n$. Then

$$P\big(f(X_1, \ldots, X_n) = y \mid X_{i_k} = \hat{x}_{i_k} \text{ for } 1 \leq k \leq t\big) = \frac{\lambda}{q^{n-t}}.$$

If f is correlation immune of order t, this probability must be equal to $P(f(X_1, \ldots, X_n) = y)$ for all choices of $i_1, \ldots, i_t$ and $\hat{x}_{i_1}, \ldots, \hat{x}_{i_t} \in \mathbb{F}_q$, i.e. $f^{-1}(y)$ forms the rows of an orthogonal array of strength t.

Conversely, if $f^{-1}(y)$ forms an orthogonal array then

$$P\big(f(X_1, \ldots, X_n) = y \mid X_{i_k} = \hat{x}_{i_k} \text{ for } 1 \leq k \leq t\big) = \frac{\lambda}{q^{n-t}} = \frac{m/q^t}{q^{n-t}} = \frac{|f^{-1}(y)|}{q^n}$$

for all choices $i_1, \ldots, i_t$ and $\hat{x}_{i_1}, \ldots, \hat{x}_{i_t} \in \mathbb{F}_q$, i.e. f is correlation immune of order t. $\qquad\square$

Another characterization of correlation immunity uses the Walsh transform. In Definition 2.10 we defined the Walsh transform for functions $f : \mathbb{F}_{2^n} \to \mathbb{F}_2$. Now we need it for functions $f : \mathbb{F}_2^n \to \mathbb{F}_2$.

Definition 3.4 Let $f : \mathbb{F}_2^n \to \mathbb{F}_2$ then the *Walsh transform* $f^W : \mathbb{F}_2^n \to \mathbb{R}$ of f is defined by

$$f^W(x) = \sum_{y \in \{0,1\}^n} f(y_1, \ldots, y_n)(-1)^{x_1 y_1 + \cdots + x_n y_n}.$$

The following characterization of correlation immune functions is due to G.-Z. Xiao and J.L. Massey [285].

Lemma 3.14 Let $f : \mathbb{F}_2^n \to \mathbb{F}_2$. Then the following statements are equivalent:

- *f is correlation immune of order t.*
- $f^W(x) = 0$ for every $x \in \mathbb{F}_2^n$ with $0 < w(x) \le t$.

Proof Let x_S denote the characteristic vector of the set S. Then

$$f^W(x_s)$$
$$= \left[\sum_{T \subseteq S} (-1)^{|T|} 2 \left| \{ y : f(y) = 1 \land (i \in T \Rightarrow y_i = 1) \land (i \in S \backslash T \Rightarrow y_i = 0) \} \right| \right]$$
$$- 2^{n - |S|}. \tag{3.6}$$

If f is correlation immune of order t and $|S| \le t$, the definition of correlation immunity implies that

$$\left| \{ y : f(y) = 1 \land (i \in T \Rightarrow y_i = 1) \land (i \in S \backslash T \Rightarrow y_i = 0) \} \right|$$

is independent from T. Hence for every correlation immune function of order t, Eq. (3.6) simplifies to $f^W(x_s) = 0$.

We prove the converse implication by induction on t. For $t = 0$ there is nothing to prove. Now let $t \ge 1$. By induction we have that

$$\left| \{ y : f(y) = 1 \land (i \in T \Rightarrow y_i = 1) \land (i \in S \backslash T \Rightarrow y_i = 0) \} \right|$$

is independent from T for every $|S| = t - 1$. For $|S| = t$ and $j \in S$ and $T \subset S \backslash \{j\}$ we can write

$$\left| \{ y : f(y) = 1 \land (i \in T \Rightarrow y_i = 1) \land (i \in (S \backslash \{j\}) \backslash T \Rightarrow y_i = 0) \} \right|$$
$$= \left| \{ y : f(y) = 1 \land (i \in T \Rightarrow y_i = 1) \land (i \in S \backslash T \Rightarrow y_i = 0) \} \right|$$
$$+ \left| \{ y : f(y) = 1 \land (i \in (T \cup \{j\}) \Rightarrow y_i = 1) \land (i \in S \backslash (T \cup \{j\}) \Rightarrow y_i = 0) \} \right|.$$

This proves that $|\{y : f(y) = 1 \wedge (i \in T \Rightarrow y_i = 1) \wedge (i \in S\backslash T \Rightarrow y_i = 0)\}|$ depends only on the parity of T.

By Eq. (3.6) we conclude that $|\{y : f(y) = 1 \wedge (i \in T \Rightarrow y_i = 1) \wedge (i \in S\backslash T \Rightarrow y_i = 0)\}|$ is independent from t, i.e. f is correlation immune of order t.   □

3.6.2 Siegenthaler's Inequality

To avoid correlation attacks we want the combiner function to be highly correlation immune and to avoid algebraic attacks we want (among other things) the degree of the combiner function to be high. In 1984 Siegenthaler [248] proved a trade-off between these two parameters. A low degree is the price one has to pay for a high correlation immunity. This result was generalized to non-binary functions by Camion and Canteaut [42].

Theorem 3.15 (see [42] Theorem 7) *Let $f : \mathbb{F}_q^n \to \mathbb{F}_q$ be correlation immune of order t, then for every monomial μ in the algebraic normal form of f there exists a subset T of size t such that $\deg_{x_i} \mu \leq q - 2$ for $i \in T$.*

If, in addition, f is balanced and $n \neq t + 1$, then there exists a subset T of size $t + 1$ such that $\deg_{x_i} \mu \leq q - 2$ for $i \in T$.

Proof Let $L_\alpha(x) = \prod_{\alpha' \in \mathbb{F}_q \backslash \{\alpha\}} \frac{x - \alpha'}{\alpha - \alpha'}$ denote the Lagrange interpolation polynomial.

Fix j of the n variables. Without loss of generality we may choose $x_{n-j+1} = \hat{x}_{n-j+1}, \ldots, x_n = \hat{x}_n$.

We denote the set $\{a \in \mathbb{F}_q^{n-j} \mid f(a_1, \ldots, a_{n-j}, x_{n-j+1}, \ldots, \hat{x}_n) = \alpha\}$ by $I_j(\alpha)$. Since f is correlation immune of order t, we have

$$\left| I_j(\alpha) \right| = \frac{|f^{-1}(\alpha)|}{q^j}$$

for all $j \leq t$. Hence $|I_j(\alpha)| = q^{t-j}|I_t(\alpha)|$ is divisible by q for $j < t$.

The algebraic normal form of $f(\cdot, \hat{x}_{n-j+1}, \ldots, \hat{x}_n)$ is

$$f(x_1, \ldots, x_{n-j}, \hat{x}_{n-j+1}, \ldots, \hat{x}_n) = \sum_{\alpha \in \mathbb{F}_q} \alpha \sum_{a \in I_j(\alpha)} \prod_{j=1}^{n} L_{a_i}(x_i).$$

Since each L_{a_i} is a monic polynomial of degree $(q - 1)$, the coefficient of degree $(q - 1)(n - j)$ of $\sum_{a \in I_j(\alpha)} \prod_{j=1}^{n} L_{a_i}(x_i)$ is $|I_j(\alpha)|$. For $j < t$ and $n \neq t + 1$ this is divisible by q, and hence the coefficient is 0 in $\mathbb{F}_q$.

So if we fix $t - 1$ variables in f we always obtain an algebraic normal form of degree less than $(q - 1)(n - t + 1)$. This is only possible if the algebraic normal form f does not contain a product of more than $n - t$ variables simultaneously having degree $q - 1$.

If, in addition, f is balanced then $|f^{-1}(\alpha)| = q^{n-1}$ and hence $|I_j(\alpha)| = q^{n-j-1}$ for $1 \le j \le t$. Thus $|I_j(\alpha)| = 0$ in $\mathbb{F}_q$, i.e. if we fix t variables we always obtain an algebraic form of degree less than $(q-1)(n-t)$. $\square$

Corollary 3.16 (Siegenthaler's inequality)   *Let $f : \mathbb{F}_q^n \to \mathbb{F}_q$ be correlation immune of order t and let d be the algebraic degree of f. Then*

$$d + t \le (q-1)n. \tag{3.7}$$

If in addition f is balanced then

$$d + t \le (q-1)n - 1. \tag{3.8}$$

Functions that achieve equality in Corollary 3.16 have *optimal non-linearity*. The next lemma describes how to construct functions with optimal non-linearity.

Lemma 3.17 *Let f_1 and f_2 be correlation immune of order t in n variables with $p(f_1 = \alpha) = p(f_2 = \alpha)$ for all α. Then*

$$f(x_1, \ldots, x_{n+1}) = x_{n+1}^{q-1} f_1(x_1, \ldots n x_n) + \left(1 - x_{n+1}^{q-1}\right) f_2$$

is also correlation immune of order t.

Proof The function f satisfies

$$P(f = \alpha) = P(x_{n+1} \ne 0)P(f_1 = \alpha) + P(x_{n+1} = 0)P(f_2 = \alpha)$$
$$= P(f_1 = \alpha) = P(f_2 = \alpha).$$

Since f_1 is correlation immune of order m, we have for $\hat{x}_{n+1} \ne 0$

$$P(f = \alpha | \hat{x}_{i_1}, \ldots, x_{i_m} = \hat{x}_{i_{m-1}}, x_{n+1} = \hat{x}_{n+1})$$
$$= P(f_1 = \alpha | \hat{x}_{i_1}, \ldots, x_{i_m} = \hat{x}_{i_{m-1}})$$
$$= P(f_1 = \alpha) = P(f = \alpha).$$

Similarly, for $\hat{x}_{n+1} = 0$ the equation follows from the correlation immunity of f_2. For $1 \le \hat{x}_{i_1}, \ldots, x_{i_m} \le n$ we have

$$P(f = \alpha \mid x_{i_1} = \hat{x}_{i_1}, \ldots, x_{i_m} = \hat{x}_{i_m})$$
$$= P(x_{n+1} \ne 0)P(f_1 = \alpha, x_{i_1} = \hat{x}_{i_1}, \ldots, x_{i_m} = \hat{x}_{i_m})$$
$$+ P(x_{n+1} = 0)P(f_2 = \alpha, x_{i_1} = \hat{x}_{i_1}, \ldots, x_{i_m} = \hat{x}_{i_m})$$
$$= P(x_{n+1} \ne 0)P(f_1 = \alpha) + P(x_{n+1} = 0)P(f_2 = \alpha)$$
$$= P(f = \alpha).$$

Thus f is correlation immune. $\square$

If both functions f_1 and f_2 have optimal non-linearity and $\deg(f_1) = \deg(f_2) = \deg(f_1 - f_2)$, the newly constructed function f also has optimal non-linearity.

So all we need to construct arbitrary functions of optimal non-linearity are $n - 1$ correlation immune functions in n variables of high degree as a starting point for a recursive application of Lemma 3.17.

For $q = 2$, linear functions will do the job. For example, we can start with

$$f_1(x_1, x_2, x_3, x_4) = x_1 + x_2 + x_3$$
$$f_2(x_1, x_2, x_3, x_4) = x_1 + x_2 + x_4$$

and apply Lemma 3.17 to obtain the function

$$f(x_1, \ldots, x_5) = x_5 f_1(x_1, x_2, x_3, x_4) + (1 + x_5) f_2(x_1, x_2, x_3, x_4)$$
$$= x_1 + x_2 + x_4 + x_3 x_5 + x_4 x_5$$

which is 3-resilient and of degree 2. Permuting the variables of f we get a second function f' for the next application of Lemma 3.17. For example the cyclic permutation gives us

$$f'(x_1, \ldots, x_5) = x_2 + x_3 + x_5 + x_4 x_1 + x_5 x_1.$$

By Lemma 3.17

$$\hat{f}(x_1, \ldots, x_5, x_6) = x_6 f(x_1, \ldots, x_5) + (1 + x_6) f'(x_1, \ldots, x_4)$$
$$= x_2 + x_3 + x_5 + x_1 x_4 + x_1 x_5 + x_1 x_6 + x_3 x_6 + x_4 x_6$$
$$+ x_5 x_6 + x_1 x_4 x_6 + x_1 x_5 x_6 + x_3 x_5 x_6 + x_4 x_5 x_6$$

is 4-resilient.

This recursive construction leads to:

Corollary 3.18 *For every n and every degree d there exists an $(n - d)$-resilient binary function f of degree d in n variables.*

Remark: In [42] an analogous result is proved for all $q \equiv 2 \mod 3$.

3.6.3 Asymptotic Enumeration of Correlation Immune Functions

Lemma 3.17 gives us a way of explicitly constructing a correlation immune function of order t. However, we may also ask: what is the probability that a random Boolean function is correlation immune of order t? To answer this question we must determine the number $N(n, t)$ of correlation immune functions of order t in n variables.

We will find an asymptotic expression for the number $N(n, t, \lambda)$ of correlation immune functions of order t in n variables and weight $\lambda 2^n$. The case of balanced functions $\lambda = \frac{1}{2}$ is especially important for cryptography.

The special case $t = 1$ has also been studied under the name *balanced colorings of a hypercube*. These are placements of equal weight to some vertices of a hypercube such that the centroid is at the center of the hypercube. Upper and lower bounds are given in [12, 193, 287]. There also exist exact enumerations for this case [202, 288].

Bounds for the general case are given in [44, 45, 235].

The first person to solve the asymptotic enumeration of correlation immune Boolean functions was O.V. Denisov [76]. However his arguments were quite complicated. In a later article [77] he withdraw his result and presented a "corrected" value. Unfortunately the correction contained a mistake and in fact Denisov's first result was correct. In the following we will follow E.R. Canfield, Z. Gao, C. Greenhill, B.D. McKay and R.W. Robinson [43] who found a more elegant proof of Denisov's Theorem.

The complement of a correlation immune function of order t is also correlation immune of order t. Hence $N(n, t, \lambda) = N(n, t, 1 - \lambda)$. In the following we will therefore restrict ourselves to the case $\lambda \leq 1/2$.

Theorem 3.19 (Theorem 1.2 of [43]) *Let*

$$M_{t,n} = \sum_{j=0}^{t} \binom{n}{j} \quad and \quad Q_{t,n} = \sum_{j=0}^{t} j \binom{n}{j}.$$

Consider a sequence of triples (n, t, λ) with $n \to \infty$ and

$$\omega\left(2^{6t-n} n^{6t+3} M_{t,n}^3\right) \leq \lambda \leq \frac{1}{2}. \tag{3.9}$$

Then

$$N(n, t, \lambda) = 2^{Q_{n,t}} \left(\lambda^{\lambda} (1 - \lambda)^{1-\lambda}\right)^{-2^n} \left(\pi \lambda (1 - \lambda) 2^{n+1}\right)^{-M_{t,n}/2} \left(1 + O\left(\eta(n, t, \lambda)\right)\right) \tag{3.10}$$

where

$$\eta(n, t, \lambda) = 2^{-n/2+3t} n^{3t+3/2} M_{n,t}^{3/2} \lambda^{-1/2} (1 - \lambda)^{-1/2} = o(1). \tag{3.11}$$

We will divide the proof of Theorem 3.19 into several lemmas.

Lemma 3.20 *Let I_t be the set of all subsets of $\{1, \ldots, n\}$ of size at most t. Let $x = (x_S)_{S \in I_t}$ be a vector of $M_{t,n}$ variables.*
Let $F : \mathbb{C}^{M_{t,n}} \to \mathbb{C}$ be defined by

$$F(x) = \prod_{\alpha \in \{\pm 1\}^n} \left(1 + D \prod_{S \in I_t} x_S^{\alpha_S}\right) \tag{3.12}$$

where

$$\alpha_s = \prod_{j \in S} \alpha_j \tag{3.13}$$

and D is an arbitrary constant.

  Then

$$N(n, t, \lambda) = \frac{1}{(2\pi i)^{M_{t,n}} D^{2^n \lambda}} \oint \cdots \oint \frac{F(x)}{x_{\emptyset}^{2^n \lambda} \prod_{S \in I_t} x_S} \, dx \tag{3.14}$$

where each x_S is integrated anticlockwise around a circle of radius 1 *centered at the origin.*

Proof For $\alpha \in \{\pm 1\}^n$ let $\bar{\alpha}$ denote the vector in $\{0, 1\}^n$ that arises from α by changing the 1 entries into 0s and the -1 entries into 1s. The map $\alpha \mapsto \bar{\alpha}$ is an isomorphism from $\{\pm 1\}^n$ with pointwise multiplication to the Boolean vector space $\mathbb{F}_2^n$.

  With this notation we can write $\alpha_S = \prod_{j \in S} \alpha_j = (-1)^{\bar{\alpha} w_S}$ where w_S is the incidence vector of the set S. Note that $(-1)^{\bar{\alpha} w_S}$ is precisely the term that occurred in the Walsh transform.

  When expanded, $F(x)$ is a sum of 2^{2^n} terms. Each summand corresponds to one Boolean function. For a Boolean Function g we have the term

$$t_g = \prod_{\substack{\alpha \in \{\pm 1\}^n \\ g(\bar{\alpha})=1}} \left(D \prod_{S \in I_t} x_S^{\alpha_S} \right)$$

$$= D^{|\{\alpha | g(\bar{\alpha})=1\}|} \prod_{S \in I_t} \prod_{\substack{\alpha \in \{\pm 1\}^n \\ g(\bar{\alpha})=1}} x_S^{(-1)^{\bar{\alpha} w_S}}$$

$$= D^{\hat{g}(0)} \prod_{S \in I_t} x_S^{\sum_{\bar{\alpha} \in \{0,1\}^n} g(\bar{\alpha})(-1)^{\bar{\alpha} w_S}}$$

$$= D^{\hat{g}(0)} \prod_{S \in I_t} x_S^{\hat{g}(w_S)}.$$

  By Lemma 3.14 the Boolean function g is correlation immune of order t if and only if its Walsh transform $\hat{g}$ vanishes for all w_S with $\emptyset \neq S \in I_t$. Therefore the coefficient of $x_{\emptyset}^{\lambda 2^n}$ in $F(x)$ is the number $N(n, t, \lambda)$ of all correlation immune functions of order t with $\hat{g}(0) = \lambda 2^n$.

  Consider the function $F(x)/(x_{\emptyset}^{\lambda 2^n} \prod_{S \in I_t} x_S)$. It has a pole at 0 and the residue of the pole is $D^{\lambda 2^n} N(n, t, \lambda)$. By Cauchy's integration formula we obtain the residue by integrating anticlockwise around the pole, which is exactly the statement of the lemma.   □

  Using the parametrization $e^{i\theta}$, $\theta \in [0, 2\pi)$, for the unit circle in the complex plane, we can change the complex integral (3.14) into a real integral. We set $x_S =$

$e^{i\theta_S}$, then $dx_S = ie^{i\theta_S}d\theta_S$. A change of variables gives:

$$N(n, t, \lambda) = \frac{1}{(2\pi)^{M_{t,n}} D^{2^n \lambda}} \int_0^{2\pi} \cdots \int_0^{2\pi} e^{-i2^n \lambda \theta_\emptyset} \prod_{\alpha \in \{\pm 1\}^n} \left(1 + De^{if_\alpha(\theta)}\right) d\theta$$

$$= \frac{(1 + D)^{2^n}}{(2\pi)^{M_{t,n}} D^{2^n \lambda}} \int_0^{2\pi} \cdots \int_0^{2\pi} e^{-i2^n \lambda \theta_\emptyset} \prod_{\alpha \in \{\pm 1\}^n} \frac{1 + De^{if_\alpha(\theta)}}{1 + D} d\theta$$

(3.15)

where

$$f_\alpha(\theta) = \sum_{S \in I_t} \alpha_S \theta_S.$$

(3.16)

We denote the integrand of the integral in (3.15) by $G(\theta)$. Note that $G(\theta)$ is a product of terms of the form $\frac{1 + De^{if_\alpha(\theta)}}{1+D}$. For $D > 0$ we have $|1 + De^{if_\alpha(\theta)}| \leq 1 + D$ with equality if and only if $f_\alpha(\theta) \equiv 0 \mod 2\pi$. Thus $G(\theta) \leq 1$.

We will prove that $G(\theta) = 1$ only at $2^{Q_{t,n}}$ points. Outside a critical region around these points $G(\theta)$ is a product of many terms strictly less than 1, i.e. $G(\theta)$ is small outside the critical region and this part of the integral will be hidden in the O-term of Theorem 3.19. The only significant contribution will come from the integral inside the critical region.

Let us start by determining all points with $G(\theta) = 1$.

Lemma 3.21 *Let*

$$C = \{\theta \in [0, 2\pi)^{M_{t,n}} \mid |G(\theta)| = 1\}.$$

(3.17)

Then

$$C = \left\{\theta \in [0, 2\pi)^{M_{t,n}} \mid 2^{|S|} \sum_{T \in I_t, T \supseteq S} \theta_T \equiv 0 \mod 2\pi \text{ for each } S \in I_t\right\}$$

(3.18)

and $|C| = 2^{Q_{t,n}}$.

Proof We already know that a point θ with $G(\theta) = 1$ corresponds to a solution of $f_\alpha(\theta) \equiv 0 \mod 2\pi$ for all $\alpha \in \{\pm 1\}^n$. All we have to do is to find a linear transformation that brings this system into the form of Eq. (3.18).

For a set S, let A_S denote the set of all $\alpha \in \{\pm 1\}^n$ with $\alpha_i = 1$ for $i \notin S$. Let $\sigma(\alpha) = \prod_{i=1}^n \alpha_i$.

Then

$$\sum_{\alpha \in A_S} \sigma(\alpha) f_\alpha(\theta) = \sum_{\alpha \in A_S} \sum_{T \in I_t} \alpha_T \theta_T$$

$$= \sum_{T \in I_t} \theta_T \sum_{\alpha \in A_S} \sigma(\alpha) \alpha_T$$

and for $S \subseteq T$ we have $\sigma(\alpha) = \prod_{i=1}^{n} \alpha_i = \prod_{i \in T} \alpha_i = \alpha_T$ since $\alpha_i = 1$ for $i \notin S \subset T$. So the inner sum simplifies to $\sum_{\alpha \in A_S} \alpha_T^2 = \sum_{\alpha \in A_S} 1 = 2^{|S|}$.

If $S \nsubseteq T$, let $j \in S \backslash T$. Let α' differ from α by changing the sign of α_j. We have $\sigma(\alpha) = -\sigma(\alpha')$ but since $j \notin T$ we have $\alpha_T = \alpha'_T$. The inner sum consists of $2^{|S|-1}$ pairs of the form $\sigma(\alpha)\alpha_T + \sigma(\alpha')\alpha'_T$ which are equal to 0. Hence the coefficient variable θ_T vanishes.

Thus

$$\sum_{\alpha \in A_S} \sigma(\alpha) f_\alpha(\theta) = 2^{|S|} \sum_{T \in I_t, T \supseteq S} \theta_T$$

which proves that

$$\mathcal{C} \subseteq \left\{ \theta \in [0, 2\pi)^{M_{t,n}} \;\middle|\; 2^{|S|} \sum_{T \in I_t, T \supseteq S} \theta_T \equiv 0 \mod 2\pi \text{ for each } S \in I_t \right\}$$

but the linear system given in Eq. (3.18) is upper triangular and hence it has full rank, which proves the equality.

(Note that we used in our argument only that the elements of A_s agree on every coordinate not in S. The only difference is that the sign of the factor $2^{|S|}$ may swap.)

Since the system in (3.18) is upper triangular we can solve it by successively assigning values to θ_S, starting with the largest sets first. We work modulo 2π and the coefficient of θ_S is $2^{|S|}$, so we have $2^{|S|}$ possible choices for θ_S (which differ by an integral multiple of $\frac{2\pi}{2^{|S|}}$). Altogether we have $\sum_{S \in I_t} 2^{|S|} = 2^{Q_{t,n}}$ solutions of (3.18). $\square$

Now we define a small rectangular region $\mathcal{R}$ around the origin. Let

$$\mathcal{R} = \left\{ \theta \in R^{M_{t,n}} \mid |\theta_S| \leq \Delta (2n)^{-|S|} \text{ for all } S \in I_t \right\} \tag{3.19}$$

where Δ is a constant factor which will be determined later in the proof.

We split the domain of integration into the critical region $\mathcal{C} + \mathcal{R}$ (the addition must be done modulo 2π) and its complement $\overline{\mathcal{C} + \mathcal{R}}$.

Lemma 3.22 *For $\theta \notin \mathcal{C} + \mathcal{R}$ we have*

$$|G| < \exp\left(-\frac{4D}{\pi^2(1+D)^2} 2^{n-2|S|} \left(2 - e^{-1/2}\right) \Delta n^{-|S|} \right). \tag{3.20}$$

Furthermore

$$\int_{\overline{\mathcal{C}+\mathcal{R}}} G(\theta) d\theta < (2\pi)^{M_{t,n}} \exp\left(-\frac{4D}{\pi^2(1+D)^2} 2^{n-2|S|} \left(2 - e^{-1/2}\right) \Delta n^{-|S|} \right). \tag{3.21}$$

Proof Let $d(x) = \min\{|x - k2\pi| \mid k \in \mathbb{Z}\}$. d defines a metric on the set of real numbers modulo 2π.

We claim that

$$d\left(2^{|S|} \sum_{T \in I_t, T \supseteq S} \theta_T\right) > \Delta n^{-|S|}$$

for some S.

Choose $\theta' \in C$ as close as possible to θ with respect to the metric d. By the definition of $\mathcal{R}$ we can find an S with the following properties:

- $d(\theta_S - \theta'_S) > \Delta(2n)^{-|S|}$.
- $d(\theta_T - \theta'_T) \leq \Delta(2n)^{-|T|}$ for all $T \supset S$.

Then

$$d\left(2^{|S|} \sum_{T \in I_t, T \supseteq S} \theta_T\right)$$

$$= d\left(2^{|S|} \sum_{T \in I_t, T \supseteq S} \theta_T\right) + d\left(2^{|S|} \sum_{T \in I_t, T \supseteq S} \theta'_T\right) \quad \text{since } \theta' \in C$$

$$\geq 2^{|S|}\left[d\left(\theta_S - \theta'_S\right) - \sum_{T \in I_t, T \supset S} d\left(\theta_T - \theta'_T\right)\right] \quad \text{triangle inequality}$$

$$> 2^{|S|}\left[(2\Delta)^{-|S|} - \sum_{j=1}^{t-|S|} \binom{n-|S|}{j} \Delta(2n)^{-|S|-j}\right] \quad \text{by choice of } S$$

$$> \Delta n^{-|S|}\left[1 - \sum_{j=1}^{t-|S|} \frac{2^{-j}}{j!}\right]$$

$$> \Delta n^{-|S|}\left(2 - e^{-1/2}\right).$$

Partition the set $\{\pm 1\}^n$ into $2^{n-|S|}$ sets $A_S^{(k)}$ with the property that the elements of each part $A_S^{(k)}$ agree on all coordinates not in S. As we have seen in the proof of Lemma 3.21:

$$\sum_{\alpha \in A_S^{(k)}} \sigma(\alpha) f_\alpha(\theta) = \pm 2^{|S|} \sum_{T \in I_t, T \supseteq S} \theta_T.$$

Hence

$$\sum_{\alpha \in A_S^{(k)}} d(f_\alpha(\theta)) \geq d\left(\sum_{\alpha \in A_S^{(k)}} d(f_\alpha(\theta))\right) = d\left(2^{|S|} \sum_{T \in I_t, T \supseteq S} \theta_T\right) > \left(2 - e^{-1/2}\right)\Delta n^{-|S|}.$$

$$(3.22)$$

We now estimate a part of $G(\theta)$. The following sequence of transformations is long and difficult to find, but the individual steps are simple and quite standard. The only unusual bound we use is $1 - \cos(x) \geq \frac{2}{\pi^2}x^2$ for $x \in [-\pi, \pi]$, which we

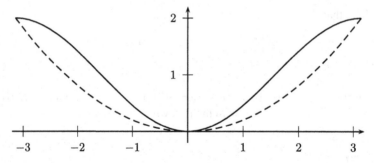

Fig. 3.4 $1 - \cos(x) \geq \frac{2}{\pi^2}x^2$

illustrate in Fig. 3.4. It is easily proved by determining the extrema of $1 - \cos(x) - \frac{2}{\pi^2}x^2$.

$$\prod_{\alpha \in A_S^{(k)}} \left| \frac{1 + De^{if_\alpha(\theta)}}{1 + D} \right|^2 = \prod_{\alpha \in A_S^{(k)}} \left(\frac{(1 + D\cos f_\alpha(\theta))^2 + (\sin f_\alpha(\theta))^2}{(1 + D)^2} \right)$$

$$= \prod_{\alpha \in A_S^{(k)}} \left(1 - \frac{2D(1 - \cos f_\alpha(\theta))}{(1 + D)^2} \right)$$

$$= \prod_{\alpha \in A_S^{(k)}} \left(1 - \frac{2D(1 - \cos d(f_\alpha(\theta)))}{(1 + D)^2} \right)$$

$$\leq \prod_{\alpha \in A_S^{(k)}} \left(1 - \frac{4Dd(f_\alpha(\theta))^2}{\pi^2(1 + D)^2} \right)$$

since $1 - \cos(x) \geq \dfrac{2}{\pi^2}x^2$ for $x \in [-\pi, \pi]$

$$\leq \prod_{\alpha \in A_S^{(k)}} \exp\left(-\frac{4Dd(f_\alpha(\theta))^2}{\pi^2(1 + D)^2} \right)$$

since $1 - x \leq e^{-x}$ (Taylor series)

$$= \exp\left(-\frac{4D}{\pi^2(1 + D)^2} \sum_{\alpha \in A_S^{(k)}} d\big(f_\alpha(\theta)\big)^2 \right)$$

$$\leq \exp\left(-\frac{4D}{\pi^2(1 + D)^2} \frac{1}{|A_S^{(k)}|} \left(\sum_{\alpha \in A_S^{(k)}} d\big(f_\alpha(\theta)\big) \right)^2 \right)$$

by Cauchy-Schwarz inequality

$$< \exp\left(-\frac{4D}{\pi^2(1+D)^2}2^{-|S|}\left(2-e^{-1/2}\right)\Delta n^{-|S|}\right) \quad \text{by (3.22).}$$

Since $G(\theta)$ is the product of $2^{n-|S|}$ parts we get:

$$G(\theta) < \exp\left(-\frac{4D}{\pi^2(1+D)^2}2^{n-2|S|}\left(2-e^{-1/2}\right)\Delta n^{-|S|}\right)$$

for all $\theta \notin C + R$.

Thus

$$\int_{\overline{C+R}} G(\theta)\,d\theta < |\overline{C+R}| \exp\left(-\frac{4D}{\pi^2(1+D)^2}2^{n-2|S|}\Delta n^{-|S|}\right)$$

$$< (2\pi)^{M_{t,n}} \exp\left(-\frac{4D}{\pi^2(1+D)^2}2^{n-2|S|}\left(2-e^{-1/2}\right)\Delta n^{-|S|}\right)$$

since the domain of integration $\overline{C+R}$ lies in the hypercube $[0, 2\pi]^{M_{t,n}}$.   □

Lemma 3.22 bounds the integral outside the critical region $C + R$. We will later choose the factor Δ such that this part of the integral will be hidden in the O-term of Theorem 3.19.

Next we look at the integral inside the critical region.

Lemma 3.23 *Let* $D = \frac{\lambda}{1-\lambda}$ *and* $\Delta = 2^{-n/2+t+3}\lambda^{-1/2}n^{t+1/2}M_{t,n}^{1/2}$, *then*

$$\int_R G(\theta)\,d\theta = \left(\frac{2\pi}{\lambda(1-\lambda)2^n}\right)^{M_{t,n}/2}\left[1 + O\left(2^n\frac{D(1-D)\Delta^3}{3(1+D)^3}\right)\right]. \qquad (3.23)$$

Proof First we write $G(\theta) = e^{g(\theta)}$. This gives

$$g(\theta) = -i2^n\lambda\theta_\emptyset + \sum_{\alpha\in\{\pm1\}^n}\log\left(\frac{1+De^{if_\alpha(\theta)}}{1+D}\right).$$

Using the Taylor series

$$h(x) = \log\left(\frac{1+De^{ix}}{1+D}\right) = i\frac{D}{1+D}x - \frac{1}{2}\frac{D}{(1+D)^2}x^2 + R(x)$$

where

$$R(x) = \int_0^x \frac{1}{2}h'''(t)(x-t)^2\,dt \qquad (3.24)$$

we get

$$g(\theta) = -i2^n\lambda\theta_\emptyset + \sum_{\alpha\in\{\pm1\}^n}\left(i\frac{D}{1+D}f_\alpha(\theta) - \frac{1}{2}\frac{D}{(1+D)^2}f_\alpha(\theta)^2 + R\big(f_\alpha(\theta)\big)\right).$$

$$(3.25)$$

Now remember that $f_\alpha(\theta) = \sum_{S \in I}, \alpha_S \theta_S$ and hence the coefficient of θ_S ($S \neq \emptyset$) is $\sum_{\alpha \in \{\pm 1\}^n} \alpha_S = 0$.

The coefficient of $\theta_\emptyset$ is

$$-i2^n\lambda + i\frac{D}{1-D} \quad \sum_{\alpha \in \{\pm 1\}^n} \alpha_\emptyset = -i2^n\lambda + i\frac{D}{1-D}2^n = 0$$

for $D = \frac{\lambda}{1-\lambda}$, which explains the choice of D.

Hence

$$g(\theta) = \sum_{\alpha \in \{\pm 1\}^n} \left(-\frac{1}{2}\frac{D}{(1+D)^2}f_\alpha(\theta)^2 + R\big(f_\alpha(\theta)\big)\right). \tag{3.26}$$

The next step used to bound the error term $R(x)$ is Eq. (3.24). We have

$$h'''(x) = \frac{iD^2(e^{ix})^2 - iDe^{ix}}{(D + e^{ix})^3}.$$

For $x = o(1)$ we get $|h'''(x)| = \frac{D(1-D)}{(1+D)^3} + o(1)$. Thus

$$|R(x)| \leq \left(\frac{D(1-D)}{(1+D)^3} + o(1)\right)\int_0^x (x-t)^2\,dt = \frac{D(1-D)x^3}{3(1+D)^3} + o\big(x^3\big).$$

Note that $\mathcal{R}$ is symmetric about the origin and the reflection $\theta \mapsto -\theta$ maps $G(\theta)$ to its complex conjugate. Hence $\int_{\mathcal{R}} G(\theta)d\theta$ is real and the integral is equal to the integral over the real part of the integrand.

Let $r(\theta) = \sum_{\alpha \in \{\pm 1\}^n} R(f_\alpha(\theta))$ then

$$\int_{\mathcal{R}} G(\theta)d\theta = \int_{\mathcal{R}} \Re(G(\theta))d\theta$$

$$= \int_{\mathcal{R}} \Re(e^{g(\theta)})d\theta$$

$$= \int_{\mathcal{R}} \Re(e^{r(\theta)})\exp\left(-\frac{1}{2}\frac{D}{(1+D)^2}\sum_{\alpha \in \{\pm 1\}^n} f_\alpha(\theta)^2\right)$$

$$= \Re(e^{r(\theta_0)})\int_{\mathcal{R}} \exp\left(-\frac{1}{2}\frac{D}{(1+D)^2}\sum_{\alpha \in \{\pm 1\}^n} f_\alpha(\theta)^2\right)$$

for some θ_0 by the intermediate value theorem.

Since

$$r(\theta_0) = \sum_{\alpha \in \{\pm 1\}^n} R(\theta_0) \leq 2^n\frac{D(1-D)\Delta^3}{3(1+D)^3} + o\big(2^n\Delta^3\big) = o(1)$$

for $\Delta = o(2^{-n})$ we get by Taylor's expansion

$$\Re\left(e^{r(\theta_0)}\right) = 1 + O\left(2^n \frac{D(1-D)\Delta^3}{3(1+D)^3}\right).$$

Now check that

$$\sum_{\alpha \in \{\pm 1\}^n} f_\alpha(\theta)^2 = 2^n \sum_{S \in I_t} \theta_S^2.$$

(The square terms in f_α^2 always have coefficient 1. The mixed terms have coefficients ± 1 and they vanish since the number of odd subsets of a non-empty set is always equal to the number of even subsets.)

Thus we must calculate

$$\int_{\mathcal{R}} \exp\left(-\frac{1}{2}\frac{D}{(1+D)^2} \sum_{\alpha \in \{\pm 1\}^n} f_\alpha(\theta)^2\right) = \int_{\mathcal{R}} \exp\left(\frac{1}{2}\lambda(1-\lambda)2^2 \sum_{S \in I_t} \theta_S^2\right) d\theta$$

$$= \prod_{S \in I_t} \int_{-\Delta(2n)^{-|S|}}^{\Delta(2n)^{-|S|}} \exp\left(\frac{1}{2}\lambda(1-\lambda)2^n \theta_S^2\right).$$

Using the bound $\int_{-x\sigma}^{x\sigma} e^{-t^2}2\sigma^2 \, dt = \sigma\sqrt{2\pi}(1 + e^{-x^2/2})$ for $x \to \infty$ with $\sigma = \sqrt{\lambda(1-\lambda)2^n}$ and $x = \Delta(2n)^{-|S|}\sigma^{-1} > \sqrt{32nM_{t,n}}$ we get

$$\int_{\mathcal{R}} G(\theta) = \left(\frac{2\pi}{\lambda(1-\lambda)2^n}\right)^{M_{t,n}/2}\left[1 + O\left(2^n \frac{D(1-D)\Delta^3}{3(1+D)^3}\right) + O\left(M_{t,n}^{-16nM_{t,n}}\right)\right].$$

The lemma follows by noting that the first O-term dominates the second. $\qquad\square$

Proof of Theorem 3.19 By Lemma 3.20 and Eq. (3.15) we have

$$N(n,t,\lambda) = \frac{(1+D)^{2^n}}{(2\pi)^{M_{t,n}}D^{2^n\lambda}} \int_0^{2\pi} \cdots \int_0^{2\pi} e^{-i2^n\lambda q} \prod_{\alpha \in \{\pm 1\}^n} \frac{1 + De^{if_\alpha(\theta)}}{1+D} \, d\theta$$

$$= \frac{(1+D)^{2^n}}{(2\pi)^{M_{t,n}}D^{2^n\lambda}} \left(\int_{\mathcal{C}+\mathcal{R}} G(\theta) \, d\theta + \int_{\overline{\mathcal{C}+\mathcal{R}}} G(\theta) \, d\theta\right).$$

The second integral is bounded by Lemma 3.22 and for $D = \frac{\lambda}{1-\lambda}$ and $\Delta = 2^{-n/2+t+3}\lambda^{-1/2}n^{t+1/2}M_{t,n}^{1/2}$ the bound is lower than the O-term in Theorem 3.19.

The first integral is $2^{Q_{t,n}} \int_{\mathcal{R}} G(\theta) \, d\theta$ and by Lemma 3.23 we know the value of the integral. This yields the value for $N(n,t,\lambda)$ given by Theorem 3.19. $\qquad\square$

Chapter 4
Correlation Attacks

4.1 CJS-Attacks

Under the name CJS-attacks we collect a class of correlation attacks that extend the approach described by V. Chepyzhov, T. Johansson and B. Smeets in [52].

4.1.1 The Basic Version

In this subsection we describe the basic version of the attack which can be found in [52].

Represent the LFSR as a linear code and let

$$G = \begin{pmatrix} 1 & & 0 & c_{n,0} & \cdots & c_{N-1,0} \\ & \ddots & & \vdots & & \vdots \\ 0 & & 1 & c_{n,n-1} & \cdots & c_{N-1,n-1} \end{pmatrix} \tag{4.1}$$

be the systematic generator matrix associated with the LFSR (see Sect. 2.2.4). Since the LFSR register produces a pseudo-random sequence the column vectors $(c_{k,0}, \ldots, c_{k,n-1})^t$, $n \leq k < N$, are also pseudo-random. For our analysis of the fast correlation attacks described in this section, we model the columns by independent uniformly random vectors.

Choose a constant B. We make the following definition.

Definition 4.1 A relation of *weight* 2 is a pair (r, r'), $n \leq r < r < N'$, with $c_{r,j} = c_{r',j}$ for all $j \in \{B, \ldots, n-1\}$.

By Ω we denote the set of all relations (of weight 2). The expected size of Ω is given by Theorem 4.1.

A. Klein, *Stream Ciphers*, DOI 10.1007/978-1-4471-5079-4_4,
© Springer-Verlag London 2013

Theorem 4.1

$$E\left(|\Omega|\right) = \binom{N-n}{2} 2^{-n+B} \approx N^2 2^{-n+B-1}.$$

Proof The probability that a pair (r, r') is a relation is 2^{-n-B}. Since there exist $\binom{N-n}{2}$ pairs (r, r') with $n \leq r < r < N'$ and the expected value is additive, we have $E(|\Omega|) = \binom{N-n}{2} 2^{-n+B}$. $\square$

The fast correlation attack is based on the following observation:

Theorem 4.2 *Let $x_0, \ldots, x_{N-1}$ be the true LFSR sequence and let $z_0, \ldots, z_{N-1}$ be the sequence observed by the attacker. Assume that the sequences (x_i) and (z_i) are correlated by $p = P(x_i = z_i) > 1/2$ and let (r, r') be a relation, then:*

$$P\left(z_r + z_{r'} = \sum_{j=0}^{B-1}(c_{r,j} + c_{r',j})x_j\right) = p^2 + (1-p)^2 \tag{4.2}$$

and for $(\hat{x}_0, \ldots, \hat{x}_{B-1}) \neq (x_0, \ldots, x_B)$

$$P\left(z_r + z_{r'} = \sum_{j=0}^{B-1}(c_{r,j} + c_{r',j})\hat{x}_j\right) = \frac{1}{2}. \tag{4.3}$$

Proof By definition

$$x_r = \sum_{j=0}^{n-1} c_{r,j}x_j \quad \text{and} \quad x_{r'} = \sum_{j=0}^{n-1} c_{r',j}x_j$$

and hence

$$x_r + x_r = \sum_{j=0}^{n-1}(c_{r,j} + c_{r',j})x_j$$

$$= \sum_{j=0}^{B-1}(c_{r,j} + c_{r',j})x_j \quad \text{since } c_{r,j} = c_{r',j} \text{ for } B \leq j < n.$$

Now

$$P\left(z_r + z_{r'} = \sum_{j=0}^{B-1}(c_{r,j} + c_{r',j})x_j\right)$$

$$= P(z_r + z_{r'} = x_r + x_{r'})$$

$$= P(z_r = x_r, z_{r'} = x_{r'}) + P(z_r \neq x_r, z_{r'} \neq x_{r'})$$

$$= p^2 + (1 - p)^2.$$

On the other hand, $\delta_c = (c_{r,j} + c_{r',j})_{j=0,\dots,B-1}$ is uniformly random in $\mathbb{Z}_2^B$, since c_r and $c_{r'}$ are independent random vectors. (Remember our model for the generator matrix!)

Hence, for $\hat{x} \neq x \in \mathbb{Z}_2^B$ the scalar product $\delta = \delta_c \cdot (\hat{x} - x)$ is a random variable independent from $\hat{x}$ and x with $P(\delta = 1) = P(\delta = 0) = 1/2$.

Therefore

$$P\left(z_r + z_{r'} = \sum_{j=0}^{B-1}(c_{r,j} + c_{r',j})\hat{x}_j\right) = P\left(z_r + z_{r'} = \delta = \sum_{j=0}^{B-1}(c_{r,j} + c_{r',j})\hat{x}_j\right) = \frac{1}{2}.$$

$\square$

Theorem 4.2 shows that the relation distinguishes between the correct vector $(x_0, \dots, x_{B-1})$ and the incorrect guess $\hat{x} = (\hat{x}_0, \dots, \hat{x}_{B-1})$. This leads us to the attack described by Algorithm 4.1.

Algorithm 4.1 Simple fast correlation attack (CJS)

1: **for** $\hat{x} \in \mathbb{Z}_2^B$ **do**
2:    Count the number $k_{\hat{x}}$ of relations with $z_r + z_{r'} = \sum_{j=0}^{B-1}(c_{r,j} + c_{r',j})\hat{x}_j$.
3: **end for**
4: Return $\hat{x}$ with $k_{\hat{x}} = \max$ as the most likely guess for $x = (x_0, \dots, x_{B-1})$.

For the moment we do not say how we count the number $k_{\hat{x}}$ of satisfied relations (see Sect. 4.1.6).

To finish the analysis of the simple fast correlation attack we must determine for which parameters N, n and B the success probability of Algorithm 4.1 is reasonable large. Let us recall Shannon's fundamental theorem of coding.

Theorem 4.3 (Shannon's coding theorem)   *Denote by* $C(p) = p \log p + (1 - p) \log(1 - p)$ *the capacity of a binary symmetric channel with error probability* p. *Let* $0 < R < C$ *and let* $\log(M_n)/n < R$ *for all* $n \in \mathbb{N}$. *Then there exists a sequence of codes* C_n *such that:*

- C_n contains M_n code words of length n; and
- the probability $P_{err}(C_n)$ of a decoding error satisfies $P_{err}(C_n) \leq 4e^{-NA}$ for some constant $A > 0$.

Proof Shannon proves that a random code C_n almost surely satisfies the bound of Theorem 4.3. See, for example, [170]. $\qquad\square$

Theorem 4.2 states that $C' = \{(\sum_{j=0}^{B-1}(c_{r,j} + c_{r',j})\hat{x}_j)_{(r,r')\in\Omega} \mid \hat{x} \in \mathbb{Z}_2^B\}$ and $(z_r + z_{r'})_{(r,r')\in\Omega}$ behaves like a random code and codewords are transmitted over a binary symmetric channel with error probability $q = 1 - p_2$, where $p_2 = p^2 + (1 - p)^2$.

Hence by Shannon's coding theorem we expect that the maximum likelihood decoding algorithm (Algorithm 4.1) will work if $|\Omega|C(p_2) > B$. Substituting the expected size of Ω from Theorem 4.1 we get that the decoding Algorithm 4.1 succeeds with high probability if the attacker can observe at least $1/2\sqrt{B\ln 2}(p - 1/2)^{-2}2^{(n-B)/2}$ bits (see [52] Theorem 2).

4.1.2 Using Relations of Different Size

The first possible extension of the CJS-attack is to use relations of size larger than 2. This extension increases the number of available relations at the price of a weaker correlation. The effect is that the attack, and especially the pre-processing, becomes slower, but the attacker needs less data.

This extension was described in the original paper [52].

The principle is very simple. We define a relation of weight w as a w-tuple $(r_0, \ldots, r_{w-1})$ with $n \leq r_0 < \cdots < r_{w-1} < N$ and $x_{r_0,j} + \cdots + x_{r_{w-1},j} = 0$ for all $j \in \{B, \ldots, n-1\}$. By analogy with Theorem 4.1 we find that there exist about $\binom{N-n}{w}2^{-n+B}$ relations of weight w. So the number of available relations increases quickly if we allow a higher weight. The price is that a relation of weight w is satisfied only with probability

$$p_w = \sum_{j=0}^{\lfloor w/2\rfloor}\binom{w}{2j}(1-p)^{2j}p^{w-2j}. \tag{4.4}$$

p_w quickly approaches $1/2$ as w increases, so the number of necessary relations for a successful attack increases quickly.

In [171] we find the following theorem.

Theorem 4.4 *Let* $p = 1/2 + \delta = P(z_i = x_i)$. *Then the success probability of the simple fast correlation attack (Algorithm 4.1) with weight $w + 1$ relations is better than the success probability of the same algorithm with weight w relations if and only if*

$$2\delta\sqrt{\frac{N}{w+1}} > 1.$$

Proof See [171] Theorem 4.10.   □

In practice we will not use relations only of weight exactly w, but instead we will use relations of weight up to w. This limits the application of Theorem 4.4. In fact the success probability always increases if we increase the admissible weight, since extra information can never hurt.

In the literature there seems to be some confusion over how to use relations of different weight. For a maximum likelihood algorithm we cannot simply count the number of satisfied relations, but we must use weighted sums. Here we follow the presentation of [83].

Let n_i be the number of relations of weight i and k_i be the number of satisfied relations of weight i. Under the assumption that our guess is correct, the probability of observing this pattern of satisfied relations is

$$p_{guess} = \prod_{i=1}^{w} p_i^{k_i} (1 - p_i)^{n_i - k_i}, \tag{4.5}$$

where

$$p_i = \sum_{j=0}^{\lfloor i/2 \rfloor} \binom{i}{2j} p^{2j} (1 - p)^{i - 2j}$$

is the probability that a relation of weight i is satisfied.

By the maximum likelihood principle we have to find the guess for which p_{guess} becomes maximal. Multiplying small floating point numbers is unfavorable, so we take the logarithm.

$$\log p_{guess} = \sum_{i=1}^{w} k_i \log(p_i) + (n_i - k_i) \log(1 - p_i).$$

Subtracting the constant $\sum_{i=1}^{w} n_i \log(1 - p_i)$ we see that we must maximize

$$\sum_{i=1}^{w} k_i \big(\log(p_i) - \log(1 - p_i) \big).$$

Finally we divide by $\log(p_w) - \log(1 - p_w)$, which leaves us the goal to maximize

$$\sum_{i=1}^{w} k_i w_i,$$

with

$$w_i = \frac{\log(p_i) - \log(1 - p_i)}{\log(p_w) - \log(1 - p_w)}.$$

Some textbooks (such as [78]) recommend replacing the real weights by integer approximations. This advice comes from the "good old times" when floating point

units were very slow. On modern hardware floating point units are highly optimized
and are often faster than integer units, so we can work directly with floating point
weights.

4.1.3 How to Search Relations

At this point we still need to explain how to find the relations used by the CJS-
attack. For weight 2 relations the solution is simple. We just sort the columns in the
generator matrix to find all pairs (r, r') with $c_{r,j} = c_{r',j}$ for $B \leq j < n$. For larger
weights the solution is less trivial.

The following theorem is in its main part due to [53] with some corrections and
improvements due to [83].

Theorem 4.5 *It is possible to compute all relations of weight up to w in $O(N^{\lceil w/2 \rceil})$
time and $O(N^{\lceil (w-1)/4 \rceil})$ memory.*

Proof First we deal with the case where w is even.

We generate a table which contains for all k-tuples with $k \leq \lceil w/4 \rceil$ the values
$c_{i_1} + \cdots + c_{i_k}$ and sort these partial relations by their characteristic (hash based
sorting). Building the table requires $O(N^{\lceil w/4 \rceil})$ time and memory.

Next we choose r such that $2^r = \Theta(N^{\lfloor w/2 \rfloor - \lceil w/4 \rceil})$. Let π denote the projection
to the first r coordinates.

For each $a \in \mathbb{F}_2^r$ enumerate the partial relations $(i_1, \ldots, i_s)$ of size $\lceil w/4 \rceil + 1 \leq
s \leq w/2$ with the property $\pi(c_{i_1} + \cdots + c_{i_s}) = a$.

For the enumeration we run through all partial relations $(i_1, \ldots, i_{\lceil s/2 \rceil})$ of size
$\lceil s/2 \rceil$ and search the pre-computed table for relations $(i_{\lceil s/2 \rceil + 1}, \ldots, i_s)$ such that
$\pi(c_{i_1} + \cdots + c_{i_{\lfloor s/2 \rfloor}}) = a + \pi(c_{i_{\lfloor s/2 \rfloor + 1}} + \cdots + c_{i_s})$. Thus for fixed a the time needed
to enumerate the partial relations $(i_1, \ldots, i_s)$ of size $\lceil w/4 \rceil + 1 \leq s \leq w/2$ with
$\pi(c_{i_1} + \cdots + c_{i_s}) = a$ is the number N_a of such relations plus $O(N^{\lceil w/4 \rceil})$ time
steps for the algorithm describe above. The memory required is $N_a = \Theta(N^s/2^r) =
\Theta(N^{\lceil w/4 \rceil})$.

Once we have all partial relations with $\pi(c_{i_1} + \cdots + c_{i_s}) = a$, we can find all
relations by searching for solutions of $c_{i_1} + \cdots + c_{i_{\lfloor w/2 \rfloor}} = c_{j_1} + \cdots + c_{j_{\lceil w/2 \rceil}}$.

Altogether the algorithm needs

$$2^r O(N^{\lfloor w/4 \rfloor}) + \sum_a N_a = O((N^{\lfloor w/2 \rfloor - \lceil w/4 \rceil}) O(N^{\lfloor w/4 \rfloor}) + O(N^{w/2})) = O(N^{w/2})$$

time steps.

Now we explain how to deal with the case of odd w. In principle we would like
to run the same algorithm as for $w - 1$. However, if we change nothing we need
$\Theta(N^{\lceil w/2 \rceil}/2^r) = \Theta(N^{\lceil w/4 \rceil + 1})$ space to store all partial relations of size $\lceil w/2 \rceil$
with $\pi(c_{i_1} + \cdots + c_{i_{\lceil w/2 \rceil}}) = a$.

But since w is odd, we will use a partial relation $(j_1, \ldots, j_{\lceil w/2 \rceil})$ of size $\lceil w/2 \rceil$ only once, when we must search all partial relations $(i_1, \ldots, i_{\lfloor w/2 \rfloor})$ with $c_{i_1} + \cdots + c_{i_{\lfloor w/2 \rfloor}} = c_{j_1} + \cdots + c_{j_{\lceil w/2 \rceil}}$. Each solution of this equation gives us a relation of size w.

Since we use the partial relations of size $\lceil w/2 \rceil$ only once, we do not need to store them. Thus we store only the partial relations of size up to $\lfloor w/2 \rfloor$ and to store them we need a memory of only $\Theta(N^{\lfloor w/2 \rfloor}/2^r) = \Theta(N^{\lceil w/4 \rceil})$. $\qquad\square$

The proof of Theorem 4.5 gives no hint as to which data structure we should use. To obtain the bounds of the theorem, we need a data structure which allows us to store and search a partial relation with given characteristic in $O(1)$ time. In addition we should not use more than $O(N^{\lceil (w-1)/4 \rceil})$ memory to store the $O(N^{\lceil (w-1)/4 \rceil})$ partial relations with $\pi(c_{i_1} + \cdots + c_{i_s}) = a$.

This calls for a hash table. We want to avoid costly memory allocating operations during the pre-processing, so we decide to choose a closed hash. Since the characteristics of the partial relations are almost uniformly distributed we can simply use the projection of the characteristic to the last h coordinates as the hash function. We choose the size 2^h of the hash table to be about twice the expected number of partial relations, so will have only few collisions. To deal with collisions we choose a quadratic probing strategy, i.e. if we want to store a partial relation at an already occupied position n we next probe the positions $n + 1^2$, $n + 2^2$, $n + 3^3$, ... until we find an empty place. (Tests of alternative collision avoiding strategies reveal that linear probing gives poor results, while more complicated probing strategies are not worth the additional effort.)

As a small implementation trick we note that we use a table of size $2^h + 100$. The 100 extra places at the end guarantees that the quadratic probing will almost surely never go beyond the end of the table. This saves us a modulo operation.

Finally we have to describe how we clear the hash table when the loop advances to the next value of a. The trick is to store the value a of the first r coordinates of the characteristic inside the hash. This has the advantage that the algorithm can detect an empty place with the old value of a. The idea of saving space by not storing the known value a results in a serious slow down, since then one must clear the hash table in every iteration of the loop, i.e. we would have 2^r extra operations per loop.

Having explained in detail the advantages of hashing over sorting we must mention that sorting is sometimes more efficient than hashing. The reason is that sorting can make better use of the cache in modern processors, while hashing produces many cache misses. See Chap. 6 "The birthday paradox: Sorting or not?" of [138] for an extensive discussion. In general, one can expect sorting to be effective if the data is read only once and hashing to be effective if the data is read several times. For the problem of finding relations this mean that hashing is best if the weight w of the relations is odd, but the case of even w can benefit from a good sorting algorithm.

The most interesting cases of Theorem 4.5 are those with memory consumption $O(N)$, i.e. $w \leq 5$. For larger w the memory consumption grows rapidly, which often forces time-memory trade-offs in the pre-processing (see [83]). For $w \leq 5$ the following tricks speed up the pre-processing by a constant factor.

- If we want to get relations $c_1 + c_2 + c_3 + c_4 = 0$ of weight 4 we use the following special algorithm.

  Without loss of generality we can assume that c_1 and c_2 have the same first component. Hence c_3 and c_4 have the same first component too.

  Therefore it is enough to enumerate all partial relations of weight 2 where the first component of $c_{i_1} + c_{i_2}$ is 0.

  So we must generate just $\approx \binom{N}{2}/2$ partial relations instead of the $\binom{N}{2}$ partial relations required in the basic algorithm.

- For $w = 3$ we can improve the trick a bit. There are two cases. In the first case the first component of c_1 and c_2 is 1 while the first component of c_3 is 0. To deal with this case we must enumerate all pairs (c_1, c_2) with first component 1. There are $\approx \binom{N/2}{2}$ such pairs.

  The second case is that all three vectors c_1, c_2 and c_3 have first component 0. In this case we can apply the same trick for the second component.

  Altogether we enumerate only $\binom{N/2}{2} + \binom{N/4}{2} + \cdots \approx N^2/6$ pairs.

- For $w = 5$ we do the following: Look at the projection to the first three coordinates. The equation $c_1 + c_2 + c_3 + c_4 + c_5 = 0$, $c_i \in \mathbb{F}_2^3$ has only the following solutions:

  Either we have two equal vectors (without loss of generality we can assume $c_4 = c_5$). We can cover this case by enumerating all $\binom{N}{3}/2^3$ partial relations for which $c_1 + c_2 + c_3 = 0$.

  The second case is that no two vectors are equal. But this means that $c_i + c_j \neq c_k$ for $i \neq j$, $j \neq k$, $i \neq k$.

  The only possibility (up to permutation of indices) is $c_1 = 0$, $c_5 = c_2 + c_3 + c_4$. This lead us to four different solutions (up to permutation) either $c_2 = (100)$, $c_3 = (010)$, $c_4 = (001)$ or the weight of c_2 and c_3 is even. For each of these four subcases there are only $\frac{N^3}{8^3}$ partial relations of weight 3 that fall into the subcase.

  Hence we must enumerate only $N^3(\frac{1}{3! \cdot 8} + 4\frac{1}{8^3})$ of the $\binom{N}{3}$ possible partial relations. This is a speed-up factor of almost 8 in comparison to the basic algorithm.

4.1.4 Extended Relation Classes

This subsection describes a substantial improvement of the original CJS-attack. The idea is to allow extended relation sets.

Definition 4.2 We call a w-tuple $(r_0, \ldots, r_{w-1})$ with $n \le r_0 < r_2 < \cdots < r_{w-1} < N$ an *extended relation of weight* w. The vector

$$\chi = \left(\sum_{j=0}^{w-1} c_{r_j, B}, \ldots, \sum_{j=0}^{w-1} c_{r_j, n-1} \right) \in \mathbb{F}_2^{n-B}$$

is called the *characteristic* of the extended relation $(r_0, \ldots, r_{w-1})$.

Let $I \subset B, \ldots, n - 1$. By Ω_I we denote the set of all extended relations (up to weight w) with characteristic χ, where χ is the characteristic vector of I.

The relations of the original CJS-attack are the extended relations with characteristic 0. In the new attack we also allow other characteristics.

The idea of extended relation classes is due to P. Lu and L. Huang [171]. In [171] the authors use only the extended relation sets Ω_I with $|I| = 1$. They appear in full generality for the first time in this book.

For simplicity we assume in the following that all extended relations have weight 2. Let Ω_I be an extended relation set and let $(\hat{x}_0, \ldots, \hat{x}_{B-1})$ be a guess for the first B bits. We say an extended relation $(r, r') \in \Omega_i$ is satisfied if and only if

$$z_r + z_{r'} = \sum_{j=1}^{B} (c_{r,j} + c_{r,j}) \hat{x}_i.$$

We denote the number of satisfied relations by k and the size of Ω_I by n.

Similarly to Theorem 4.2 we have for an extended relation $(r, r') \in \Omega_I$ the equation

$$P\left(z_r + z_{r'} = \sum_{j=0}^{B-1} (c_{r,j} + c_{r',j}) x_j + \xi_I \right) = p_2 = p^2 + (1 - p)^2 \qquad (4.6)$$

where $\xi_I = \sum_{j=B}^{n-1} (c_{r,j} + c_{r',j}) x_j$ depends only on I. Thus the probability that k_I of the n_I relations are satisfied under the condition $x_i = \hat{x}_i$ $(0 \le i < B)$ and $\xi_I = 0$ is $P_0 = p_2^{k_I} p_2^{n_I - k_I}$. Similarly the probability that k_I of the n_I relations are satisfied under the condition $x_i = \hat{x}_i$ $(0 \le i < B)$ and $\xi_I = 1$ is $P_1 = p_2^{n_I - k_I} p_2^{k_I}$.

If all possible initial values have equal probability then ξ_I is a random variable with $p(\xi_I = 0) = p(\xi_I = 1) = 1/2$.

So in a likelihood test we have to compute

$$L_I = \log\left(\frac{1}{2} P_0 + \frac{1}{2} P_1 \right) = \log\left(p_2^{k_I} (1 - p_2)^{n_I - k_I} + p_2^{n_I - k_I} (1 - p_2)^{k_I} \right) - 1. \quad (4.7)$$

The expression in the form of Eq. (4.7) is unwieldy, but for large n_I the expected difference between n_I and k_I will be large, thus $\max\{P_0, P_1\} \gg \min\{P_0, P_1\}$ and we can safely replace Eq. (4.7) by

$$L_i \approx \log \max\{P_0, P_1\} - 1 = \hat{k} \log p_2 + (n_I - \hat{k}) \log(1 - p_2) - 1 \qquad (4.8)$$

where $\hat{k} = \max\{k_I, n_i - k_i\}$.

If n_I is small we transform Eq. (4.7) as follows:

Assume without loss of generality that $k_I > n_I - k_I$. Then

$$L_I = \log\left(p_2^{k_I} (1 - p_2)^{n_I - k_I} + p_2^{n_I - k_I} (1 - p_2)^{k_I} \right) - 1$$

$$= \log\left(p_2^{k_I} (1 - p_2)^{n_I - k_I} \right) + \log\left(1 + \left(p_i / (1 - p_i) \right)^{n_I - 2k_i} \right) - 1.$$

The first term is the approximation we used in Eq. (4.8). The second term can be interpreted as a correction value which, for large n_I, will be small and hence can be considered negligible. If n_I is small there are only a few possible values of $n_I - 2k_I$ and we can tabulate the correction values.

It is possible to use more than one extended relation set at the same time. Under the assumption that everything is independent we just add the likelihood values, i.e. we compute

$$L = L_0 + \sum_{k=1}^{e} L_{I_k}.$$

This assumption is valid if the extended relation sets do not contain overlapping relations and if the characteristic vectors of the index sets I_k are linearly independent. As already stated in the description of the simple CJS-attack we will ignore any dependency that may occur in our relation sets. So we should expect that the real attack needs more data than estimates based on the model of independent relations suggest. Experiments show that the effects that come from the dependencies are negligible.

One problem with extended relations is that one must compute many more likelihood values than in the original CJS-attack. Even if we take into account that each extended relation set will be much smaller than the relation set used in the original CJS-attack, this is an expensive operation. A way to deal with this problem is to use sequential tests (see Sect. 15.3). The idea is the following:

The likelihood values of a wrong guess are smaller on average than the likelihood values for the right guess. So instead of directly computing the full value $L = L_0 + \sum_{k=1}^{e} L_{I_k}$ for each guess and selecting the largest value as the answer for the decoding algorithm, we compute only a part of the sum: $L' = L_0 + \sum_{k=1}^{e'} L_{I_k}$, where $e' < e$. If this value is small we can guess that the full likelihood value L is also small and stop the computation.

Turning this heuristic argument into a precise statistic with proved error bound needs some work. Section 15.3 gives an introduction to sequential tests. In the present case we can use a slight modification of Wald's sequential ratio test in which we formally set the error probability of the first kind to 0.

The main advantage of extended relation sets is that we can choose I so that Ω_I contains as many relations as possible. Therefore extended relation sets are of most use if we can use low weight relations and small relation sets.

For example consider an LFSR of length 50 with feedback polynomial

$$f(z) = z^{50} + z^{49} + z^{48} + z^{47} + z^{46} + z^{39} + z^{34} + z^{32} + z^{30} + z^{29} + z^{27} + z^{24}$$
$$+ z^{23} + z^{19} + z^{18} + z^{17} + z^{15} + z^{14} + z^{13} + z^{12} + z^{8} + z^{7} + z^{6} + z^{5} + z^{4}$$
$$+ z^{3} + z^{2} + z + 1$$

and the moderately strong correlation of $1 - p = 0.65$. We want to use relations of weight 2. As parameter B we choose $B = 20$. The standard relation set which could be used by the simple CJS-attack has only 8 relations, but one can find 1500 extended relations with sizes between 35 and 51.

It remains to explain how we find large sets of extended relations.

For each vector $x \in \mathbb{F}_2^{l-B}$ let N_x be the number of columns in the generator matrix for which the projection to the last $l - B$ coordinates is x. Then the number of k-tuples $(i_1, \ldots, i_k)$ for which sum of the k columns $c_{i_1}, \ldots, c_{i_k}$ of the generator matrix has only zeros in the last $l - B$ coordinates is

$$N_x^{(k)} = \sum_{x_1 + \cdots + x_k = x} N_{x_1} \cdots N_{x_k}. \tag{4.9}$$

Equation (4.9) describes a convolution. So we can compute all $N_x^{(k)}$ efficiently by computing the Fourier transform $\hat{N} = \theta(N)$, raising each element of $\hat{N}$ to the k power, and then applying the inverse Fourier transform.

For $l - B < 35$ this is no problem. The number $N_x^{(k)}$ is almost the number of extended relations in Ω_x of size k. We just count every relation $k!$ times and we also count pseudo-relations such as (x_2, x_2). The correction is very simple. Let $R_x^{(k)}$ be the number of relations of weight k in Ω_x. In the extreme case $k = 0$ we have $R_0^{(0)} = 1$ and $R_x^{(0)} = 0$ for $x \neq 0$.

Then

$$N_x^k = \sum_{j=0}^{\lfloor k/2 \rfloor} R_x^{k-2j} \binom{N + j - 1}{N - 1}.$$

($\binom{N+j-1}{n+1}$ is the number of ways of choosing, with repetition, j columns from the N columns in the generator matrix. Each relation of weight $k - 2j$ can be extended to a pseudo-relation of weight k by adding j pairs of equal columns.)

For small $l - B$ this allows us to determine the exact size of all extended relation sets and we can choose the largest one for the attack. For large $l - B$ we switch to a heuristic method. We can compute the exact size of the extended relation sets for a $B' > B$. Then we choose a large extended relation set with respect to B'. This splits into $2^{B'-B}$ extended relation sets with respect to B which are most likely large.

4.1.5 Twice Step Decoding

Twice step decoding is a technique to improve the success rate (or the speed) of fast correlation attacks. In the basic form of the fast correlation attack (Algorithm 4.1) we loop over all $\hat{x} \in \mathbb{Z}_2^B$ and search for the value $\hat{x}$ with the maximal likelihood value.

In twice step decoding we compute the likelihood values for all $\hat{x} \in \mathbb{Z}_2^B$. Instead of stopping at this point, we run Algorithm 4.2.

Algorithm 4.2 Twice step decoding

1: Compute likelihood values for all $\hat{x} \in \mathbb{Z}_2^B$ and sort the vectors $\hat{x}$ by the likelihood values
2: **loop**
3:      Get the $\hat{x} = (\hat{x}_0, \ldots, \hat{x}_{B-1})$ with the highest likelihood value.
4:      Assume that $(x_0, \ldots, x_{B-1}) = \hat{x}$.
5:      Under this assumption decode a linear $[N, n - B]$. Denote by $(\hat{x}_B, \ldots, \hat{x}_{n-1})$
        the output of the decoding algorithm.
6:      Generate the LFSR sequence $\hat{x}_0, \ldots, \hat{x}_{N-1}$.
7:      **if** $|\{i \mid \hat{x}_i = \hat{z}_i\}| > T$ **then**
8:          **return** $\hat{x}$ as the most likely guess for the LFRS sequence x.
9:      **else**
10:         Delete $(\hat{x}_0, \ldots, \hat{x}_{B-1})$ from the list of possible B-tuples.
11:     **end if**
12: **end loop**

The idea of Algorithm 4.2 is to interpret the likelihood values computed by the simple correlation attack as a hint of the order in which we should test the vectors $\hat{x}$ (line 1).

Next we take an $\hat{x}$ with highest *a posteriori* probability (line 3) and try to reconstruct the LFSR sequence under the assumption that $\hat{x}$ is the true x. This leaves us with the problem of decoding an $[N, n - B]$-code. We will use the simple fast correlation attack (Algorithm 4.1) for this purpose. So we must choose a new parameter B' and compute the corresponding relations. Since now we assume the first B values of the LFSR as known, the number of relations is much larger, so in the second decoding step even a small B' gives us a high success probability. This means that each of the internal decoding steps is very fast.

After line 5 of Algorithm 4.2 we have reconstructed a possible initial state of the LFSR. We know that this internal state is correct with high probability if the initial assumption $x_0 = \hat{x}_0, \ldots, x_{B-1} = \hat{x}_{B-1}$ was correct. To test if this is the case we generate the LFSR sequence $\hat{x}_0, \ldots, \hat{x}_{N-1}$ (line 6). If all our assumptions were correct we have $x_i = \hat{x}_i$ and hence $P(\hat{x}_i = z_i) = p > 1/2$. If our assumptions are wrong we get $P(\hat{x}_i = z_i) = 1/2$.

To distinguish these cases we use a classical one-sided test, i.e. we choose the smallest T such that $P(\text{Bin}(N, p) \geq T) > 1 - \epsilon$ for some small ϵ and accept the hypothesis $H_0 : \hat{x}_i = x_i$ if and only if $\hat{x}_i = z_i$ for at least T indices $0 \leq i < N$. Otherwise we reject H_0. This test is done in line 7 of Algorithm 4.2.

If we reject the guess $\hat{x}$ we go back to the beginning of the loop and test the $\hat{x}$ with the second highest *a posteriori* probability and so on, until we find the right $\hat{x}$.

One way to interpret twice step decoding is that we run the simple Algorithm 4.1 with the large parameter $B + B'$ and test the $\hat{x}$ in the order that is determined by the first step.

So in twice step decoding we must choose $B + B'$ so that the simple attack with that search window succeeds with a probability of almost 1. If B is too small, the first step of the twice step decoding will produce a random order of the candidates $\hat{x}$. In this case twice step decoding will be not faster than the simple algorithm and we will even lose speed because we must test the $\hat{x}$ in some artificial order. If, on the other hand, B is too large, the first step will most likely assign the highest *a posteriori* probability to $\hat{x} = x$. In this case Algorithm 4.2 will run the loop for the second step only once (or at least only a few times). The effect is that Algorithm 4.2 spends its whole time in line 1, i.e. in the first step. Hence it is no better than the simple algorithm. We must choose B between these two extreme cases to benefit from twice step decoding.

As mentioned above, one normally uses the simple Algorithm 4.1 as a decoding algorithm for the second step in twice step decoding (line 5 of Algorithm 4.2). However, it is possible to iterate twice step decoding. Ultimately, each step of a multi-step decoding algorithm adds only one bit. We will discuss this variant when we come to the application of convolutional decoding algorithms in cryptography (see Sect. 4.2.3.2).

4.1.6 Evaluation of the Relations

In the previous sections we just stated that the attacker has to compute for all 2^B possible values of $(x_0, \ldots, x_{B-1})$ the quality $\sum_{j=1}^{w} w_j e_j$ where e_j denotes the number of satisfied relations of weight j. In this section we want to answer the question how the attacker should do this. There are basically two methods, both having their pros and cons.

4.1.6.1 Simple Counting

For the first method we use a bit field b to indicate, for each relation, whether it is satisfied or not. To get the number of satisfied relations we must perform a sideway addition (see Sect. 12.1.2). In fact the sideway addition is the most time-consuming part of the whole algorithm, so it should be carefully optimized.

We use a gray code to loop over the 2^B possible values of $(x_0, \ldots, x_{B-1})$, i.e. in every step we flip only one x_i. In the pre-processing step we can compute the bit field b_i, in which all relations that include x_i are marked by a 1. The update of the bit field b of satisfied relations is then performed by $b \leftarrow b \oplus b_i$, i.e. we need only a simple XOR of two bit fields.

This algorithm for the evaluation of the relations needs $O(2^B R)$ steps where R denotes the number of relations.

The greatest pro for the simple counting algorithm is that it opens the door for the use of sequential statistics. To understand why, we look at the simple case in which all relations are of the same quality. Let $p > 1/2$ be the probability that a relation is satisfied for the correct guess $(x_0, \ldots, x_{B-1})$. Let R_k be the random variable that is 1 if the kth relation is satisfied and 0 if it is unsatisfied.

Our goal is to distinguish the hypothesis $H_0 : P(R_k = 1) = p$ (i.e. the guess $(x_0, \ldots, x_{B-1})$ is correct) from the alternative $H_1 : P(R_k = 1) = 1/2$ (i.e. the guess $(x_0, \ldots, x_{B-1})$ is wrong).

By the Neyman-Pearson lemma the optimal test to distinguish H_0 and H_1 is to compute

$$L_R = \prod_{k=0}^{R-1} \frac{P_0(R_k)}{P_1(R_k)}$$

and reject H_0 in favor of H_i if L_R lies below a critical value c_α. (The value c_α depends on the chosen error probability α.)

One can show (see Lemma 15.2) that under H_0, $L_R \rightarrow \infty$ with probability 1 and that under H_1, $L_R \rightarrow 0$ with probability 1. Now suppose we have a large number of relations. It would be a waste of time to successively compute $L_0, \ldots, L_{R-1}$ to see that L_{R-1} is small. We should be able to distinguish a sequence that tends to 0 from a sequence that tends to ∞ by looking only at the first few elements. Sequential statistics (see Sect. 15.3) formalizes this intuition. Especially when using extended relation sets, sequential tests should be used to skip unnecessary relation sets. Sequential statistics can easily speed up the simple counting by a factor of 2 or even higher.

4.1.6.2 Fourier Transformation

The second way to evaluate the relations uses a Fourier transform.

For $c = (c_0, \ldots, c_{B-1}) \in \mathbb{Z}_2^B$ denote by n_c the total quality of relations with characteristic c. By k_c we denote the total quality of relations with characteristic c, which satisfy in addition the equation $x_{r_1} + \cdots + x_{r_w} = 0$. Note that the values n_c can be determined in the pre-processing phase, while the values k_c must be calculated during the attack.

Now consider the guess $x = (x_0, \ldots, x_{B-1})$ for the first B bits. The total quality of all relations satisfied by this guess is

$$Q(x) = \sum_{x \cdot c \equiv 0 \bmod 2} k_c + \sum_{x \cdot c \equiv 1 \bmod 2} (n_c - k_c).$$

Since the value of n_c is known after the pre-processing, one can pre-compute $\sum_{x \cdot c \equiv 1 \bmod 2} n_c$, thus the attacker need only compute

$$Q'(x) = \sum_{x \cdot c \equiv 0 \bmod 2} k_c - \sum_{x \cdot c \equiv 1 \bmod 2} k_c = \sum_{c \in \mathbb{Z}_2^B} (-1)^{x \cdot c} k_c$$

for all $x \in \mathbb{Z}_2^B$.

Computing $Q'(x)$ is a Fourier transformation with respect to the group $\mathbb{Z}_2^n$. As for the ordinary Fourier transform there exists a very efficient divide and conquer algorithm (see Algorithm 4.3).

Algorithm 4.3 Fast Fourier transform over the group $\mathbb{Z}_2$

1: {**Input:** $Q'(c) = k_c$ for all $c \in \mathbb{Z}_2^B$.}
2: **for** i **from** 0 **to** $B - 1$ **do**
3:    **for** $x \in \mathbb{Z}_2^B$ with $x_i = 0$ **do**
4:       $(Q'(x), Q'(x + e_i)) \leftarrow (Q'(x) + Q'(x + e_i), Q'(x) - Q'(x + e_i))$
5:    **end for**
6: **end for**
7: {**Output:** $Q'(c) = \sum_c (-1)^{x \cdot c} k_c$ for all $c \in \mathbb{Z}_2^B$.}

Fig. 4.1 A simple convolutional code

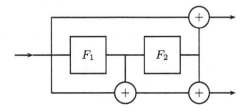

If $R \ll 2^B$ the effort of calculating the k_c is negligible. The Fourier transform needs $O(2^B B)$ steps, i.e. the complexity is independent of the number of relations. Therefore the Fourier transform is superior to simple counting if one uses many relations of low quality.

A hybrid algorithm that combines the simple counting technique with the Fourier transform is also possible (see [53]).

4.2 Attacks Based on Convolutional Codes

4.2.1 Introduction to Convolutional Codes

When mathematicians speak of error-correcting codes they usually mean block codes, but there is another class of codes, the convolutional codes, that are to block codes as stream ciphers are to block ciphers in cryptology. In this section we will give a quick introduction to convolutional codes and show how to use them in correlation attacks.

An (n, k) *convolutional encoder* maps k infinite (binary) streams to n output streams. Normally the encoding is done by a finite state machine like the one shown in Fig. 4.1.

If we associate with a sequence $(a_i)_{i \in \mathbb{N}}$ its generating function $\sum_{i=0}^{\infty} a_i D^i$, we can define an (n, k) *convolutional code* as a k-dimensional subspace of $F((D))^n$ in complete analogy to the definition of block codes. Now we can transfer the notions of generator matrix and parity check matrix from block codes to convolution codes. For example, the encoder shown in Fig. 4.1 is represented by the generator matrix $(1 + D^2 \ 1 + D + D^2)$.

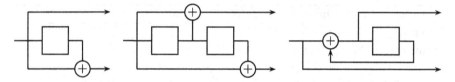

Fig. 4.2 Three different encoders of the same code

One big difference between block codes and convolutional codes is that, since a convolutional code has infinitely many codewords, the encoder as a finite description of the code becomes more important. For example, decoding algorithms for convolutional codes (e.g. Viterbi decoding and sequential decoding) depend on the used encoder, while many block decoding algorithms (e.g. syndrome decoding, BCH decoding, etc.) attempt to reconstruct the codeword and are independent of the encoder. A consequence of this fact is that many measures for the performance of convolutional codes are defined as properties of an encoder and, in a slight abuse of terminology, we use "the convolutional code has property X" as an abbreviation for "the convolutional code has an encoder with property X".

Figure 4.2 exhibits three different encoders of the same code.

The corresponding generator matrices are

$$G_1 = \begin{pmatrix} 1 & 1+D \end{pmatrix}, \qquad G_2 = \begin{pmatrix} 1+D & 1+D^2 \end{pmatrix}, \qquad G_3 = \begin{pmatrix} \dfrac{1}{1+D} & 1 \end{pmatrix} \qquad (4.10)$$

As one can check, G_2 and G_2 are scalar multiples of G_1, i.e. all matrices generate the same code.

G_2 is an example of a *catastrophic encoder*. It maps the infinite sequence $111\ldots$ to the sequence $11010000\ldots$, so only three errors can flip the output to $0000\ldots$, which would be decoded as $000\ldots$. If a finite number of errors in the received codeword results in an infinite number of decoding errors, we speak of a catastrophic error propagation. So G_2 should not be used as an encoder.

G_3 is an example that shows how feedback can be used to construct finite state machines that belong to non-polynomial generator matrices.

For convolutional codes the minimal distance between two codewords is called the *free distance* d_{free} of the code. The free distance is the most important quality measure for convolutional codes.

Another important distance measure is the column distance and distance profile, which is a property of the encoder. Let $(c_0, \ldots) = C(e_0, \ldots)$ and $(c_0', \ldots) = C(e_0', \ldots)$ be the two codewords with $e_0 \neq e_0'$. The *column distance* d_i measures the distance between the first i outputs of the encoder, i.e. d_i is the minimal distance between $(c_0, \ldots, c_{i-1})$ and $(c_0', \ldots, c_{i-1}')$ under the restriction that $e_0 \neq e_0'$. For non-catastrophic encoders, $d_i = d_{free}$ for large i.

The sequence $d_1, d_2, \ldots$ is called the *distance profile* of the convolutional encoder.

At this point we stop our introduction and direct the reader to the further references (see [78] and the references given there). The handbook of coding theory also contains a good introduction [184], but it does not cover what are, for us, the important decoding algorithms, which we will cover in the next section.

4.2.2 Decoding Convolutional Codes

The decoding algorithms for convolutional codes that are used in practice are mostly the convolutional code equivalent of the simple decoding of a block code that compares the received word with all code words. Since convolutional codes have infinitely many code words, even this simple algorithm must truncate the words, and the decoding algorithms differ in their truncation strategies. Here we describe only two of the most popular variants.

4.2.2.1 Viterbi Decoding

Viterbi proposed his decoding algorithm in 1967 [271], which was later proved to be optimal in the maximum likelihood sense by Forney [96].

We describe the Viterbi algorithm in the more general setting of discrete state Markov processes.

Definition 4.3 A *discrete state Markov process* consists of a finite state set $\mathcal{S}$ and output alphabet $\mathcal{A}$ and a set of transitions $\mathcal{T} \subset \mathcal{S} \times \mathcal{S} \times \mathcal{A} \times [0,1]$.

For each state s we have a finite number of transitions (s, s_k, a_k, p_k), where:

- s_k is the successor state;
- a_k is the output associated to the transition; and
- p_k is the probability of the transition.

The Viterbi algorithm solves the following problem: Given a discrete state Markov process and sequences of observations A find the state sequence X with the maximal *a posteriori* probability $P(X|A)$.

Viterbi's idea is quite simple. The transition probabilities

$$P(x_t \mid x_{t-1}, \ldots, x_0, a_t, \ldots, a_0) = P(x_t \mid x_{t-1}, a_t)$$

are known in advance. By

$$P(x_t, \ldots, x_0 | a_t, \ldots, a_0) = P(x_{t-1}, \ldots, x_0 | a_{t-1}, \ldots, a_0) P(x_t \mid x_{t-1}, a_t) \quad (4.11)$$

one can compute the *a posteriori* probabilities recursively. A consequence of Eq. (4.11) is that $P(x_{t-1}, \ldots, x_0 | a_t, \ldots, a_0) > P(x'_{t-1}, \ldots, x'_0 | a_t, \ldots, a_0)$ and $x_{t-1} = x'_{t-1}$ implies $P(x_t, \ldots, x_0 | a_t, \ldots, a_0) > P(x_t, x'_{t-1}, \ldots, x'_0 | a_t, \ldots, a_0)$. Since the decoder needs to find only the sequence $x_T, \ldots, x_0$ for which $P(x_T, \ldots, x_0 | a_t, \ldots, a_0)$ becomes maximal, it need only store, for every $t \in \{0, \ldots, T\}$ and every state s, the probabilities

$$P(t, s) = \max_{x_{t-1}, \ldots, x_0} P(x_t = s, x_{t-1}, \ldots, x_0 | a_t, \ldots, a_0).$$

The Viterbi algorithm uses Eq. (4.11) to compute these probabilities recursively. Algorithm 4.4 shows the Viterbi algorithm in pseudo-code.

Algorithm 4.4 Viterbi decoding

1: $P(-1, x_{\text{initial state}}) \leftarrow 1$, $P(-1, x) \leftarrow 0$ for $x \neq x_{\text{initial state}}$ {The Markov chain states in $x_{\text{initial state}}$}
2: **for** t **from** 0 **to** T **do**
3:     **for** all states s **do**
4:         $P(t, s) = \max_{s'} P(t - 1, s') P(x_t = s \mid x_{t-1} = s', a_t)$
5:     **end for**
6: **end for**
7: Find s with $P(T, s) = \max$ and output the corresponding sequence $x_T, \ldots, x_0$ as the result.

Fig. 4.3 An example for the Viterbi algorithm

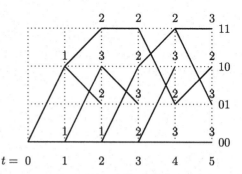

The best way to understand the algorithm is to look at an example. Consider the convolution encoder from Fig. 4.1. It starts with all flip-flops set to zero. The input 11101... is mapped to the codeword 1110011000.... Assume that the transmission adds two errors and the receiver receives 1000011000 instead of 1110011000.

The run of the Viterbi algorithm is illustrated by the trellis diagram in Fig. 4.3.

We loop over all possible inputs and remember the number of corresponding errors. For example the entry 1 in the row 00 at time step 1 stands for the input sequence 00 that sends the encoder to the state in which both flip-flops are 0 and shows that the corresponding output sequence 0000 differs in one place from the received sequence 1000. At the time step $t = 3$ there are two possible input sequences 000 and 100 that send the encoder to the internal state 00. If 000 is the input, the number of errors would be 2 and if 100 is the input, the number of errors would be 3. We do not know if the internal state of the encoder at time step 2 is 00, but it is more likely that the input starts with 000 than with 100. We indicate this in the trellis diagram by drawing the line that connects the state 00 at time step 2 with the state 00 at time step 3 and not drawing the line that connects the state 01 at time step 2 with state 00 at time step 3.

At the end we decode the input sequence that produces a code word which is closest to the received sequence (nearest neighborhood decoding). In the example this input corresponds to the internal state 10 and, as we can see from the diagram, the corresponding input sequence is 11101, sending the encoder through the states $00 - 10 - 11 - 11 - 01 - 10$.

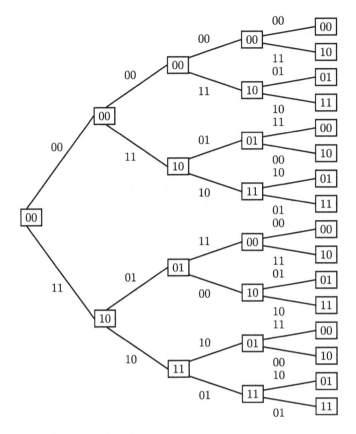

Fig. 4.4 A tree diagram for a (2, 1) encoder

The error correction capability of the Viterbi algorithm depends mostly on the free distance of the convolutional code and its complexity depends mostly on the number of flip-flops used in the encoder.

4.2.2.2 Sequential Decoding

Sequential decoding was proposed in 1961 by J.M. Wozencraft and B. Rieffen [283] and was the first practical method for decoding convolutional codes. There are several interesting variations of the algorithm, the most important being the Fano algorithm [86] and the ZJ-algorithm [135, 289].

Sequential decoding is a tree search algorithm. We represent the computation of an (n, k) convolutional encoder as a tree where each node has 2^k outgoing branches, one for each possible k-bit input word. For example, the (2, 1)-encoder of Fig. 4.1 has the tree diagram shown in Fig. 4.4.

In the diagram the nodes of the tree are labeled by the internal state of the encoder. The branches are labeled by the output produced by the encoder in the cor-

Table 4.1 A Fano metric for a (2, 1) convolutional code and a BSC with $p = 0.25$

	00	01	10	11
00	0.17	−1.42	−1.42	−3.00
01	−1.42	0.17	−3.00	−1.42
10	−1.42	−3.00	0.17	−1.42
11	−3.00	−1.42	−1.42	0.17

responding step. Every possible computation of the encoder is represented as a path in the tree. As one can see, the diagram is highly redundant. For example the entire infinite tree is isomorphic to the subtree in the upper half. The idea of sequential decoding is to search in the tree diagram for the "best path". This is done by Algorithm 4.5.

Algorithm 4.5 Sequential decoding

1: {initialization}
2: Store the root of the tree diagram into a priority list.
3: **while** the path stored with the weight has not reached a terminal node **do**
4:     Remove the path with the smallest weight from the priority list.
5:     Compute the weights of its successors and store the successors together with their weights in the priority list.
6: **end while**
7: Return the path with the smallest weight as the decoding result.

There are different variants of how to compute the weight of a path. The following method proposed by Fano [86] is very common.

The Fano metric is essentially just the normal log likelihood metric with an adjustment term to make paths of different length comparable. The log-likelihood metric of a code word $x_0, \ldots, x_{t-1}$ given that we received $y_0, \ldots, y_{t-1}$ is

$$M(y_0 \ldots y_{t-1} | x_0 \ldots x_{t-1}) = \log P(y_0 \ldots y_{t-1} | x_0 \ldots x_{t-1}).$$

In the Fano metric we use for an (n, k) convolutional code the adjustment

$$M(y_0 \ldots y_{t-1} | x_0 \ldots x_{t-1}) = \log P(y_0 \ldots y_{t-1} | x_0 \ldots x_{t-1}) + t(n - k).$$

Algorithm 4.5 becomes clear if we look at an example. Assume again that we use the (2, 1) convolutional code of Fig. 4.1 to encode 11101..., i.e. we send the codeword 1110011000.... The channel is a binary symmetric channel with error probability $p = 0.25$. Assume that the transmission adds two errors and the receiver receives 1000011000 instead of 1110011000.

Table 4.1 shows the corresponding Fano metric.

Now we run the sequential decoding. Table 4.2 shows for each step the content of the priority list.

Table 4.2 Example of the sequential decoding algorithm

Step 1	Step 2	Step 3	Step 4	Step 5	Step 6
$0(-1.42)$	$00(-1.25)$	$1(-1.42)$	$000(-2.67)$	$001(-2.67)$	$0011(-2.50)$
$1(-1.42)$	$1(-1.42)$	$000(-2.67)$	$001(-2.67)$	$10(-2.84)$	$10(-2.84)$
	$01(-4.42)$	$001(-2.67)$	$10(-2.84)$	$11(-2.84)$	$11(-2.84)$
		$01(-4.42)$	$11(-2.84)$	$0000(-4.09)$	$0000(-4.09)$
			$01(-4.42)$	$0001(-4.09)$	$0001(-4.09)$
				$01(-4.42)$	$01(-4.42)$
					$0010(-5.67)$

Step 7	Step 8	Step 9	Step 10	Step 11
$10(-2.84)$	$11(-2.84)$	$111(-2.67)$	$1110(-2.50)$	$11101(-2.33)$
$11(-2.84)$	$00110(-3.92)$	$00110(-3.92)$	$00110(-3.92)$	$00110(-3.92)$
$00110(-3.92)$	$00111(-3.92)$	$00111(-3.92)$	$00111(-3.92)$	$00111(-3.92)$
$00111(-3.92)$	$0000(-4.09)$	$0000(-4.09)$	$0000(-4.09)$	$0000(-4.09)$
$0000(-4.09)$	$0001(-4.09)$	$0001(-4.09)$	$0001(-4.09)$	$0001(-4.09)$
$0001(-4.09)$	$100(-4.26)$	$100(-4.26)$	$100(-4.26)$	$100(-4.26)$
$01(-4.42)$	$101(-4.26)$	$101(-4.26)$	$101(-4.26)$	$101(-4.26)$
$0010(-5.67)$	$01(-4.42)$	$01(-4.42)$	$01(-4.42)$	$01(-4.42)$
	$0010(-5.67)$	$0010(-5.67)$	$0010(-5.67)$	$11100(-5.50)$
		$110(-5.84)$	$1111(-5.67)$	$0010(-5.67)$
			$110(-5.84)$	$1111(-5.67)$
				$110(-5.84)$

After the first step we have two possible paths: 0 and 1. Both paths give one error and hence they have the same Fano metric. If there is a tie, we choose one of the best paths, in our example we choose to extend the path 0. In the next few steps nothing special happens, but in step 7 we reach a terminal node with the path 00110 and the path 00111. Since the weight of these paths is not minimal we cannot stop. So we continue until we reach, in step 11, the point where the first element of the priority list is a terminal node. So we obtain 11101 as the output of the sequential decoding.

Optimal performance for sequential decoding is achieved by using encoders with a rapidly growing distance profile. In contrast to Viterbi decoding, the number of flips-flops used by the encoder does not have a significant influence on the decoding speed.

4.2.3 Application to Cryptography

We are now ready to apply ideas from convolutional coding to cryptography.

Take a relation (r, r') (of weight 2) as defined for the CJS-attack, i.e. $c_{r,j} = c_{r',j}$ for $j \geq B$. The important observation for the CJS-attack was Theorem 4.2, especially

$$P\left(z_r + z_{r'} = \sum_{j=0}^{B-1}(c_{r,j} + c_{r',j})x_j\right) = p^2 + (1-p)^2. \qquad (4.12)$$

Since the linear code generated by an LFSR is cyclic, we can shift the indices in Eq. (4.12), i.e. Eq. (4.12) implies

$$P\left(z_{r+t} + z_{r'+t} = \sum_{j=0}^{B-1}(c_{r,j} + c_{r',j})x_{j+t}\right) = p^2 + (1-p)^2 \qquad (4.13)$$

for all $t \in \mathbb{Z}$.

In the CJS attacks we used Eq. (4.12), reducing the correlation attack to the problem of decoding the $[R, B]$ block code

$$C' = \left\{\left(\sum_{j=0}^{B-1}(c_{r,j} + c_{r',j})\hat{x}_j\right)_{(r,r')\in\Omega} \;\middle|\; \hat{x} \in \mathbb{Z}_2^B\right\}.$$

In the same way we can use Eq. (4.13) to reduce the correlation attack to the problem of decoding a $(R, 1)$ convolutional code with memory size B.

$$E_{C'} : (\hat{x}_t)_{t\in N} \mapsto \left(\left(\sum_{j=0}^{B-1}(c_{r,j} + c_{r',j})\hat{x}_j\right)_{(r,r')\in\Omega}\right)_{t\in\mathbb{N}}.$$

This idea is due to Johansson and Jönsson [136] and is even older than the CSJ-attacks.

4.2.3.1 A Fast Correlation Attack Based on Viterbi's Algorithm

First we apply Viterbi's algorithm to correlation attacks. This is the idea described in [136] and analyzed in detail in [137].

In comparison to the original Viterbi algorithm, there are some necessary variations. The most important is that in convolutional coding we always assume that the encoder is initialized with the state 0. This allows us to assign the probability 1 to the initial state in the Viterbi algorithm (see line 1 of Algorithm 4.4). For the convolutional (R, B) code constructed in the previous section we have no such *a priori* information. Thus line 1 of Algorithm 4.4 must be replaced by a CSJ-attack which calculates the correct likelihood values. The work that we have to do for the initial CSJ-attack is approximately the same as we have to do for a single Viterbi step.

In [83] the following variation of the attack based on Viterbi's algorithm was developed. Consider k time steps of the LFSR as one step of the convolutional code. Thus we reduce the correlation attack to the decoding of a (R_k, k) convolutional code of memory size B.

Each Viterbi step of the (R_k, k) convolutional code of memory order B needs 2^{k+B} operations. In fact, the number of operations also depends on R_k; we must evaluate the relations. However, as we have seen in Sect. 4.1.6.2, the time needed to count the satisfied relations via a Fourier transform is almost independent of the number of relations.

For the correlation attack this means that we can process t time steps of the LFSR in about $\frac{t}{k}2^{k+B}$ operations. Since $2^1/1 = 2^2/2$, the attack using the $(R_2, 2)$ convolutional code is almost as fast as the original attack using the $(R_1, 1)$ convolutional code.

The advantage of the new attack is that $R_2 \approx 3R_1$, and hence the variation with the $(R_2, 2)$ convolutional code evaluates about $3/2$ more relations than the original attack for almost no extra work.

The evaluation of the relations via simple counting (see Sect. 4.1.6.1) needs some adjustments. As pointed out in [83], the problem of counting the satisfied relations reduces naturally to the problem of counting the 1 bits in the columns of a bit matrix. Since transposing a bit matrix is an expensive operation (see [276] Sect. 7.3), one cannot reduce this problem to the normal bit count algorithms. In [82, 83] the authors give a specialized vertical bit count algorithm that solves this problem efficiently.

The Viterbi algorithm can also benefit from the use of sequential tests. We can detect paths which have a bad likelihood value early and avoid performing further computations in these paths. The underlying statistical problem is the following.

Given a random sequence of independent random variables $X_1, X_2, \ldots$ we know $X_i \sim F$ for $i < t$ and $X_i \sim G$ for $i \geq t$. But the value t is unknown and may be infinite. The problem is to detect as early as possible that we have passed the *change point* t. We will not go into the details of the statistics of such change point problems (see Sect. 15.3 and the references given there, especially [249]).

4.2.3.2   A Fast Correlation Attack Based on Sequential Decoding

Sequential decoding applied to correlation attacks also gives an interesting algorithm. It is closely related to twice step decoding (see Sect. 4.1.5).

First note that in the sequential decoding, we always have full paths. This means if we extend a path from length $k - 1$ to length k, we can use relations where the columns of the generator matrix satisfies $c_{i_1} + \cdots + c_{i_r} = c$ with $c_{k-1} = 1$ and $c_j = 0$ for $j \geq k$. Thus in the pre-processing phase we generate for every k an extra relation set $\Omega^{(k)}$. Since for each relation in $\Omega^{(k)}$ we have 2^k degrees of freedom in the sum c, the size of the relation sets Ω_k grows exponentially.

In comparison to classical sequential decoding this has the effect that we reach the point where a path is either detected as definitively true or false very quickly. One may interpret the sequential decoding as a variation of the twice step decoding

(Sect. 4.1.5) where we take multiple steps and increase the search window by only one bit at each step.

Sequential decoding also works very well with extended relation sets. If our current search window has width k we can interpret the relations from the set $\Omega^{(k+1)}$ as extended relations with characteristic set $\{k\}$. So it is very attractive to use this extended relation set. If we must extend the path, we fix one more bit and the extended relations by normal relations, so we can reuse the previously determined number of satisfied (extended) relations.

To speed up the decoding one can start by computing all paths of length B with a simple CSJ-attack for some suitably chosen B. If B is small enough the sequential decoding algorithm would generate these paths anyway and we save some memory operations when we build the stack in one step. If B is too large, the CSJ-attack will generate many useless paths and we will lose the speed-up from sequential decoding.

4.3 Attacking LFSRs with Sparse Feedback Polynomials

The attacks described in the previous sections work independently of the feedback polynomial of the LFSR under attack. If the feedback polynomial is sparse, we can use a very efficient attack due to W. Meier and O. Staffelbach. In their paper [187] from 1989 they begin the development of correlation attacks with two attacks. Their algorithm A is somewhat similar to CJS-attacks and is now outdated by its modern descendants.

Their algorithm B works perfectly for LFSRs with sparse feedback polynomials. The key idea is the following. In the correlation attacks of the previous sections we always assumed that the correlations were independent. Some correlations in the relations could be safely ignored, but they always led to slightly worse success probabilities. If the feedback is sparse it is easy to generate several relations which have a common column. The attack of Meier and Staffelbach explores the dependency between such relations.

Let $f = x^{n_0} + \cdots + x^{n_k}$ be the sparse feedback polynomial of the LFSR under attack. By $(x_i)_{i \in \mathbb{N}}$ we denote the true LFSR sequence and by $(z_i)_{i \in \mathbb{N}}$ we denote the observed sequence. We know that these sequences are correlated, i.e. we know $P(x_i = z_i) = p$.

The feedback polynomial gives us a relation between the x_i. We know

$$0 = x_{n_0+(i-n_j)} + \cdots + x_{n_k+(i-n_j)}$$

and hence

$$P\left(x_i = z_{n_0+(i-n_j)} + \cdots + z_{n_{j-1}+(i-n_j)} + z_{n_{j+1}+(i-n_j)} + \cdots + z_{n_k+(i-n_j)}\right)$$

$$= p_k = \sum_{j=0}^{\lfloor k/2 \rfloor} \binom{k}{2j} (1-p)^{2j} p^{k-2j}. \tag{4.14}$$

We start with the "*a priori* probability" $P(x_i = z_i) = p$ and use the relations of Eq. (4.14) to get a better "*a posteriori* probability".

Example 4.1 Consider the LFSR with feedback polynomial $x^n + x^k + 1$. Let $p = 0.75$ be the strength of the correlation between the true LFSR and the observed sequence.

Assume $z_i = 1$, i.e. we know $P(x_i = 1) = 0.75$. In addition (4.14) gives us the relations

$$P(x_i = z_{i+n} + z_{i+k}) = p_2 = 0.625,$$

$$P(x_i = z_{i+(n-k)} + z_{i-k}) = p_2 = 0.625,$$

$$P(x_i = z_{i+(k-n)} + z_{i-n}) = p_2 = 0.625.$$

Suppose that two of the three observed relations are satisfied, e.g. let us assume that $z_{i+n} + z_{i+k} = z_{i+(n-k)} + z_{i-k} = 1$ and $z_{i+(k-n)} + z_{i-n} = 1$.

The *a posteriori* probability is computed as:

$$P(x_i = 1|\text{observed relations}) = \frac{pp_2^2(1 - p_2)}{pp_2^2(1 - p_2) + (1 - p)(1 - p_2)^2 p_2}.$$

($p_2^2(1 - p_2)$ is the conditional probability that two of three relations are satisfied under the condition that $x_i = 1$. The factor p is for the *a priori* probability that $x_i = 1$. Similarly $(1 - p)(1 - p_2)^2 p_2$ is the probability that we observe two satisfied relations and that $x_i = 0$.)

For the concrete values we get $P(x_i = 1|\text{observed relations}) = 0.833$. The *a posteriori* probability is much better than the *a priori* probability 0.75.

This leads to Algorithm 4.6.

Algorithm 4.6 Meier's and Staffelbach's attack against LFSRs with sparse feedback polynomials

---

(1) Allocate an array of N floating points to store the probabilities $P(x_i = 1)$ for $i = 1, \ldots, N$. Initialize the array with p or $1 - p$ according to the observed sequences.
(2) For all $i = 1, \ldots, N$ compute the *a posteriori* probability using the relations from Eq. (4.14).
(3) Repeat step (2) until at least n probabilities are below ϵ or above $1 - \epsilon$.
(4) At this point we have determined n outputs of the LFSR with a probability of at least $1 - \epsilon$. Use these n outputs of the LFSR to reconstruct the initial values by solving a linear system of equations.

The correlation of Eq. (4.14) becomes very weak when the feedback polynomial is not sparse. Even a feedback polynomial of weight 10 is sufficient to guarantee that the probabilities in Algorithm 4.6 will not converge to 0 or 1. However, for trinomials Algorithm 4.6 will terminate after a few iterations with the correct result, even if the LFSR has size 1000 and we observe only 100000 bits.

Chapter 5
BDD-Based Attacks

Binary decision diagram-based attacks, introduced in 2002 by M. Krause [166], are a type of time-memory trade-off attack based on a remarkably simple idea. The set of internal states that is consistent with the observed output sequence describes a Boolean function. The attack is successful if only few solutions are left. It is more efficient to compute the function than to perform a complete search over all possible keys, but the binary decision diagram used to store the Boolean function requires a lot of memory.

5.1 Binary Decision Diagrams

A *binary decision diagram* (*BDD*) is a method of representing and manipulating Boolean functions. The diagram consists of nodes labeled with numbers $\{1, \ldots, n\}$ for variables $\{x_1, \ldots, x_n\}$ and sinks labeled with $\bot$ for false and $\top$ for true. Each node has two successors. One successor represents the case that the corresponding variable is false (drawn as a dashed line) and the other represents the case that corresponding variable is true (drawn as a straight line).

Figure 5.1 shows a simple example that represents the majority function $x_1x_2 + x_1x_3 + x_2x_3$ in three variables.

The important point that makes binary decision diagrams an effective representation of Boolean functions is that nodes can share a successor. A BDD is called *reduced* if for each node the two successors are different and if no two nodes have the same label and successors. The diagram in Fig. 5.1 is not reduced. Figure 5.2 shows a reduced BDD.

Reducing a BDD simply requires the repeated application of the two reduction steps shown in Fig. 5.3. The reduction process is confluent, so one always reaches the same reduced form regardless of the order in which the reduction steps are applied. See also Algorithm 5.2 for an effective implementation of the reduction process. In the following we will always assume that the BDD is reduced.

There will still exist many different BDDs describing the same Boolean function. One can choose at each node which variable to test. The problem of finding the

A. Klein, *Stream Ciphers*, DOI 10.1007/978-1-4471-5079-4_5,
© Springer-Verlag London 2013

Fig. 5.1 A non-reduced
binary decision diagram

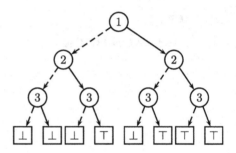

Fig. 5.2 A reduced binary
decision diagram

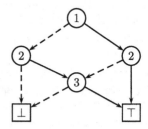

smallest BDD which represents a given Boolean function is NP-complete, so we are
interested in subclasses of BDDs for which the computational part is efficient.

5.1.1 Ordered BDDs

The most common type of binary decision diagram is the ordered binary decision
diagram. A binary decision diagram is *ordered* if the label of each node is greater
than the label of its father, i.e. the variables are read from x_1 to x_n when evaluating
the diagram. Prescribing the order of the variables has the advantage that there exists
a unique reduced ordered binary decision for each function.

Many authors use the term 'binary decision diagram' as a synonym for 'reduced
ordered binary decision diagram'. To emphasize the distinction from free binary
decision diagrams (see Sect. 5.1.2) we will always speak of ordered binary decision
diagrams.

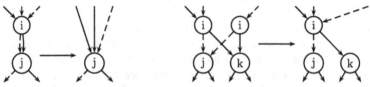

(a) Deleting unnecessary tests   (b) Identifying nodes with the same successors

Fig. 5.3 Reducing a binary decision diagram

For the following algorithms we assume that the nodes of a BDD are stored as a list of branch instructions $I_s, \ldots, I_0$. Each branch instruction I_k is a triple (v_k, L_k, H_k), where v_k is the value of the node and L_k and H_k are the indices of the branch nodes for the false and true case, respectively. By convention we require that I_v is the root, $I_1 = (n + 1, 1, 1)$ represents the sink $\top$ and $I_0 = (n + 1, 0, 0)$ represents the sink $\bot$.

Furthermore, we assume that the instructions are ordered according to the values v_k, which makes some algorithms easier to describe.

Counting all solutions of an ordered binary decision diagram is a simple bottom up algorithm that assigns to each node the number of paths that lead from the node to $\top$ (see Algorithm 5.1).

Algorithm 5.1 Counting solutions of an ordered BDD

1: $c_0 \leftarrow 0, c_1 \leftarrow 1$
2: **for** k **from** 2 **to** v **do**
3:     $l \leftarrow L_k, h \leftarrow H_k$
4:     $c_k \leftarrow 2^{v_k - v_l - 1} c_l + 2^{v_k - v_h - 1} c_h$
5: **end for**
6: **return** $2^{v_s - 1} c_s$

Enumerating all solutions of an ordered binary decision diagram is even simpler. Since every variable occurs at most once per path, every path from the root to $\top$ describes a solution. A deep-first search yields all solutions of the BDD in lexicographic order.

One way to build a reduced ordered BDD is to start with a non-reduced BDD and reduce it. Since BDD algorithms usually run short of memory rather than time, we want to do the reduction without too much extra storage. The following algorithm is due to D. Sieling and I. Wegener [251]. It uses an additional pointer AUX per node and two extra bits.

In the following we assume that the extra bits are the sign bits of the AUX and the LO pointer of the node. If we want to test if the extra bit is set we simply check AUX < 0. If we want to set the extra bit and store the value a we set AUX $\leftarrow \overline{a}$ where $\overline{a}$ denotes the bitwise complement of a. We cannot use AUX $\leftarrow -a$ for this purpose since a could be 0.

Algorithm 5.2 contains a lot of low level pointer operations. The AUX pointers are used to link nodes at the same level, while the algorithm searches for duplicated nodes. When a node is marked for deletion the low-pointer becomes negative and points to the replacement. All deleted nodes form a stack and are linked by their high fields. The pointer AVAIL denotes the top of the stack.

To help our understanding of the algorithm we follow the pointer operations by hand in a small example. Figure 5.4 shows some important intermediate steps of Algorithm 5.2. The AUX fields are indicated by dotted lines.

Algorithm 5.2 Reducing an ordered BDD

1: $AUX(0) \leftarrow AUX(1) \leftarrow AUX(ROOT) \leftarrow -1$ {Initialize}
2: $HEAD(v) \leftarrow -1$ for $1 \leq v \leq v_{max}$.
3: $s \leftarrow ROOT$
4: **while** $s \neq 0$ **do**
5:     $p \leftarrow s, s \leftarrow \overline{AUX(p)}, AUX(p) \leftarrow HEAD(V(p)), HEAD(V(p)) \leftarrow \overline{p}$
6:     **if** $AUX(LO(p)) \geq 0$ **then**
7:         $AUX(LO(p)) \leftarrow \overline{s}, s \leftarrow LO(p)$
8:     **end if** $AUX(0) \leftarrow AUX(1) \leftarrow 0$
9:     **if** $AUX(LO(p)) \geq 0$ **then**
10:         $AUX(LO(p)) \leftarrow \overline{s}, s \leftarrow LO(p)$
11:     **end if**
12: **end while**
13: {Now nodes with the same value are linked by their AUX fields.}
14: **for** v **from** v_{max} **downto** $V(HEAD)$ **do**
15:     $p \leftarrow \overline{HEAD(v)}, s \leftarrow 0$
16:     **while** $p \neq 0$ **do**
17:         $p' \leftarrow \overline{AUX(p)}$ {This will be the next value of p.}
18:         $q \leftarrow HI(p)$
19:         **if** $LO(q) < 0$ **then**
20:             $HI(p) \leftarrow \overline{LO(q)}$ {If $HI(p)$ was deleted, set the pointer to the new value.}
21:         **end if**
22:         $q \leftarrow LO(p)$
23:         **if** $LO(q) < 0$ **then**
24:             $q \leftarrow LO(p) \leftarrow \overline{LO(q)}$ {If $LO(p)$ was deleted, set the pointer to the new value.}
25:         **end if**
26:         **if** $q = HI(p)$ **then**
27:             $LO(p) = \overline{q}, HI(p) \leftarrow AVAIL, AUX(p) \leftarrow 0, AVAIL \leftarrow p$ {Both successors are the same, delete the node, see Fig. 5.3 (a).}
28:         **else if** $AUX(q) \geq 0$ **then**
29:             $AUX(p) \leftarrow s, s \leftarrow \overline{q}, AUX(q) \leftarrow \overline{p}$
30:         **else**
31:             $AUX(p) \leftarrow AUX(\overline{AUX(q)}), AUX(\overline{AUX(q)}) \leftarrow \overline{p}$
32:         **end if**
33:         $p \leftarrow p'$
34:     **end while**
35:     {Nodes with $LO(p) = x \neq HI(p)$ are now linked together via their AUX field.}
36:     $r \leftarrow \overline{s}, s \leftarrow 0$
37:     **while** $r \geq 0$ **do**
38:         $q \leftarrow \overline{AUX(r)}, AUX(r) \leftarrow 0$
39:         **if** $s = 0$ **then**

Algorithm 5.2 (*Continued*)

40:          $s \leftarrow q$
41:      **else**
42:          $\text{AUX}(p) \leftarrow q$
43:      **end if**
44:      $p \leftarrow q$
45:      **while** $\text{AUX}(p) > 0$ **do**
46:          $p \leftarrow \text{AUX}(p)$
47:      **end while**
48:      $r \leftarrow \overline{\text{AUX}(p)}$
49: **end while**
50: $q \leftarrow p \leftarrow s$
51: **while** $p \neq 0$ **do**
52:      $s \leftarrow \text{LO}(p)$
53:      **repeat**
54:          $r \leftarrow \text{HI}(q)$
55:          **if** $\text{AUX}(r) \geq 0$ **then**
56:              $\text{AUX}(r) \leftarrow \overline{q}$
57:          **else**
58:              $\text{LO}(q) \leftarrow \text{AUX}(r), \text{HI}(q) \leftarrow \text{AVAIL}, \text{AVAIL} \leftarrow q$
59:          **end if**
60:          $q \leftarrow \text{AUX}(q)$
61:      **until** $q = 0$ or $\text{LO}(q) \neq s$
62:      **repeat**
63:          **if** $\text{LO}(p) \geq 0$ **then**
64:              $\text{AUX}(\text{HI}(p)) \leftarrow 0$
65:          **end if**
66:          $p \leftarrow \text{AUX}(p)$
67:      **until** $p = q$
68:      **end while**
69: **end for**
70: **if** $\text{LO}(\text{ROOT}) < 0$ **then**
71:      $\text{ROOT} < \overline{\text{LO}(\text{ROOT})}$
72: **end if**

We must also be able to build binary decision diagrams step by step. For example we might have the binary decision diagrams for f and g and want to find the diagram for $f \wedge g$.

Let $\circ$ be any Boolean operator. If either f or g is constant then $f \circ g$ has an obvious value (constant, f, g or the complement of f or g).

Otherwise we write $f = x_1 ? f_H : f_L$ and $g = x_1 ? g_H : g_L$. (Here $X ? Y : Z = (X \wedge Y) \vee (\bar{X} \wedge Z)$ denotes the C-style if-then-else operator.) Then

$$f \circ g = x_1 ?(f_H \circ g_H):(f_L \circ g_L). \tag{5.1}$$

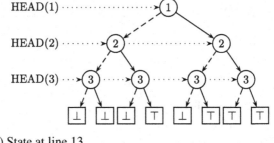

(a) State at line 13

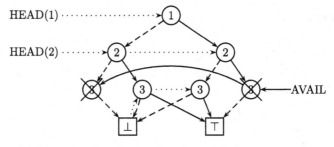

(b) State after line 49, first iteration of the loop

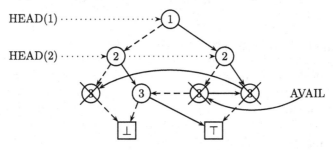

(c) State after line 68, first iteration of the loop

Fig. 5.4 Algorithm 5.2 applied to the diagram of Fig. 5.1

This suggest a deep-first recursive algorithm to compute the binary decision diagram for $f \circ g$. To ensure that the new BDD is reduced one uses memoization. Each time we want to create a node, we first consult a look-up table to check if an equivalent node already exists. In addition one must memoize the calls of the synthesis function to avoid unnecessary calls.

This is the algorithm of choice if we want to build a so-called BDD-base, which represents f, g and $f \circ g$ and several other BDDs simultaneously. The disadvantage is that this algorithm needs dynamic storage allocation.

An alternative is a breadth-first algorithm based on the idea of melting two BDDs. Let $a = (v, l, h)$ and $a' = (v', l', h')$ be two BDD nodes. Then the melt $a \diamond a'$ is

Fig. 5.5 The melt of two
BDDs

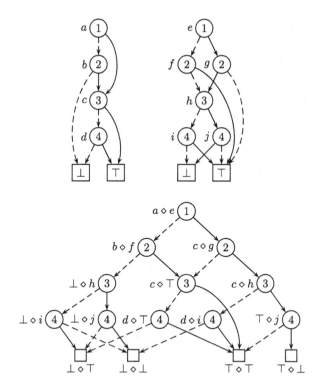

defined by

$$a \diamond a' = \begin{cases} (v, l \diamond l', h \diamond h') & \text{if } v = v', \\ (v, l \diamond a', h \diamond a') & \text{if } v < v', \\ (v, a \diamond l', a \diamond h') & \text{if } v > v'. \end{cases} \tag{5.2}$$

Figure 5.5 shows an example of the melting procedure. The melted diagram has
for sinks $\perp \diamond \perp$, $\perp \diamond \top$, $\top \diamond \perp$ and $\top \diamond \top$.

The melted diagram contains all the information needed to compute any Boolean
function $f \circ g$, one just replaces the sink $\perp \diamond \perp$ by $\perp \circ \perp$, and so on.

The breadth-first algorithm constructs the melt of the two BDDs and reduces it
afterwards by a variation of Algorithm 5.2. With some careful optimizations one
can make this very efficient with respect to memory consumption and cache access. However the pointer operations are very involved. We will skip the detailed
description of this algorithm (see the fully documented reference implementation
that comes with this book or [156] Algorithm S).

Algorithm 5.3 Check that a given BDD is free

Require: BDD nodes are ordered, each node comes after its successors
 1: used-variables[$\bot$] $\leftarrow$ used-variables[$\top$] $\leftarrow \emptyset$
 2: **for all** nodes $a = (v, l, h)$ **do**
 3:     used-variables[a] $\leftarrow$ used-variables[l] $\cup$ used-variables[h]
 4:     **if** $v \in$ used-variables[a] **then**
 5:         **return** false
 6:     **else**
 7:         used-variables[a] $\leftarrow$ used-variables[a] $\cup \{v\}$
 8:     **end if**
 9: **end for**
10: **return** true

The above algorithms are the most important BDD algorithms. We mention just two more common extensions, which are very helpful in many applications but are not so important for BDD-based attacks.[1]

- Many applications have a natural variable order. For the case where there is no natural order, dynamic variable reordering can be used to find small BDDs. In BDD-based attacks against stream ciphers we have a natural variable order, the order in which the variables are computed by the cipher.
- In many applications many knots have a high pointer to the false sink. Zero suppressed BDDs are a variation which stores such knots implicitly. This can reduce the memory requirements, however in cryptography we find that the low pointer points to a sink just as often as the high pointer, so this variant is of less interest.

5.1.2 Free BDDs

Free binary decision diagrams are a generalization of ordered binary decision diagrams. For a free BDD we require that on each path from the root to a sink every variable occurs at most once, but the order of the variables may be different from path to path.

Free BDDs share many properties with ordered BDDs. One can still enumerate all solutions efficiently, and with a little modification Algorithm 5.1 counts the number of solutions of a free BDD. The reduction process described in Fig. 5.3 is still confluent.

Furthermore it is easy to check that a given BDD is free (Algorithm 5.3).

With a free BDD we associate a *control graph*. One obtains the control graph from a free BDD by identifying the two sinks and reducing the resulting graph

---

[1] BDD packages are usually optimized in equal parts for speed, memory efficiency and flexibility. In cryptography we can scarify most of the flexibility for memory efficiency (see also Sect. 12.2).

Fig. 5.6 A free binary
decision diagram

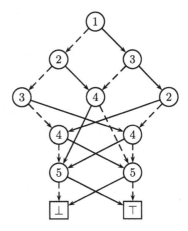

Fig. 5.7 The control graph of
the free BDD in Fig. 5.6

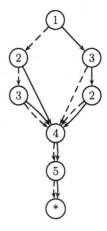

using only rule (b) of Fig. 5.3. Figure 5.7 shows the control graph of the free BDD
given in Fig. 5.6.

The control graph of an ordered BDD is just a line of nodes. Free BDDs with
prescribed control graphs behave nearly identically to ordered BDDs. For a given
control graph each Boolean function has a unique minimal BDD. If all enlisted
free BDDs have the same control graph, the synthesis algorithm from the previous
subsection can be translated to free BDDs.

The size of the control graph appears as a linear factor in the time and space
complexity.

At this point we close our short introduction and refer the reader to the literature,
see [156, 189, 278].

5.2   An Example of a BDD-Based Attack

The idea of a BDD-based attack is to generate a BDD that describes all internal states consistent with the observed output. If the BDD has only one solution then we have found the internal state. BDD-based attacks are a special variant of time-memory trade-off attacks. We illustrate the idea with the example of the cipher E_0 that is used in Bluetooth applications.

5.2.1   The Cipher E_0

Bluetooth is a technology used for short range communication. It has been developed since 1998 by the Bluetooth Special Interest Group, which has more than 5000 member companies.

At several points cryptography plays a role in the Bluetooth protocol (authentication, key exchange, etc.). The payload is encrypted by the stream cipher E_0 described below. We will focus on the cipher and skip the key-scheduling algorithm as well as the reinitialization protocol used in Bluetooth. The reader may consult the Bluetooth specification [31] for the other parts of the protocol.

E_0 consist of four linear feedback shift registers with lengths 25, 31, 33 and 39 (in total 128 bits). In addition it has a non-linear combiner with 4 internal bits. The feedback polynomials of the four LFSRs are

$$f_1 = x^{25} + x^{20} + x^{12} + x^8 + 1,$$
$$f_2 = x^{31} + x^{24} + x^{16} + x^{12} + 1,$$
$$f_3 = x^{33} + x^{28} + x^{24} + x^4 + 1$$

and

$$f_4 = x^{39} + x^{36} + x^{28} + x^4 + 1.$$

The bluetooth protocol specifies that the output of LFSR 1 is the bit at position 1. The outputs of the LFSRs 2, 3 and 4 are the bits at position 7, 1 and 7, respectively. This is only important if one uses the original key scheduling protocol, so in the following we assume that the four LFSRs are in canonical form and always use the bit at position 0 as output.

Figure 5.8 illustrates the E_0 algorithm.

Denote the sequences generated by the four LFSRs by $s_t^{(i)}$ ($i \in \{1, 2, 3, 4\}$). The combiner has four internal bits denoted by $c_t^{(0)}$, $c_t^{(1)}$, $c_{t-1}^{(0)}$ and $c_{t-1}^{(1)}$. Let

Fig. 5.8 The cipher E_0

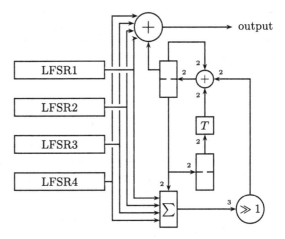

$c_t = (c_t^{(1)} c_t^{(0)})_2 = 2c_t^1 + c_t^0$. The register c_t is updated by the following equations:

$$z_{t+1} = \left\lfloor \frac{s_t^{(1)} + s_t^{(2)} + s_t^{(3)} + s_t^{(4)} + c_t}{2} \right\rfloor$$

$$c_{t+1} = z_{t+1} \oplus c_t \oplus T(c_{t-1})$$

$$T\left(c^{(1)}, c^{(0)}\right) = \left(c^{(0)}, c^{(0)} + c^{(1)}\right)$$

and the output of E_0 is defined as

$$x_t = s_t^{(1)} \oplus s_t^{(2)} \oplus s_t^{(3)} \oplus s_t^{(4)} \oplus c_t^{(0)}.$$

In Fig. 5.8 the small numbers at the wires indicate the number of bits transported on that wire.

5.2.2 Attacking E_0

At the moment E_0 is still practically secure. The known attacks fall into two classes: either they need a very long key stream which is not available in the Bluetooth protocol or they need too much time for a practical attack.

The best attack of the first kind is a correlation attack (see [172]) which needs $\approx 2^{29}$ bits of keystream and $\approx 2^{38}$ operations.

In the second class we find guess and check attacks. The simplest of these attacks (see [134]) simply guesses the initial bits of the LFSRs 1, 2 and 3 and the internal state of the combiner and computes the 35 bits of LFSR 4 from the output. This attack needs about 2^{93} operations. Improved versions (see [94]) lower the complexity to about 2^{84} operations.

In the following we describe a BDD-based attack due to Y. Shaked and A. Wool [239].

For each time step we use 4 Boolean variables $s_t^{(1)}$, $s_t^{(2)}$, $s_t^{(3)}$ and $s_t^{(4)}$ that represent the output of the four LFSRs. The two variables $c_t^{(0)}$ and $c_t^{(1)}$ represent the internal state of the compressor. In addition we introduce variables $z_t^{(0)}$ and $z_t^{(1)}$ to represent the two bits of

$$z = \left\lfloor \frac{\sum_{i=1}^{4} s_t^{(i)} + c_t}{2} \right\rfloor.$$

We will construct a BDD that represents all allocations which are consistent with the observed output. As variable order we choose $s_t^{(1)}, s_t^{(2)}, s_t^{(3)}, s_t^{(4)}, c_t^{(0)}, c_t^{(1)}, z_t^{(1)}$, $z_t^{(0)}$ ($t = 1, 2, 3, \ldots$). (This is just the order in which the bits are generated.)

The BDD is synthesized from small BDDs that describe the LFSR feedback polynomials and the compressor.

The elementary BDDs are:

A BDD that determines whether the XOR of some variables is true (or false). We use this BDD to describe the output $s_t^{(1)} \oplus s_t^{(2)} \oplus s_t^{(3)} \oplus s_t^{(4)} \oplus c_t^{(0)}$, the feedback polynomials of the four LFSRs and the update function of the compressor $0 = c_{t+1} \oplus z_{t+1} \oplus c_t \oplus T(c_{t-1})$. Figure 5.9 (a) shows a BDD of this type.

In addition we need a BDD for the defining relation of z_t (shown in Fig. 5.9 (b)).

The attack against E_0 requires us to compute the AND of all these BDDs. Since we are only interested in the combined BDD, we do not build a BDD base. We will use the breadth-first algorithm described in Sect. 5.1.1.

To determine the complexity of the attack we must estimate the size of the intermediate BDDs. When we look at the definition of the melt of two BDDs (see Eq. (5.2)) we see that the BDD for $B_1 \wedge B_2$ needs at most all the nodes of the form $a \diamond a'$, i.e. its size is bounded by $|B_1| \cdot |B_2|$. This is the worst case upper bound on the size of the melt (and indeed there are BDDs for which the size of the melt grows quadratically). However, if we look a bit closer at the definition of the melt, we see that if B_2 is a BDD of the form shown in Fig. 5.9 (a), then every node is at most doubled.

Similarly, if we look at the melt of all BDDs that deal with the compressor, we see that we need at most 160 nodes per time step to describe the behavior of the compressor. The BDD consists of 16 groups of nodes, one for each possible compressor state, and each group has 10 nodes to describe the sum $s_t^{(1)} + s_t^{(2)} + s_t^{(3)} + s_t^{(4)}$, like the first 10 nodes in the BDD of Fig. 5.9 (b)). We do not need any nodes to describe the variables c_t and z_t because we can apply an existential quantifier to those variables after the melt. Thus this part results in a BDD of size $160t$.

Each recursion equation of an LFSR is described by a BDD like the one in Fig. 5.9 (a). Since the BDD contains only two nodes per level, melting it with another BDD at most doubles the nodes. We need $t - 25$ recursion equations to describe the first LFSR, $t - 31$ equations for second LFSR, $t - 33$ for the third LFSR and $t - 39$ for the fourth LFSR. Thus for $t \geq 39$ we get $2^{4t-128} \cdot (160t)$ as an upper bound for the size of the BDD that describes the first t time steps of E_0.

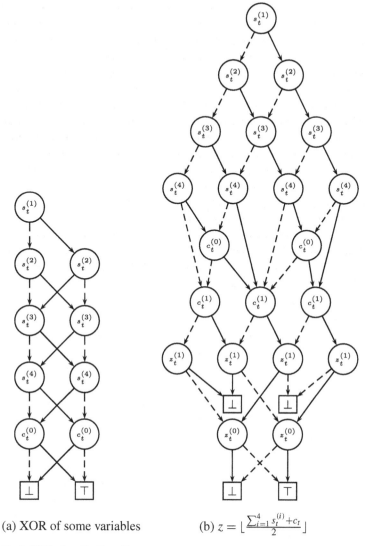

(a) XOR of some variables

(b) $z = \lfloor \frac{\sum_{i=1}^{4} s_t^{(i)} + c_t}{2} \rfloor$

Fig. 5.9 Basic BDDs for attacking E_0

Another bound on the size of the BDD comes from the number of solutions. A BDD with n variables and only s solutions consists of s (partly overlapping) paths of length n. Thus it has at most ns nodes. In the case of the BDD describing E_0 we obtain the following: E_0 has 128 degrees of freedom and each time step yields one equation, so after t steps we expect about 2^{128-t} possible solutions. This bounds the size of the BDD to $4t \cdot 2^{128-t}$.

Putting both bounds for the BDD together, we see that the BDD reaches its largest size approximately at time step 50. At this point it has about 2^{86} nodes which is also a good estimate for the time complexity of the attack. As we can see, the simple BDD-based attack already reaches a time complexity comparable with the best guess and check attacks. The problem is that the space complexity of 2^{86} is clearly unacceptable. So we need some time-memory trade-off.

The natural approach is to guess some of the initial bits and apply the BDD-based attack to a reduced version of E_0. We guess all 25 bits of the first LFSR. That has the advantage that we can compute the whole output of the first LFSR which removes some BDDs from the melt. Furthermore, we guess the first 14 bits of the second and third LFSRs. Then we can compute the first 14 bits of the fourth LFRS directly. For the size of the BDD at time step t we get the following bounds $(16 \cdot 8 \cdot t)2^{3t-128+25}$ (from melting the single BDDs) and $16 \cdot 2^{128-25-2 \cdot 14-t}$ (from the number of paths). This bounds the size of the BDD to $\approx 2^{33}$ which fits into the main memory of a well-equipped computer. Computing the BDD takes only 2^{33} steps.

Thus the time complexity of the attack is about $2^{25+2 \cdot 14}2^{33} = 2^{86}$.

The time complexity of this attack is still quite high, but all other attacks against E_0 that need only a few output bits have a similar complexity (in Sect. 6.2.2 we will present an algebraic attack against E_0). So, at the moment, E_0 still provides an acceptable level of security.

Chapter 6
Algebraic Attacks

6.1 Tools for Solving Non-linear Equations

Let us recall the basic idea behind algebraic attacks (see Sect. 3.3.2). The output of a non-linear combination of LFSRs can be described by a non-linear function applied to the initial state. The goal of the attacker is to solve a system of non-linear equations to recover the initial state.

Therefore our study of algebraic attacks begins with a collection of tools for solving non-linear systems of equations.

We will present two methods for solving systems of non-linear equations. Gröbner bases are a general method and linearization is a specialized method for the over-defined systems common to cryptography.

6.1.1 Gröbner Bases

6.1.1.1 Ordering on Monomials

Gröbner bases are used to solve the ideal membership problem. Let us examine two cases in which we know how to solve this problem.

In the univariate case we solve the ideal membership problem by computing the greatest common divisor with the Euclidean algorithm and the main step is the division with remainder. The key idea is to eliminate high degree terms.

The second case is a system of linear equations which we solve by Gaussian elimination. Here the key idea is to eliminate variables step by step.

Both cases involve the ordering of terms in polynomials and a common key element is the elimination of high order terms.

We start with the definition of monomial orderings in the multivariate case.

A *monomial in* $F[x_1, \ldots, x_n]$ is an expression of the form $x_1^{\alpha_1} \cdots x_n^{\alpha_n}$. For short, we write x^α with $\alpha = (\alpha_1, \ldots, \alpha_n)$ for the monomial $x_1^{\alpha_1} \cdots x_n^{\alpha_n}$. The *degree* of x^α is $|\alpha| = \alpha_1 + \cdots + \alpha_n$.

A. Klein, *Stream Ciphers*, DOI 10.1007/978-1-4471-5079-4_6,
© Springer-Verlag London 2013

Definition 6.1 A *monomial order* on $F[x_1, \ldots, x_n]$ is a relation $\prec$ on $\mathbb{N}^n$ or equivalently a relation on the monomials x^α with $\alpha \in \mathbb{N}^n$ which satisfies the following properties:

(1) $\prec$ is a total ordering on $\mathbb{N}^n$.
(2) If $\alpha \prec \beta$ then $\alpha + \gamma \prec \beta + \gamma$ for all $\gamma \in \mathbb{N}^n$.
(3) $\prec$ is a well-ordering of $\mathbb{N}^n$, i.e. every non-empty subset of $\mathbb{N}^n$ has a smallest element with respect to $\prec$.

An alternative formulation of the well-ordering condition is that $\prec$ has no infinite descending chain, i.e. there exists no sequence $(\alpha_i)_{i \in \mathbb{N}}$ with $a_{i+1} \prec \alpha_i$.

For the univariate case $F[x]$, the only monomial order is the normal order $<$ on the non-negative integers.

Example 6.1 Some important monomial orderings are:

(1) The *lexicographic ordering* $\prec_{\text{lex}}$, which is defined by $(\alpha_1, \ldots, \alpha_n) \prec_{\text{lex}} (\beta_1, \ldots, \beta_n)$ if and only if there exists a $k \in \{1, \ldots, n\}$ with $\alpha_k < \beta_k$ and $\alpha_j = \beta_j$ for all $j < k$.
(2) The *graded lexicographic ordering* $\prec_{\text{grlex}}$, which is defined by $\alpha \prec_{\text{grlex}} \beta$ if either

$$|\alpha| = \sum_{j=1}^{n} \alpha_i < |\beta| = \sum_{j=1}^{n} \beta_i$$

or

$$|\alpha| = |\beta| \quad \text{and} \quad \alpha \prec_{\text{lex}} \beta.$$

(3) The *graded reverse lexicographic ordering* $\prec_{\text{grevlex}}$, which is defined by $a \prec_{\text{grevlex}} \beta$ if and only if either $|\alpha| < |\beta|$ or $|\alpha| = |\beta|$ and $\alpha \succ_{\text{lex}'} \beta$ where $\prec_{\text{lex}'}$ is the lexicographic ordering with $x_n \prec x_{n-1} \prec \cdots \prec x_1$.

Theorem 6.1 $\prec_{\text{lex}}$, $\prec_{\text{grlex}}$ *and* $\prec_{\text{grevlex}}$ *are monomial orderings.*

Proof Note that $\alpha \prec_{\text{lex}} \beta$ if and only if the first non-zero coordinate of $\alpha - \beta \in \mathbb{Z}^n$ is negative. From this we see immediately that $\prec_{\text{lex}}$ is a total order since for $\alpha \neq \beta$, $\alpha - \beta \neq 0$ and hence the first element is either negative or positive.

Furthermore, since $\alpha - \beta = (\alpha + \gamma) - (\beta + \gamma)$ we have $\alpha \prec_{\text{lex}} \beta$ if and only if $\alpha + \gamma \prec_{\text{lex}} \beta + \gamma$.

Assume that $\prec_{\text{lex}}$ is not a well-ordering, i.e. assume that there exists an infinite descending chain

$$\alpha^{(1)} \succ_{\text{lex}} \alpha^{(2)} \succ_{\text{lex}} a^{(3)} \succ_{\text{lex}} \cdots.$$

By definition of $\prec_{\text{lex}}$ we find that the first coordinates of the $\alpha^{(i)}$ must be monotonically decreasing, i.e. $\alpha_1^{(1)} \geq \alpha_1^{(2)} \geq \cdots$. Since $<$ is a well-ordering on $\mathbb{N}$, each monotonically decreasing sequence must become stationary, i.e. there exists an $N_1 \in \mathbb{N}$ and an α_1 with $\alpha_1^{(i)} = \alpha_1$ for $i \geq N_1$.

By the same argument we see that $\alpha_2^{(N_1)} \geq \alpha_2^{(N_1+1)} \geq \cdots$ is a monotonically decreasing sequence and hence it becomes stationary, i.e. $\alpha_2^{(i)} = \alpha_2$ for some α_2 and $i \geq N_2$. Since $N_2 \geq N_1$ we also have $\alpha_1^{(i)} = \alpha_1$ for all $i \geq N_2$.

Repeating these steps we find a bound N_n such that $a_k^{(i)} = \alpha_k$ for all $i > N_2$, but this means that the sequence $\alpha^{(i)}$ becomes stationary, which contradicts the assumption, i.e. $\prec_{\text{lex}}$ is a well-ordering.

Both orders $\prec_{\text{grlex}}$ and $\prec_{\text{grevlex}}$ sort the monomials first by their degree. The only difference is their tie breaking strategies. Since we use a total order to break the ties, $\prec_{\text{grlex}}$ and $\prec_{\text{grevlex}}$ are total orders.

Since $|\alpha + \gamma| = |\alpha| + |\gamma|$ we have $|\alpha| > |\beta|$ if and only if $|\alpha + \gamma| > |\alpha + \gamma|$, hence $\prec_{\text{grlex}}$ and $\prec_{\text{grevlex}}$ satisfy condition (b) of Definition 6.1.

For every $\alpha \in \mathbb{N}^n$ the number of elements $\beta \in \mathbb{N}^n$ with $|\beta| \leq |\alpha|$ is finite (to be exact, the number is $\binom{n+|\alpha|-1}{|\alpha|}$). Hence every descending chain in a graded order must become stationary.   □

A uniform way to describe all possible monomial orderings is via matrix orders.

Definition 6.2 For a regular matrix $M \in \mathbb{R}^{n \times n}$ with only non-negative entries the *matrix order* $\prec_M$ on $\mathbb{N}^n$ is defined by $\alpha \prec_M \beta$ if and only if $M\alpha \prec_{\text{lex}} M\beta$. (Here $\prec_{\text{lex}}$ denotes the lexicographic order on $R^{\mathbb{N}}$.)

Theorem 6.2 (Robbiano [224]) *Every matrix order is a monomial ordering. For every monomial ordering $\prec$ there exists a regular matrix $M \in \mathbb{R}^{n \times n}$ with only non-negative entries with $\alpha \prec \beta \iff \alpha \prec_M \beta$.*

Proof Let $M \in \mathbb{R}^{n \times n}$ be a regular matrix with only non-negative entries. We prove that $>_M$ is a monomial ordering.

Since M is regular, $M\alpha \neq M\beta$ and hence either $M\alpha \prec_{\text{lex}} M\beta$ or $M\alpha \succ_{\text{lex}} M\beta$.

If $\alpha \prec_M B$, i.e. if $M\alpha \prec_{\text{lex}} M\beta$ then $M\alpha + M\gamma \prec_{\text{lex}} M\beta + M\gamma$, i.e. $\alpha + \gamma \prec_M \beta + \gamma$.

Assume that $\prec_M$ is not a well-ordering, i.e. assume that there exists an infinite sequence α_i with $\alpha_i \succ_M \alpha_{i+1}$. Let $M = (m_1, \ldots, m_n)$ be the rows of M. Since the set $\{m_1 x \mid x \in Z^n$ and $m_1 x < c\}$ is finite for every $c \in \mathbb{R}$, the sequence $m_1\alpha_i$ must become stationary, i.e. there exists an N_1 with $m_1\alpha_i = c_1$ for all $i > N_1$. By the same argument we find an $N_2 > N_1$ with $m_2\alpha_i = c_2$, and so on.

However, for $i > N_n$ this means $M\alpha_i = c$ and since M is regular the sequence α_i must be stationary.

For the opposite implication see [224].   □

Example 6.2 The identity matrix defines the lexicographic order.

The graded lexicographic order is defined by the matrix

$$M_{\text{grlex}} = \begin{pmatrix} 1 & \cdots & 1 & 1 \\ 1 & & 0 & 0 \\ & \ddots & & \vdots \\ 0 & & 1 & 0 \end{pmatrix}$$

and the graded reverse lexicographic order is defined by

$$M_{\text{grevlex}} = \begin{pmatrix} 1 & 1 & \cdots & 1 \\ 1 & \cdots & 1 & 0 \\ \vdots & \cdots & \cdots & \vdots \\ 1 & 0 & \cdots & 0 \end{pmatrix}.$$

With a monomial order we can extend the notion of leading terms from univariate to multivariate polynomials.

Definition 6.3 Let $f \in \sum_{\alpha \in \mathbb{N}^n} c_\alpha x^\alpha \in F[x_1, \ldots, x_n]$ be a non-zero polynomial and let $\prec$ be a monomial order.

(a) Each $c_\alpha x^\alpha$ with $c_\alpha \neq 0$ is a *term* of f.
(b) The *multidegree* of f is

$$\text{mdeg}(f) = \max{}_\prec \{\alpha \in \mathbb{N}^n \mid c_\alpha \neq 0\},$$

where $\max_\prec$ is the maximum with respect to $\prec$.
(c) The *leading coefficient* of f is $\text{lc}(f) = c_{\text{mdeg}(f)}$.
(d) The *leading monomial* of f is $\text{lm}(f) = x^{\text{mdeg}(f)}$.
(e) The *leading term* of f is $\text{lt}(f) = \text{lc}(f)\text{lm}(f)$.

Now the transfer of the division algorithm to multivariate polynomials is straightforward (see Algorithm 6.1).

Note that the remainder is not necessary unique, since there could be several possible values of i in line 4.

Example 6.3 Let $f = 3x^2y + 2$, $f_1 = xy - 1$ and $f_2 = x^2 - x$ and choose $\prec_{\text{lex}}$ as the monomial order.

Then $\text{lt}(f_1) = xy$ divides $\text{lt}(f) = 2x^2y$ and $\text{lt}(f_2) = x^2$ also divides $2x^2y$.

If we choose $i = 1$ in line 4 of Algorithm 6.1 we obtain, after the first step, $p = 3x + 2$ and the division algorithm terminates immediately with $q_1 = 3x$, $q_2 = 0$ and $r = 3x + 2$.

If, on the other hand, we choose $i = 2$ in line 4 of Algorithm 6.1, we obtain $p = 3xy + 2$ after the first step and the division algorithm needs one more step before ending with the result $q_1 = 3$, $q_2 = 3y$ and $r = 5$.

As one can see, neither q_1, q_2, r nor the number of reduction steps are uniquely determined.

Algorithm 6.1 Multivariate division with remainder

Require: $f, f_1, \ldots, f_k \in F[x_1, \ldots, x_n]$ are non-zero polynomials, f is a field and $\prec$ is a monomial order on $F[x_1, \ldots, x_n]$.

Ensure: $q_1, \ldots, q_k, r \in F[x_1, \ldots, x_n]$ satisfy $f = q_1 f_1 + \cdots + q_k f_k + r$ and no monomial in r is divisible by any $\mathrm{lt}(f_1), \ldots, \mathrm{lt}(f_k)$.

1: $r \leftarrow 0, \ p \leftarrow 0$
2: $q_i \leftarrow 0$ for $i \in \{1, \ldots, k\}$
3: **while** $p \neq 0$ **do**
4:     **if** $\mathrm{lt}(f_i)$ divides $\mathrm{lt}(p)$ for some i **then**
5:         $q_i \leftarrow q_i + \frac{\mathrm{lt}(p)}{\mathrm{lt}(f_i)}, \ p \leftarrow p - \frac{\mathrm{lt}(p)}{\mathrm{lt}(f_i)} f_i$
6:     **else**
7:         $r \leftarrow r + \mathrm{lt}(p), \ p \leftarrow p - \mathrm{lt}(p)$
8:     **end if**
9: **end while**
10: **return** $q_1, \ldots, q_k, r$

We will make Algorithm 6.1 deterministic by always choosing the smallest possible value of i in line 4. We will write

$$r = f \operatorname{rem}(f_1, \ldots, f_k)$$

for the remainder, which is now uniquely defined.

For $k = 1$ the remainder solves the ideal membership problem, $f \in \langle f_1 \rangle$ if and only if $f \operatorname{rem}(f_1) = 0$. For $k \geq 2$ this is not the case as the following example shows.

Example 6.4 Let $f = xy^2 + 2x$, $f_1 = xy + x$ and $f_2 = y^2 + 2$. Then with the lexicographic order $\prec_{\mathrm{lex}}$ the algorithm gives us $q_1 = y$, $q_2 = 0$ and $r = -xy + 2x$, i.e. $f \operatorname{rem}(f_1, f_2) \neq 0$. But $f = xf_2$ and hence $f \in \langle f_1, f_2 \rangle$.

Next we want to define special ideal bases for which the result of the multivariate division with remainder is unique (no matter which i is chosen in line 4) and hence give the correct answer for the ideal membership problem.

6.1.1.2 Monomial Ideals and the Hilbert Basis Theorem

Definition 6.4 A *monomial ideal* I is an ideal generated by monomials, i.e. there exists a set $A \subseteq \mathbb{N}^n$ with

$$I = \langle x^A \rangle = \langle x^\alpha \mid \alpha \in A \rangle.$$

Lemma 6.3 *Let* $I = \langle x^\alpha \mid \alpha \in A \rangle$ *be a monomial ideal. Then* $x^\beta \in I$ *if and only if* x^β *is a multiple of* x^α *for some* $\alpha \in A$.

Proof Together with x^α, I also contains every multiple x^β.

For the opposite direction let $x^\beta = \sum q_i x^{\alpha_i}$ with $\alpha_i \in A$. Then x^β occurs in at least one term $q_i x^{\alpha_i}$ and so x^β is a multiple of x^{α_i}.   □

The next lemma states that one can determine if f lies in a monomial ideal I by looking only at the monomials in I.

Lemma 6.4 *Let I be a monomial ideal and let $f \in F[x_1, \ldots, x_n]$. Then the following are equivalent*:

(a) $f \in I$.
(b) f *is a linear combination of monomials in I*.

Proof Let $f = \sum q_i x^{\alpha_i}$. Then every monomial that occurs in f must also occur on the right hand side, i.e. it must be a multiple of x^{α_i} for some i. By Lemma 6.3 this means that every monomial of f lies in I.

The implication (b) $\implies$ (a) is trivial and holds for every ideal I.   □

Corollary 6.5 *Two monomial ideals are the same if they contain the same monomials.*

The next lemma states that a monomial ideal is finitely generated. This is a big step towards Hilbert's basis theorem.

Lemma 6.6 (Dickson's lemma) *Every monomial ideal is generated by a finite set of monomials.*

Proof Let $A \in \mathbb{N}^n$ be the set of all α with $x^\alpha \in I$. The set $\mathbb{N}^n$ is partial ordered by $\leq$ with

$$\alpha \leq \beta \quad \Longleftrightarrow \quad a_i \leq \beta_i \quad \text{for all } i \in \{1, \ldots, n\}.$$

Let B be the set of minimal elements in A, i.e. $B = \{\beta \in A \mid \nexists \alpha \in A : \alpha < \beta\}$. By the preceding lemmas, the monomials x^β with $\beta \in B$ generate I.

We claim that B is finite.

For $n = 1$ the partial order $<$ is a well-ordering and B contains just the unique minimal element of A. Let $A_i = \{(a_1, \ldots, \alpha_{i-1}, \alpha_{i+1}, \ldots, \alpha_n) \mid \alpha \in A\}$ be the projection of A onto all but the ith coordinate. By induction, the set $B_i = \{\bar{\beta}^{(1)}, \ldots, \bar{\beta}^{k_i}\}$ of minimal elements of A_i is finite. For each $\bar{\beta}^{(j)} \in B_i$ choose a $\beta^{(j)} \in A$ which projects to $\bar{\beta}^{(j)}$.

Let $b_i = \max_{1 \leq j \leq k_i}\{\beta_i^{(j)}\}$. We claim that for every $\alpha = (a_1, \ldots, \alpha_n) \in A$ there exists a $\beta \leq \alpha$ with $\beta_i \leq b_i$. By definition of B_i, there exists a $\bar{\beta}^{(j)}$ with $\bar{\beta}_{i'}^{(j)} \leq a_{i'}$ for $i' \neq i$. If $\beta_i^{(j)} \leq \alpha_i$ then $\beta^{(j)}$ is the element we are searching for. If $\alpha_i < \beta_i^{(j)}$ then $\alpha_i \leq b_i$ and $\alpha \leq \alpha$ is the element we seek.

This proves that the ith coordinate of a minimal element of A is bounded by b_i. Since we have such a bound for all $i \in \{1, \ldots, n\}$, there are only finitely many minimal elements.   □

Monomial ideals give a very elegant proof of Hilbert's basis theorem.

Theorem 6.7 (Hilbert's basis theorem) *Every ideal of $F[x_1, \ldots, x_n]$ is finitely generated.*

Proof Let $I' = \langle \mathrm{lt}(I) \rangle$ be the monomial ideal corresponding to I. Let $B' = \{m_1, \ldots, m_k\}$ be a final set of monomials generating I', which exists by Lemma 6.6. For $i \in \{1, \ldots, k\}$ choose $f_i \in I$ with $\mathrm{lt}(f_i) = m_i$. We claim that $B = \{f_1, \ldots, f_k\}$ is a basis for I.

Let $f \in I$. The division with remainder algorithm (Algorithm 6.1) gives us polynomials $g_1, \ldots, g_k$ and r with $f = g_1 f_1 + \cdots + g_k f_k + r$ and either $r = 0$ or no term of r is divisible by a leading term of any f_i.

Since $r = f - g_1 f_1 - \cdots - g_k f_k \in I$ we have $\mathrm{lt}(r) \in I' = \langle \mathrm{lt}(f_1), \ldots, \mathrm{lt}(f_k) \rangle$ and hence $r = 0$. This proves $I = \langle f_1, \ldots, f_n \rangle$.   □

An important consequence of Hilbert's basis theorem is:

Corollary 6.8 (Ascending chain condition) *The ring $F[x_1, \ldots, x_n]$ is Noetherian, i.e. any ascending chain $I_1 \subseteq I_2 \subseteq I_3 \subseteq \cdots$ stabilizes, that is $I_N = I_{N+1} = \cdots = I$ for N large enough.*

Proof Let $I = \bigcup_{j=1}^{\infty} I_j$. By Hilbert's basis theorem I has a finite basis $I = \langle g_1, \ldots, g_k \rangle$. Set $N = \min\{j \in \mathbb{N} \mid g_1, \ldots, g_k \in I_j\}$, then $I_N = I_{N+1} = \cdots = I$. □

6.1.1.3 Gröbner Bases and Buchberger's Algorithm

The proof of Theorem 6.7 suggests that an ideal basis $\{f_1, \ldots, f_k\}$ with

$$\langle \mathrm{lt}(f_1), \ldots, \mathrm{lt}(f_k) \rangle = \langle \mathrm{lt}(I) \rangle$$

is something special. We give such a basis a name.

Definition 6.5 Let $\prec$ be a monomial order and the $I \subseteq F[x_1, \ldots, x_n]$ be an ideal. A finite set $G \subseteq I$ is a *Gröbner basis* for I with respect to $\prec$ if $\langle \mathrm{lt}(G) \rangle = \langle \mathrm{lt}(I) \rangle$.

The name Gröbner basis was given to such a basis in 1965 by Bruno Buchberger in honor of his advisor Wolfgang Gröbner.

As we have already seen in the proof of Hilbert's basis theorem, every ideal of $F[x_1, \ldots, x_2]$ has a finite Gröbner basis.

One special property of Gröbner bases is that, for these bases, the division algorithm produces a unique remainder.

Theorem 6.9 *Let G be a Gröbner basis for the ideal I and let $f \in F[x_1, \ldots, x_n]$ be an arbitrary polynomial. Then there exists a unique polynomial $r \in R$ such that:*

(a) $f - r \in I$; and
(b) no term of r is divisible by any monomial in $\mathrm{lt}(G)$.

Proof The existence of r follows from the multivariate division algorithm (Algorithm 6.1).

Assume that $f = h_1 + r_1 = h_2 + r_2$ with $h_1, h_2 \in I$ and no term of r_1 or r_2 is divisible by any leading monomial in $\mathrm{lt}(G)$.

Then $r_1 - r_2 = h_2 - h_1 \in I$ and since G is a Gröbner basis $\mathrm{lt}(r_1 - r_2) \in \langle \mathrm{lt}(G) \rangle$. By Lemma 6.3, $\mathrm{lt}(r_1 - r_2)$ is divisible by some monomial in $\mathrm{lt}(G)$, i.e. $r_1 - r_2 = 0$. $\square$

As a consequence, we conclude that for a Gröbner basis the division algorithm (Algorithm 6.1) produces the same remainder r no matter which i we choose in line 4. Note that the quotients $q_1, \ldots, q_k$ still depend on the order of reduction steps in Algorithm 6.1. An immediate consequence is that Gröbner bases solve the ideal membership problem.

Corollary 6.10 *Let G be a Gröbner basis for the ideal I. Then $f \in I$ if and only if $f \operatorname{rem} G = 0$.*

Now we consider the problem of constructing a Gröbner basis. One reason why not all bases are Gröbner bases is that the leading term of the linear combination $ax^\alpha f + bx^\beta g$ may not be divisible by $\mathrm{lt}(f)$ or $\mathrm{lt}(g)$, since the leading terms of $ax^\alpha f$ and $bx^\beta g$ cancel. We will look at such cancellations in detail, which leads to the definition of S-polynomials.

Definition 6.6 Let $g, h \in F[x_1, \ldots, x_n]$.

(a) If $\mathrm{mdeg}(f) = \alpha$ and $\mathrm{mdeg}(g) = \beta$, let $\gamma = (\gamma_1, \ldots, \gamma_n)$ with $\gamma_i = \max\{\alpha_i, \beta_i\}$. We call x^γ the *least common multiple* of $\mathrm{lm}(f)$ and $\mathrm{lm}(g)$ and write $x^\gamma = \mathrm{lcm}(\mathrm{lm}(f), \mathrm{lm}(g))$.
(b) The *Syzygien polynomial* or *S-polynomial* is

$$S(g, h) = \frac{x^\gamma}{\mathrm{lt}(f)} f - \frac{x^\gamma}{\mathrm{lt}(g)} g.$$

Lemma 6.11 *A finite set $G = \{g_1, \ldots, g_k\}$ is a Gröbner basis of the ideal $I = \langle G \rangle$ if and only if*

$$S(g_i, g_j) \operatorname{rem} G = 0$$

for $1 \leq i < j \leq k$.

Proof $S(g_i, g_j) \in I$ and by Corollary 6.10 this implies $S(g_i, g_j) \operatorname{rem} G = 0$ if G is a Gröbner basis.

Now assume that $S(g_i, g_j) \operatorname{rem} G = 0$ for $1 \leq i < j \leq k$. We have to prove that for $f \in I$, the leading term satisfies $\mathrm{lt}(f) \in \langle \mathrm{lt}(G) \rangle$.

Let $f = \sum_{j=1}^{k} c_i g_i$ be a representation of f with the property that $\delta = \max_{\prec}\{\text{mdeg}(c_i g_i) \mid 1 \leq i \leq k\}$ is minimal under all possible representations of f. Such a minimum exists, since $\prec$ is a well-order.

If $\delta = \text{mdeg } f$ then

$$\text{lt}(f) = \sum_{\substack{1 \leq i \leq k \\ \text{mdeg}(c_i g_i) = \delta}} \text{lt}(c_i g_i)$$

and hence $\text{lt}(f) \in \langle \text{lt}(G) \rangle$.

Now assume that $\text{mdeg } f \prec \delta$, i.e. the leading terms cancel. Let j be the largest index for which $\text{mdeg}(c_i g_i) = \delta$. Then

$$\sum_{\substack{1 \leq i \leq k \\ \text{mdeg}(c_i g_i) = \delta}} \text{lt}(c_i) g_i = \sum_{\substack{1 \leq i < j \\ \text{mdeg}(c_i g_i) = \delta}} \frac{\text{lt}(c_i)}{\text{lcm}(\text{lm } g_i, \text{lm } g_k) / \text{lm } g_i} S(g_i, g_j)$$

because otherwise the leading terms do not cancel.

By assumption $S(g_i, g_j) \text{ rem } G = 0$, i.e. the multivariate division with remainder (Algorithm 6.1) gives us a representation $S(g_i, g_j) = \sum_{l=1}^{k} c_l^{(i,j)} g_l$ with $\text{mdeg}(c_l^{(i,j)} g_l) \leq \text{mdeg } S(g_i, g_j)$.

Then

$$f = \sum_{j=1}^{k} c_i g_i - \sum_{\substack{1 \leq i \leq k \\ \text{mdeg}(c_i g_i) = \delta}} \text{lt}(c_i) g_i + \sum_{\substack{1 \leq i < j \\ \text{mdeg}(c_i g_i) = \delta}} \left(\frac{\text{lt}(c_i)}{\text{lcm}(\text{lm } g_i, \text{lm } g_k) / \text{lm } g_i} \sum_{l=1}^{k} c_l^{(i,j)} g_l \right)$$

is a representation of f with a smaller δ, contradicting the assumption that the original representation was already minimal. □

Lemma 6.11 suggest the following way to compute a Gröbner basis. Compute $S(p, q) \text{ rem } G$ for all Syzygien polynomial and add the remainders to the basis if necessary. This leads to Algorithm 6.2.

Theorem 6.12 *Algorithm 6.2 terminates after finitely many steps and returns a Gröbner basis G of $I = \langle f_1, \ldots, f_k \rangle$.*

Proof First note that at every step $\langle G \rangle = I$. If the algorithm terminates then $S(g_i, g_j) \text{ rem } G = 0$ for all $g_1, g_j \in G$ and, by Lemma 6.11, G is a Gröbner basis.

So we need only show that Algorithm 6.2 terminates. Every time we add an element to G in line 7 the leading monomial of the new element does not lie in $\langle \text{lt}(G) \rangle$. So Algorithm 6.2 produces a chain $\langle \text{lt}(G_1) \rangle \supset \langle \text{lt}(G_1) \rangle \supset \cdots$. By the ascending chain condition (Corollary 6.8) this chain cannot be infinite, i.e. Algorithm 6.2 enlarges G only a finite number of times. □

The presentation of Algorithm 6.2 was chosen for simplicity. An efficient implementation must contain several improvements. First, each S-polynomial must be

---
Algorithm 6.2 Buchberger's algorithm

Ensure: G is a Gröbner basis of $I = \langle f_1, \ldots, f_k \rangle$
  1: $G \leftarrow \{f_1, \ldots, f_k\}$
  2: **repeat**
  3:    $G' \leftarrow G$
  4:    **for** each pair (p, q), $p \neq q$ in G' **do**
  5:       Compute $S = S(p, q) \operatorname{rem} G$
  6:       **if** $S \neq 0$ **then**
  7:         $G \leftarrow G \cup \{S\}$
  8:       **end if**
  9:    **end for**
10: **until** $G = G'$
11: **return** G

---

evaluated at most once. Secondly, there are many criteria to detect S-polynomial that reduce to zero quickly, and the speed of implementation of Buchberger's algorithm depends mainly on how these criteria are applied (see for example [39, 103]).

Finally, the more recent algorithms of Jean-Charles Faugère (F_4 [87] and F_5 [88]) are more efficient than the original Buchberger algorithm. In Sect. 6.1.2.4 we will sketch the F_4 algorithm to highlight its similarities with the XL-algorithm.

There are many good implementations (commercial and open source) of Gröbner basis algorithms. To give just one example, the open source computer algebra system Singular [253] contains effective implementations of Gröbner basis algorithms.

6.1.1.4 The Gröbner Walk

One problem with Gröbner bases is that elimination orders such as the lexicographic order usually perform badly. The reason is that eliminating x_1 can increase the degree of all other variables arbitrarily. Graded orders are more efficient and normally graded reverse lexicographic orders are the best. The Gröbner walk (see [6, 55]) is a technique that allows one to convert a Gröbner basis $G_\prec$ with respect to an order $\prec$ to a Gröbner basis $G_{\prec'}$ with respect to another order $\prec'$ very quickly. It is often better to compute the Gröbner basis with respect to a good order such as the graded reverse lexicographic order and then convert it into the order needed for the current application.

Consider two matrix orders $\prec_A$ and $\prec_B$. For simplicity we assume that only the first rows of A and B are different (see [55] for general matrices). The walk consists of computing the Gröbner bases with respect to $\prec_{(1-t)A+tB}$ and letting t range from 0 to 1. The important part is that one can do this in discrete steps and the update process is much cheaper than computing the Gröbner basis from the start.

To formulate the Gröbner walk we need some notation and terminology.

Definition 6.7 For each matrix order M the first row ω of M is called the *leading form* of M.

ω defines an ω-*degree* on $K[x_1, \ldots, x_n]$. A monomial $x^e = \prod_{i=1}^{n} x_i^{e_i}$ has ω-degree

$$\deg_\omega(x^e) = (e, \omega) = \sum_{i=1}^{n} e_i \omega_i.$$

The ω-degree of a polynomial $f \in K[x_1, \ldots, x_n]$ is the largest ω-degree of its monomials. The *initial form* $\text{in}_\omega(f)$ of f is the part of all monomials with maximal ω-degree.

A polynomial f is ω-*homogeneous* if all its monomials have the same ω-degree, i.e. $f = \text{in}_\omega(f)$.

Let $<$ be a partial order on the monomials and let $\prec'$ be a monomial order. The refinement of $<$ by $\prec'$ is defined by

$$\alpha \prec \beta \quad \Longleftrightarrow \quad (\alpha < \beta) \vee \left((\alpha \not< \beta) \wedge (\beta \not< \alpha) \wedge (\alpha \prec' \beta)\right).$$

Lemma 6.13 *Let A and B be two regular matrices that differ only in the first row. Let ω_A and ω_B be the first row of A and B respectively.*

Let G be a Gröbner basis of I with respect to $\prec_A$. Then G is a Gröbner basis with respect to $\prec_B$ if $\text{in}_{\omega_A}(g) = \text{in}_{\omega_B}(g)$ for all $g \in G$.

Proof Since $\text{in}_{\omega_A}(g) = \text{in}_{\omega_B}(g)$ for all $g \in G$ and the matrix orders differ only in the first row, we get $\text{lt}_A(g) = \text{lt}_B(g)$ for all g. Hence G is either a Gröbner basis for both orders or for none. $\qquad\qquad\square$

Lemma 6.13 assures us that the Gröbner walk proceeds in discrete steps. Let G_t be a Gröbner basis with respect to the matrix order $\prec_{(1-t)A+tB}$. Let t' be defined as

$$t' = \min\{\tau \in (t, 1] \mid \text{in}_{(1-\tau)\omega_A + \tau\omega_B}(g) \neq \text{in}_{(1-t)\omega_A + t\omega_B}(g) \text{ for one } g \in G\}. \quad (6.1)$$

By Lemma 6.13, G_t is a Gröbner basis for all $\prec_{(1-\tau)A+\tau B}$ for all $\tau \in (t, t')$. $\prec_1 = \prec_{(1-t')A+t'B}$ refines $\omega = (1-t')\omega_A + t'\omega_B$, i.e. $\text{lt}_{\prec_1}(\text{in}_\omega(f)) = \text{lt}(f)$. Hence

$$\left\langle \text{lt}_{\prec_1}\left(\langle \text{in}_\omega(I) \mid \rangle\right)\right\rangle = \left\langle \text{lt}_{prec_1}(I)\right\rangle$$

$$= \left\langle \text{lt}_{\prec_1}(G_t)\right\rangle \quad \text{since } G_t \text{ is a Gröbner basis for } I$$

$$\text{with respect to } \prec_1$$

$$= \left\langle \text{lt}_{\prec_1}\left(\text{in}_\omega(G_t)\right)\right\rangle$$

which proves that $\text{in}_\omega(G_t)$ is a Gröbner basis of $\langle \text{in}_\omega(I)\rangle$.

Now we convert $\text{in}_\omega(G_t)$ to a Gröbner basis $G'_{t'}$ of $\langle \text{in}_\omega(I)\rangle$ with respect to $\prec_2 = \prec_{(1-t')A+t'B}$. The crucial point is that this conversion is much simpler that converting G_t directly into a Gröbner basis $G_{t'}$ with respect to $\prec_2$. One point which makes the conversion efficient is that $\langle \text{in}_\omega(I)\rangle$ is much simpler than I. The other point is that $\text{in}_\omega(G_t)$ consists only of ω-homogeneous polynomials. For such ideals we can replace the complex Buchberger algorithm by a very fast method described by Traverso [266].

Let $G_t = \{g_1, \ldots, g_k\}$ and $G'_{t'} = \{m_1, \ldots, m_h\}$. Since $m_1, \ldots, m_h$ are ω-homogeneous we can determine ω-homogeneous polynomials p_{ij} with

$$m_i = \sum_{j=1}^{k} p_{ij} \, \mathrm{in}_\omega(g_j). \tag{6.2}$$

Replacing $\mathrm{in}_\omega(g_j)$ by g_j we get

$$f_i = \sum_{j=1}^{k} p_{ij} g_j \tag{6.3}$$

and we set $G_{t'} = \{f_1, \ldots, f_h\}$. By construction $\mathrm{in}_\omega(f_i) = m_i$ and hence

$$
\begin{aligned}
\langle \mathrm{lt}_{\prec_2}(I) \rangle &= \langle \mathrm{lt}_{\prec_2}\big(\mathrm{in}_\omega(I)\big) \rangle \\
&= \langle \mathrm{lt}_{\prec_2}\big(G'_{t'}\big) \rangle \quad \text{since } G' \text{ is a Gröbner basis of } \langle \mathrm{lt}_{\prec_2}\big(\mathrm{in}_\omega(I)\big) \rangle \\
&= \langle \mathrm{lt}_{\prec_2}\big(\mathrm{in}(G_{t'})\big) \rangle \quad \text{by (6.3)} \\
&= \langle \mathrm{lt}_{\prec_2}(G_{t'}) \rangle \quad \text{since } \prec_2 \text{ refines } \omega.
\end{aligned}
$$

Thus $G_{t'}$ is a Gröbner basis of I with respect to ω'.

This leads to Algorithm 6.3, which transforms a Gröbner basis from one monomial order to another.

Algorithm 6.3 The Gröbner walk

Require: G_A is a Gröbner basis of I with respect to $\prec_A$.
Ensure: G_B is a Gröbner basis of I with respect to $\prec_B$.
1: $t = 0$, $G_0 = G_A$
2: **while** $t < 1$ **do**
3:      Find t' (see (6.1))
4:      Convert the Gröbner basis $\mathrm{in}_{(1-t')\omega_A + t'\omega_B}(G_t)$ of $\langle \mathrm{in}_{(1-t')\omega_A + t'\omega_B}(I) \rangle$ to a
        Gröbner basis with respect to $\prec_{(1-t')A + t'B}$.
5:      Compute the conversion polynomials (see (6.2)) and the new Gröbner basis
        $G_{t'}$ as in (6.3).
6:      $t \leftarrow t'$
7: **end while**

A good implementation of Algorithm 6.3 requires a lot of work in addition to the basic form described above. The choice of the path is especially important, since the number of steps and the complexity of the steps heavily depends on it. We do not go into details here and instead refer to [6, 55].

For ideals of dimension 0, i.e. ideals with only a finite number of solutions, there is a specialized method for the change of the monomial ordering due to Faugère, Gianni, Lazard and Mora [89] which is more efficient than the general Gröbner walk.

6.1.2 Linearization

Linearization is a technique to solve massive over-defined systems of non-linear equations. We start our description with the simple original linearization method and then treat modern variants such as relinearization and the XL-Algorithm. Since the following algorithms involve a lot of linear algebra, efficient matrix operations become essential (see also Sect. 12.5).

6.1.2.1 Ordinary Linearization

A very simple way to solve an over-defined system of non-linear equations is to replace all monomials by new variables and solve the resulting system of linear equations.

For example, consider the Geffe generator (see Sects. 3.2 and 3.3.2).

Let x_i be the output of the Geffe Generator and a_i, b_i and c_i be the output of the three LFSRs. Then

$$x_i = a_i b_i + c_i (b_i + 1).\tag{6.4}$$

Writing a_i, b_i and c_i as linear combinations of the initial values of the LFSRs we get

$$x_i = \sum_{j=0}^{L_a-1}\sum_{k=0}^{L_b-1} \alpha_{j,k}^{(i)} a_j b_k + \sum_{j=0}^{L_b-1}\sum_{k=0}^{L_c-1} \beta_{j,k}^{(i)} b_j c_k + \sum_{j=0}^{L_c-1} \gamma_j^{(i)} c_j \tag{6.5}$$

where the coefficients $\alpha_{j,k}^{(i)}$, $\beta_{j,k}^{(i)}$ and $\gamma_j^{(i)}$ can be computed from the feedback polynomials f_a, f_b and f_c of the three LFSRs.

Now we substitute $a_j b_k$ by $d_{j,k}$ and $b_j c_k$ by $e_{j,k}$ and get the linear system

$$x_i = \sum_{j=0}^{L_a-1}\sum_{k=0}^{L_b-1} \alpha_{j,k}^{(i)} d_{j,k} + \sum_{j=0}^{L_b-1}\sum_{k=0}^{L_c-1} \beta_{j,k}^{(i)} e_{j,k} + \sum_{j=0}^{L_c-1} \gamma_j^{(i)} c_j. \tag{6.6}$$

In Eq. (6.6) we have $L = L_a L_b + L_b L_c + L_c$ unknowns ($L_a L_b$ variables $d_{j,k}$, $L_b L_c$ variables $e_{j,k}$ and L_c variables c_j). Provided that the attacker can observe enough output bits of the Geffe generator, he can solve the system of linear equations (6.6) by Gaussian elimination in $O(L^3)$ steps.

If L is small, it is no problem to obtain L independent equations and the cost of the Gaussian elimination is feasible.

Once we have the solution of the linear system it is easy to obtain the solution of the original non-linear system. In the example the linear system contains the variables c_i and at least one of these variables will be non-zero. By investigating $e_{j,i} = b_j c_i$ for a $c_i \neq 0$ we find the values of b_j, and so on.

Linearization is a very effective method. Surprisingly many stream ciphers are vulnerable to this kind of attack. In the following we will see how to improve the basic idea further.

6.1.2.2 Relinearization

A drawback of the linearization method is that it needs a lot of equations. Relinearization was introduced in 1999 by Kipnis and Shamir [147] with the aim of lowering the number of required equations.

The technique can best be described by an example. Consider a system of homogeneous quadratic equations

$$\sum_{1 \leq j \leq k \leq n} \alpha_{j,k}^{(i)} x_j x_j = b^{(i)} \quad i \in \{1, \dots, m\}. \tag{6.7}$$

Such equations arise, for example, from the simple non-linear filter generator shown in Fig. 3.3.

Using linearization we transform it into a system of m linear equations in $\binom{n}{2} + n$ variables $x_{j,k}$:

$$\sum_{1 \leq j < k \leq n} \alpha_{j,k}^{(i)} x_{j,k} = b^{(i)} \quad i \in \{1, \dots, m\}. \tag{6.8}$$

If $m < \binom{n}{2} + n$, the linear system is under-defined. We apply Gaussian elimination to express each $x_{j,k}$ as a linear combination of l parameters $t_1, \dots, t_l$. (If the m equations of (6.8) are linearly independent, $l = \binom{n}{2} + n - m$. In practice almost all equations will be linearly independent.)

Next consider the equations

$$x_{j,k} x_{l,h} = (x_j x_k)(x_l x_h) = (x_j x_l)(x_k x_h) = x_{j,l} x_{k,h}. \tag{6.9}$$

This can be viewed as a quadratic equation in the $x_{j,k}$ variables and hence as a quadratic equation in the parameters $t_1, \dots, t_l$ expressing them.

This system of quadratic equations is now solved by ordinary linearization.

The new system has $2\binom{n}{4} + 3\binom{n}{3} + \binom{n}{2}$ equations. We have $\binom{n}{4}$ equations of the form $x_{j,k} x_{l,h} = x_{j,l} x_{k,h} = x_{j,h} x_{k,h}$ where j, k, l, h are all different, each of them delivering 2 equations to the system. In addition we have $3\binom{n}{3}$ equations of the form $x_{j,j} x_{k,h} = x_{j,k} x_{j,h}$ and $\binom{n}{2}$ equations of the form $x_{j,j} x_{k,k} = x_{j,k}^2$.

Experiments show that almost all of these equations are linearly independent. The number of variables in the second linear system is $\binom{l}{2} + l$.

Thus relinearization will solve the system (6.8) if

$$2\binom{n}{4} + 3\binom{n}{3} + \binom{n}{2} \geq \binom{l}{2} + l,$$

$$2\binom{n}{4} + 3\binom{n}{3} + \binom{n}{2} \geq \frac{1}{2}\left(\frac{1}{2}n(n+1) - m\right)\left(\frac{1}{2}n(n+1) - m + 1\right),$$

$$m \geq -\frac{\sqrt{3}\sqrt{2n^4 - 2n^2 + 27} - 3n^2 - 3n - 9}{6},$$

$$m \geq \left(\frac{1}{2} - \frac{1}{\sqrt{6}}\right)n + O(\sqrt{n}).$$

This is a significant improvement over ordinary linearization, which needs $m \geq n^2$. The price is that one has to solve a larger system of linear equations.

The idea of relinearization can of course also be applied to higher degree systems, but for higher degrees regularization generates a lot of redundant equations. There are no good theoretical bounds on how many equations are needed for a successful higher degree linearization. Nevertheless, in practice relinearization has proved to be quite an effective weapon for the attacker.

6.1.2.3   The XL-Algorithm

In 2000 N. Courtois, A. Klimov, J. Patarin and A. Shamir presented an improved variant of the relinearization algorithm [63] which they called the extended linearization algorithm (or XL-algorithm, for short). Again the real complexity of the algorithm is not known, but the algorithm behaves well in practice.

Let f_i, $1 \leq i \leq m$, be a set of multivariate polynomials. We assume that the system $f_i = 0$ has only a finite number of solutions (normally only one solution).

The *extended linearization algorithm* (XL-algorithm) is given by Algorithm 6.4.

We illustrate Algorithm 6.4 with a small example:

Example 6.5 Consider the problem of solving

$$x^2 + \alpha xy = \beta,$$

$$y^2 + \gamma xy = \delta$$

with $\alpha \neq 0$ and $\gamma \neq 0$.

Basic linearization does not help because we have three different monomials but only two equations.

Starting the XL-algorithm with $D = 4$ gives in the first step the following equations

$$x^4 + \alpha x^3 y = \beta x^2, \tag{6.10}$$

Algorithm 6.4 XL-algorithm

---

1. **Multiply:** Generate all products of the form $x^d f_i$ with $\deg x^d f_i \leq D$, ($x^d = \prod_{j=1}^{n} x_j^{d_j}$ may be any monomial).
2. **Linearize:** Consider each monomial of degree $\leq D$ as a new variable and perform Gaussian elimination on the equations obtained in step 1.

    Choose the variable order in such a way that all terms containing one variable (say x_1) are eliminated last.
3. **Solve:** Step 2 should result in a univariate equation in powers x_1. If this is not the case, the choice of D in step 1 was too small.

    Otherwise, solve the univariate equation in x_1 (for example by using Berlekamp's algorithm, see for example Algorithm 14.31 in [100]).
4. **Repeat:** Substitute the values of x_1 found in step 3 into the equations, simplify them and run the algorithm again to find the values of the other variables.

$$x^3 y + \alpha x^2 y^2 = \beta xy, \tag{6.11}$$

$$x^2 y^2 + \alpha x y^3 = \beta y^2, \tag{6.12}$$

$$x^2 y^2 + \gamma x^3 y = \delta x^2, \tag{6.13}$$

$$x y^3 + \gamma x^2 y^2 = \delta xy, \tag{6.14}$$

$$y^4 + \gamma x y^3 = \delta y^2. \tag{6.15}$$

Gaussian elimination applied to these 6 equation in addition to the two original equations yields the equation

$$\beta^2 + x^2 (\alpha\beta\gamma - \alpha^2\delta - 2\alpha) + x^4 (1 - \alpha\gamma) = 0.$$

Now the XL-algorithm proceeds to step 3 and finds candidates for x. Substituting these into the original equation yields the corresponding values for y.

6.1.2.4 The XL-Algorithm and Gröbner Bases

The XL-algorithm has a very strong relation to Gröbner bases and especially to the F_4 algorithm for calculating Gröbner bases, which will be sketched in this subsection.

The idea of the F_4 algorithm is to replace the polynomial operations in the Buchberger algorithm by matrix operations which allows the use of fast linear algorithms and to perform several polynomial reductions in one step.

We do not discuss F_4 in detail. In the outer loop you see that F_4 loops as the Buchberger algorithm over all pairs to check the Syzygien polynomials. The important part of the F_4 algorithm is the creation of the list L (which we simplified a bit in Algorithm 6.5). The algorithm spends most of its time in line 12 (computation of

---

Algorithm 6.5 F_4 algorithm (simplified)

Require: $I = \langle F \rangle$

Ensure: G is a Gröbner basis of I with respect to $\prec$.

1: $G \leftarrow F$

2: Pairs $\leftarrow \{(f, g) | f, g \in G, f \neq g\}$

3: **while** Pairs $\neq \emptyset$ **do**

4:      Select a pair $P = (f_i, f_j)$ from Pairs.

5:      Pairs $\leftarrow$ Pairs$\setminus\{P\}$

6:      $cm \leftarrow \mathrm{lcm}(\mathrm{lm}(f_i), \mathrm{lm}(f_j))$

7:      Add all polynomials of the form mf_i with $\mathrm{lm}(mf_i)|cm$ to the list L.

8:      Add all polynomials of the form mf_j with $\mathrm{lm}(mf_j)|cm$ to the list L.

9:      Add all polynomials of the form mg with $g \in G$ and $\mathrm{lm}(mg)|cm$ to the list L.

10:     {the full F_4 has a symbolic pre-processing to reduce the number of polynomials generated in this step}

11:     Linearize the polynomials in L, i.e. interpret L as a matrix.

12:     $\bar{L} \leftarrow$ row echelon form of L with respect to $\prec$.

13:     $\bar{L}^+ \leftarrow \{f \in \bar{L} | \mathrm{lm}(f) \notin \mathrm{lm}(L)\}$

14:     **for** $h \in \bar{L}^+$ **do**

15:         Pairs $\leftarrow$ Pairs $\cup \{(h, g) | g \in G\}$

16:         $G \leftarrow G \cup \{h\}$

17:     **end for**

18: **end while**

---

the row echelon form). But this is just Gaussian elimination (or a clever variation of it using methods from fast and sparse linear algebra).

As one can see, F_4 and the XL-algorithm are both using linearization and Gaussian elimination and there are only minor differences in way the matrix is constructed. In [9] the relation is studied in detail and the main result is that the XL-algorithm does indeed calculate a Gröbner basis.

Thus the XL-algorithm is just another way to describe Gröbner bases. It is nevertheless an interesting point of view which is worth knowing.

6.2 Pre-processing Techniques for Algebraic Attacks

6.2.1 Reducing the Degree

The use of pre-processing techniques can best be explained with an example. Suppose we want to attack a simple combination of five LFSRs in which the output of the combiner is the majority function, i.e. $z_t = f(s_t^{(1)}, s_t^{(2)}, s_t^{(3)}, s_t^{(4)}, s_t^{(5)})$ with

$$f(x_1, x_2, x_3, x_4, x_5) = \begin{cases} 1 & \text{if at least three of the five input variables are 1} \\ 0 & \text{otherwise.} \end{cases}$$

One can express f as a function of degree 4:

$$f(x_1, x_2, x_3, x_4, x_5) = \sigma_3 + \sigma_4$$

where σ_i denotes the ith elementary symmetric function in $x_1, \ldots, x_5$, i.e.

$$\sigma_3 = x_1x_2x_3 + x_1x_2x_4 + x_1x_2x_5 + x_1x_3x_4 + x_1x_3x_5 + x_1x_4x_5$$
$$+ x_2x_3x_4 + x_2x_3x_5 + x_2x_4x_5 + x_3x_4x_5,$$
$$\sigma_4 = x_1x_2x_3x_4 + x_1x_2x_3x_5 + x_1x_2x_4x_5 + x_1x_2x_4x_5 + x_2x_3x_4x_5.$$

Thus f is a function of degree 4. Suppose all LFSRs have length 20, then linearization would result in a system of $\binom{5}{3}20^3 + \binom{5}{4}20^4 = 880000$ variables.

However, there is a much cheaper solution. Assume that we observe 0 as output. Then we know that at least three of the five LFSRs have 0 as output. Hence at least one of the first three LFSRs must be 0, i.e. we know that $s_t^{(1)}s_t^{(2)}s_t^{(3)} = 0$. Applying linearization to this equation gives a system of $20^3 = 8000$ variables.

In the above example we found the low degree equation by an ad hoc argument.

In general we are searching for a polynomial g such that $\deg(fg)$ is low. Here fg is taken in the ring of Boolean polynomials

$$R = F_2[x_1, \ldots, x_n]/\langle x_1^2 - x_1, \ldots, x_n^2 - x_n \rangle.$$

Then $f = 0$ implies $fg = 0$ which can be used to lower the complexity. This is exactly what happened in our example (with $g = x_1x_2x_3$ and $fg = g$). To find such a g we can compute the Gröbner basis of $\langle f, x_1^2 - x_1, \ldots, x_n^2 - x_n \rangle$ with respect to a graded monomial order. Here we can make use of the fact that $x_1^2 - x_1, \ldots, x_n^2 - x_n$ are in the ideal. One consequence is that no exponent is larger than 1, which makes it cheap to store such Boolean polynomials. Many implementations (for example Magma) have efficient Gröbner basis algorithms for this special case.

To use the time steps in which the output of the combiner is non-zero, we search for a low degree polynomial g with $gf = 0$. Then $f(s_t^{(1)}, \ldots, s_t^{(n)}) = 1$ implies $g(s_t^{(1)}, \ldots, s_t^{(n)}) = 0$ which we can use to mount an algebraic attack.

If we are lucky, we might find two distinct low degree polynomials g and h with $gf = h$. Then we get a low degree equation for both cases $f = 0$ and $f \neq 0$.

The following theorem gives us an upper bound on the size of the equations needed in an algebraic attack.

Theorem 6.14 (Theorem 6.0.1 in [64]) *Let f be a Boolean function in n variables. Then there exists a Boolean function $g \neq 0$ of degree at most $\lceil n/2 \rceil$ and a Boolean function h of degree at most $\lceil n/2 \rceil$ such that $gf = h$.*

Proof Let A be the set of all monomials of degree at most $\lceil n/2 \rceil$ and let $B = Af$. Then $|A| + |B|$ is $2 \sum_{i=0}^{\lceil n/2 \rceil} \binom{n}{i} = \sum_{i=0}^{n} \binom{n}{i} + \binom{n}{\lceil n/2 \rceil} > 2^n$. Hence some linear dependency exists between the Boolean functions in A and B. Since the monomials

Fig. 6.1   A combiner with memory

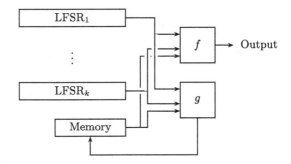

in A are linearly independent, the dependency must use elements of B, i.e. we get a $g \neq 0$ with $gf = h$.   □

The techniques described above appeared in 2003 under the name 'fast algebraic attacks' [8, 62, 64]. Our presentation is based on [64].

When implementing these kinds of algebraic attack one has to be careful when substituting the key stream into the new equations [124]. A naive approach can easy result in a high complexity (sometimes even higher than the time needed for the linearization itself). Techniques like fast Fourier transforms and the asymptotic fast Berlekamp-Massey algorithm (see Algorithm 2.2) help to avoid this problem.

A special case of degree decrease is provided by differential attacks. The operator

$$D : f(x_1, x_2, \dots, x_n) \mapsto f(x_1, x_2, \dots, x_n) + f(x_1 + 1, x_2, \dots, x_n)$$

is the binary analog of the differential operator $\frac{\partial}{\partial x_1}$. The practical use is that it reduces the degree. A natural point where it can be applied in a cryptographic attack is at the key scheduling when the attacker observes two sessions which differ only in the initialization vector bit x_1; he can learn all that is needed to compute $Df(x_1, x_2, \dots, x_n)$. The application of the differential operator in attacks is limited, since designers of cryptosystems tend to choose the algebraic functions f so that Df is still very complex. There are only very rare cases in which the differential operator yields good equations and the more general degree reduction techniques do not find these equations automatically. A special case of a differential attack is the cube-attack which we describe later in the example of a round reduced version of Trivium (see Sect. 10.1).

6.2.2 Dealing with Combiners with Memory

Combiners with memory are an improvement on the simple non-linear combiners discussed in Chap. 3. In Sect. 5.2.1 we presented E_0 as an example of a combiner with memory. In general, such combiners take the form indicated in Fig. 6.1.

In addition to k LFSRs, the combiner also has an internal memory of l bits. We denote the output of the LFSRs by $s_t^{(i)}$, ($t \in \mathbb{N}$, $i \in \{1, \ldots, k\}$). The internal state at time step t is denoted by $z_t = (z_t^{(1)}, \ldots, z_t^{(l)})$. The output of the combiner with memory is

$$x_t = f\left(s_t^{(1)}, \ldots, s_t^{(k)}, z_t^{(1)}, \ldots, z_t^{(l)}\right).$$

The internal state is updated according to

$$z_{t+1}^{(i)} = g_i\left(s_t^{(1)}, \ldots, s_t^{(k)}, z_t^{(1)}, \ldots, z_t^{(l)}\right) \qquad (6.16)$$

for $i = 1, \ldots, l$.

Linearization cannot be applied directly to combiners with memory. If one introduce new variables for all monomials containing a $z_t^{(i)}$ the number of variables explodes, and if one uses Eq. (6.16) to express the $z_t^{(i)}$ by the LFSR outputs and the initial values $z_1^{(i)}$ of the internal state, the degree of the equations explodes.

So in order to apply linearization to a combiner with memory we must search in a pre-processing step for a relation which uses only the outputs x_t of the combiner and the outputs $s_t^{(i)}$ of the LFSRs.

Once more, Gröbner bases are our tool of choice. Consider the ideal

$$I = \Big\langle x_{t'} - f\left(s_{t'}^{(1)}, \ldots, s_{t'}^{(k)}, z_{t'}^{(1)}, \ldots, z_{t'}^{(l)}\right),$$
$$z_{t'+1}^{(j)} - g_j\left(s_{t'}^{(1)}, \ldots, s_{t'}^{(k)}, z_{t'}^{(1)}, \ldots, z_{t'}^{(l)}\right),$$
$$x_{t'}^2 - x_{t'}, \left(s_{t'}^{(i)}\right)^2 - s_{t'}^{(i)}, \left(z_{t'}^{(j)}\right)^2 - z_{t'}^{(j)}\Big\rangle$$

where t' runs from t to $t + \delta_t$, i runs from 1 to k and j runs from 1 to l. Again we work in the ring of Boolean polynomials, which means that the ideal contains all polynomials of the form $x^2 - x$.

For the monomial ordering we divide the variables into three groups. Group 1 contains the variables $z_{t'}^{(j)}$, group 2 contains the variables $s_{t'}^{(i)}$ and group 3 contains the variables $x_{t'}$. Between the groups we use a lexicographic order: Every monomial containing a variable of group 1 is bigger than any monomial containing only variables of group 2 and 3 and every monomial containing a variable of group 2 is bigger than any monomial containing only variables of group 3.

Inside the groups we use a graded ordering, i.e. if two monomials contain both variables of group 1 we compare the degree in the $z_{t'}^{(j)}$ to decide which is smaller.

For our application it is not important how the graded ordering inside the groups is refined to a full order.

Now compute the Gröbner basis of I with respect to this order. The chosen monomial order guarantees that the first element of the Gröbner basis will not contain the variables $z_{t'}^{(j)}$ describing the internal state if possible, and that the degree in the variables $s_{t'}^{(i)}$ describing the output of the LFSRs is as small as possible. This means if there is any relation between the LFSRs and the output of the combiner with memory that is suitable for an attack based on linearization, the Gröbner basis will find the best relation.

In Sect. 6.3.2 we will see how this kind of attack is used in practice.

One can prove that for δ_t large enough the above approach is always successful.

Theorem 6.15 (see [8] Theorem 1) *For any combiner with memory that uses k LFSRs as input and has a memory of size l, there exists a relation using only the output of the LFSRs and the output of the combiner at $l + 1$ successive time steps. This relation has degree at most $\lceil k(l + 1)/2 \rceil$ in the $k(l + 1)$ variables describing the output of the LFSRs.*

For $l = 0$, Theorem 6.15 degenerates to Theorem 6.14. Since Theorem 6.15 makes no use of the degree of the combination function or the feedback function of the internal memory, the bound of Theorem 6.15 is in practice often much higher than the real equations. This limits the application of Theorem 6.14.

6.3 Real World Examples

We close this chapter with two examples of an algebraic attack against real ciphers (see also Sect. 8.2.2, which contains an algebraic attack against the GSM-protocol and Sect. 10.1 which contains an algebraic attack against a reduced variant of Trivium).

6.3.1 LILI-128

The cipher LILI-128 [72] was a submission for the NESSIE-Project (New European Schemes for Signatures, Integrity, and Encryption). It was submitted in 2000 by the ISRC (Information Security Research Centre). The name LILI-128 indicates that this generator is only one of a family of ciphers with different key lengths.

The cipher was withdrawn when the first security problems were reported and it is now totally broken (see [132]).

We use it as example of an algebraic attack in a more complex setting.

LILI use the idea of irregularly clocked shift registers. (We will treat this class of ciphers in the next chapter in detail.)

LILI consists of two LFSRs. The clock control LFSR L_c has length 39 and feedback polynomial $x^{39} + x^{35} + x^{31} + x^{17} + x^{15} + x^{14} + x^2 + 1$. At every time step, L_c is clocked exactly once. Then one computes $f_c = 2x_{12} + x_{20} + 1$ (the computation is done in $\mathbb{Z}$). Then the data-generation LFSR L_d is clocked f_c times. L_d is an LFSR of size 89 with feedback polynomial $x^{89} + x^{83} + x^{80} + x^{55} + x^{53} + x^{42} + x^{39} + x + 1$. A non-linear filter f_d is applied to L_d to produce the output. Let $z_0, \ldots, z_{88}$ be the internal state of L_d. The output is

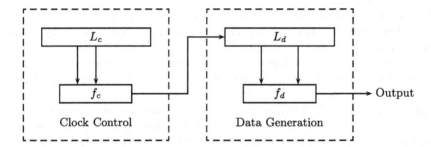

Fig. 6.2 The LILI-128 keystream generator

$$f_d = z_{12} + z_7 + z_3 + z_1 + z_{80}z_{20} + z_{80}z_7 + z_{65}z_3 + z_{65}z_0$$

$$+ z_{44}z_1 + z_{44}z_0 + z_{30}z_{20} + z_{80}z_{65}z_{12} + z_{80}z_{65}z_7 + z_{80}z_{65}z_3$$

$$+ z_{80}z_{65}z_1 + z_{80}z_{44}z_7 + z_{80}z_{44}z_3 + z_{80}z_{30}z_{20}$$

$$+ z_{80}z_{30}z_{12} + z_{80}z_{30}z_7 + z_{65}z_{44}z_{20} + z_{65}z_{44}z_3$$

$$+ z_{65}z_{30}z_{20} + z_{65}z_{30}z_7 + z_{65}z_{30}z_3 + z_{80}z_{65}z_{44}z_{20}$$

$$+ z_{80}z_{65}z_{44}z_7 + z_{80}z_{65}z_{44}z_3 + z_{80}z_{65}z_{44}z_0 + z_{80}z_{65}z_{30}z_{20} + z_{80}z_{65}z_{30}z_7$$

$$+ z_{80}z_{65}z_{30}z_1 + z_{80}z_{44}z_{30}z_{12} + z_{80}z_{44}z_{30}z_3 + z_{65}z_{44}z_{30}z_7 + z_{65}z_{44}z_{30}z_1$$

$$+ z_{65}z_{30}z_{20}z_{12} + z_{65}z_{30}z_{20}z_7 + z_{80}z_{65}z_{44}z_{30}z_7 + z_{80}z_{65}z_{44}z_{30}z_3$$

$$+ z_{80}z_{65}z_{30}z_{20}z_{12} + z_{80}z_{65}z_{30}z_{20}z_7 + z_{65}z_{44}z_{30}z_{20}z_{12} + z_{65}z_{44}z_{30}z_{20}z_7$$

$$+ z_{80}z_{65}z_{44}z_{30}z_{20}z_{12} + z_{80}z_{65}z_{44}z_{30}z_{20}z_7.$$

The designers of LILI claim that f_d was chosen according to principles given in [232] to provide optimal resistance against correlation attacks.

Figure 6.2 illustrates the algorithm.

We want to attack LILI-128 using the techniques from this chapter. First we must get rid of the irregular clock control. In case of LILI-128 we have several ways to achieve this.

- The size of the clock register is relative small. One can simply guess the 39 bits of L_c. This adds a factor of 2^{39} to the attack.
- In a pre-processing step you can compute for each of the 2^{39} possible values of the register L_c relations in the output. Storing the relations needs a lot space, but is feasible. Then you check which relations are satisfied by the output LILI and obtain the content of L_C. So you can avoid the factor 2^{39} during the attack. (See also Sect. 8.2.2, where such a technique is applied to A5/2.)
- In $2^{39} - 1$ time steps (one period of the LFSR L_c) the register L_d is clocked exactly $\Delta_d = 5 \cdot 2^{38} - 1$ times. So if we can observe output bits with difference Δ_d we can ignore the irregular clocking of L_d.

Now we attack the linear filter generator build of L_d and f_d. The degree of f_d is 6 so simple linearization would give a system of more than $\binom{89}{6} \approx 2^{29.2}$ variables. Let us try the pre-processing techniques of Sect. 6.2.1 to reduce the degree.

In case of LILI-128 we need no complicated algebraic techniques to find lower degree equations. Note the degree 5 and 6 terms of f_d. They are divisible by z_{65} and by z_{30} as a common factor, so $f_d(z_{65} + 1)$ and $f_d(z_{30} + 1)$ have at most degree 5. (The degree 5 and 6 terms of f vanish and the degree 4 terms give new degree 5 terms.) One can iterate the trick with $f_d(z_{30} + 1)$ and find that $f_d(z_{30} + 1)(z_{80} + 1)$ has degree 4.

Using this equation for linearization we get a system of only $\sum_{j=1}^{4} \binom{89}{j} \approx 2^{21.2}$ variables. Using fast linear algebra such as Strassen's matrix multiplication (see Sect. 12.5 or [100] for an overview) this results in a complexity of 2^{60} for the algebraic attack. This may not seem great, but it is much better than the complete key search (2^{128} operations) and was reason enough to abandon LILI-128.

We can also apply Gröbner basis techniques to the problem. Using the graded lexicographic order we find that the Gröbner basis of

$$I = \left\langle f_d, z_0^2 - 1, z_1^2 - 1, \dots, z_{80}^2 - 1 \right\rangle$$

contains 13 elements of degree 4.

As noted in [64] LILI-128 is extraordinarily weak against algebraic attacks. A random function f_d would have provided more resistance.

6.3.2 E_0

We use once more the cipher E_0 as an example (see Sect. 5.2.1).

Let us recall the structure of E_0:

- E_0 consists of 4 LFSRs.
- Additional Memory of E_0 has size 4 bits $c_t = (c_t^{(1)}, c_t^{(0)})$ and $c_{t-1} = (c_{t-1}^{(1)}, c_{t-1}^{(0)})$.

Thus, by Theorem 6.15 we know that there must be an equation using only 5 successive time steps and having degree at most 10. Since the combination function and feedback polynomials of E_0 are quite simple, we expect that Theorem 6.15 overestimates the complexity.

Recall the equations defining E_0:

$$z_{t+1} = \left\lfloor \frac{s_t^{(1)} + s_t^{(2)} + s_t^{(3)} + s_t^{(4)} + c_t}{2} \right\rfloor,$$

$$c_{t+1} = z_{t+1} \oplus c_t \oplus T(c_{t-1}),$$

$$T\left(c^{(1)}, c^{(0)}\right) = \left(c^{(0)}, c^{(0)} + c^{(1)}\right),$$

$$x_t = s_t^{(1)} \oplus s_t^{(2)} \oplus s_t^{(3)} \oplus s_t^{(4)} \oplus c_t^{(0)}.$$

Here $s_t^{(i)}$ denotes the output of the LLSRs, c_t the internal state and x_t the output of E_0.

We write the equation defining c_{t+1} as Boolean equations

$$c_{t+1}^{(1)} = \left[\sigma_4(t) + \sigma_3(t)c_t^{(0)} + \sigma_2(t)c_t^{(1)}\right] + c_t^{(1)} + c_{t-1}^{(0)},$$

$$c_{t+1}^{(0)} = \left[\sigma_2(t) + c_t^{(1)} + c_t^{(0)}\sigma_1(t)\right] + c_t^{(0)} + c_{t-1}^{(1)} + c_{t-1}^{(0)}$$

where $\sigma_i(t)$ denotes the elementary symmetric function of degree i in the variables $s_t^{(1)}, s_t^{(2)}, s_t^{(3)}, s_t^{(4)}$.

Now we apply the Gröbner basis technique described in Sect. 6.2.2 to find relations between the output of the LFSRs and the output of E_0. The Gröbner basis gives us the following relation of degree 4 combining 4 successive time steps.

$$
\begin{aligned}
1 = {}& x_t + x_{t+1} + x_{t+2} + x_{t+3} \\
&+ \sigma_1(t+1)\big((x_{t+1}+1)x_{t+3} + (x_{t+1}+1)x_{t+2} + (x_{t+1}+1)x_t + 1\big) \\
&+ \sigma_2(t+1)(1 + x_t + x_{t+1} + x_{t+2} + x_{t+2}) + \sigma_3(t+1)x_{t+1} + \sigma_4(t+1) \\
&+ \sigma_1(t) + \sigma_1(t)\sigma_1(t+1)(1 + x_{t+1}) + \sigma_1(t)\sigma_2(t+1) + \sigma_1(t+2)x_{t+2} \\
&+ \sigma_1(t+2)\sigma_1(t+1)x_{t+2}(1 + x_{t+1}) + \sigma_1(t+2)\sigma_2(t+1)x_{t+2} \\
&+ \sigma_2(t+2) + \sigma_2(t+2)\sigma_1(t+1)(1 + x_{t+1}) + \sigma_2(t+2)\sigma_2(t+1) \\
&+ \sigma_1(t+3) + \sigma_1(t+3)\sigma_1(t+1)(1 + x_{t+1}) + \sigma_1(t+3)\sigma_2(t).
\end{aligned}
$$

Using this equation as a basis for an algebraic attack, linearization produces a system of $\approx 2^{23}$ variables. With fast linear algebra it takes only $\approx 2^{65}$ operations to solve such a system. The applicability of this attack is limited, since the Bluetooth protocol does not allow us to observe 2^{23} variables. E_0 still provides enough security.

Chapter 7
Irregular Clocked Shift Registers

In the previous chapters we have studied non-linear combinations of linear feedback shift registers. Another way to generate a sequence with high linear complexity from an LFSR is to use an irregular clocking. An advantage of such a mechanism is that it avoids most algebraic attacks and thus allows the use of smaller LFSRs. A disadvantage is that these constructions are harder to analyze, so there is a risk that an attacker might find a trick we have missed. In addition, the speed of irregular clocked generators is lower than the speed of regular clocked ciphers and we risk side channel attacks.

7.1 The Stop-and-Go Generator and the Step-Once-Twice Generator

We begin our analysis with two simple generators.

The stop-and-go generator [25] consists of two LFSRs. The first LFSR is regular clocked and its output is the clock control for the second LFSR, i.e. R_2 is clocked at time step t if R_1 outputs 1 at this time step. The output of the stop-and-go generator is the output of R_2. It differs from a simple linear shift register sequence by duplicating some outputs. Figure 7.1 illustrates the construction.

The stop-and-go generator is not suitable for cryptography. First, it has poor statistical properties. For example, two consecutive bits of the stop-and-go generator are equal if either

- the output of R_1 at this time step is 0 and hence R_2 is not clocked (probability $1/2$); or
- the output of R_1 is 1, i.e. R_2 is clocked but the two consecutive outputs of R_2 are equal (probability $1/4$).

Thus the bigram aa occurs in the output of the stop-and-go generator with probability $3/4$, which is much higher than the probability $1/2$ for a sequence of independent and uniformly distributed bits.

A. Klein, *Stream Ciphers*, DOI 10.1007/978-1-4471-5079-4_7,
© Springer-Verlag London 2013

Fig. 7.1 The stop-and-go
generator

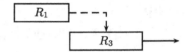

Another problem is that every switch from 1 to 0 and vice versa indicates that the output of R_1 has to be 1. Thus an attacker can reconstruct without difficulty many outputs of R_1, i.e. he can recover the initial seed of R_1. As soon as the clock control is known, the problem of breaking the stop-and-go generator is equivalent to the problem of breaking a linear shift register.

This is a typical problem for all irregular clocked stream ciphers. As soon as the clocking mechanism is known they become very weak ciphers. So one generic attack is to guess the initial state of the LFSR that controls the clock. Therefore it is important that this LFSR must be large enough to make this attack impractical (see also the attack against A5/2 in Sect. 8.2.2).

The stop-and-go generator is the basis for the better alternating step generator, thus it is of interest to find its linear complexity.

Theorem 7.1 *Consider a stop-and-go generator with the following properties*:

- *The control sequence is a de Bruijn sequence* $(\kappa_i)_{i \in \mathbb{N}}$ *of period* 2^k.
- *The generator sequence is an m-sequence* $(\mu_i)_{i \in \mathbb{N}}$ *with minimal polynomial p. Let* $m = \deg(p)$ *and let* $M = 2^m - 1$ *denote the period of the m-sequence.*

Then the period of the stop-and-go generator is $2^k M$ *and the sequence generated by the stop-and-go generator has minimal polynomial* p^j *for some j with* $2^{k-1} < j \le 2^k$.

Proof Let the coefficients of p be denoted by p_i, i.e. $p(x) = \sum_{i=0}^{m} p_i x^i$.

Let $f_t = \sum_{j=0}^{t} \kappa_j$ (the summation is done in $\mathbb{N}$) denote the number of times the generator sequence is clocked at time step t. The output of the stop-and-go generator at time step t is $\mu_{f(t)}$.

Let s be the period of the stop-and-go generator, i.e. $\mu_{f(t+s)} = \mu_{f(t)}$ for all $t \in \mathbb{N}$. This implies

$$f(t + s) \equiv f(t) \mod M \tag{7.1}$$

for all $t \in \mathbb{N}$. Subtracting the equation $f(t - 1 + s) \equiv f(t - 1) \mod M$ we get $\kappa_t = \kappa_{t+s}$. Thus $s = x2^k$. Since exactly half of the elements in a de Bruijn sequence are equal to 1, we obtain $f_{t+s} = x2^{k-1} + f_t$. Together with $\gcd(2^{k-1}, M) = 1$, Eq. (7.1) implies $x = M$, i.e. $s = 2^k M$.

Since μ is an m-sequence, it has the constancy on cyclotomic cosets property (see Theorem 2.4), i.e. p is not only the minimal polynomial of $(\mu_i)_{i \in \mathbb{N}}$ but also of $(\mu_{i2^{k-1}})_{i \in \mathbb{N}}$.

We claim that $p(x^{2^k}) = p(x)^{2^k}$ is a feedback polynomial of the sequence generated by the stop-and-go generator, i.e. we claim $\sum_{i=0}^{m} p_i \mu_{f_{t+i2^k}} = 0$ for all $t \in \mathbb{N}$.

$$\sum_{i=0}^{m} p_i \mu_{f_{t+i2^k}} = \sum_{i=0}^{m} p_i \mu_{f_t+i2^{k-1}}$$

$$\left(2^k \text{ is a full period of the de Bruijn sequence}\right),$$

$$= \sum_{i=0}^{m} p_i \mu_{f_t+\tau}$$

$$(\tau \text{ exists by the constancy on cyclotomic cosets property}),$$

$$= 0$$

$$(\text{since } p \text{ is the feedback polynomial of the LFSR sequence } \mu).$$

Since p is irreducible this proves that the minimal polynomial of the stop-and-go generator sequence is $p(x)^j$ for some $j \leq 2^k$.

Assume that $j \leq 2^{k-1}$. Then $p(x)^j | (x^M - 1)^j | (x^M - 1)^{2^{k-1}} | x^{M2^{k-1}} - 1$, i.e. the stop-and-go generator sequence has period $2^{k-1} M$, a contradiction as we have already proved that the period is $2^k M$. $\square$

Setting $p(x) = x - 1$ in the proof of Theorem 7.1 we obtain that the linear complexity of a de Bruijn sequence lies between 2^{k-1} and 2^k. Theorem 7.1 is just a variation of the theorem on de Bruijn sequences given in [49].

The stop-and-go generator shows that duplicating bits in an LFSR sequence leads to problems. A better approach is to delete bits. One variant is the step-once-twice generator. It consists of one control register and one generation register. If the output of the control register is 0 the generation register is stepped once and if the output of the control register is 1 the generation register is stepped twice.

Thus, on average, every third bit of the generation sequence is deleted. A practical consequence is that the encryption speed decreases. An analysis of the step-once-twice generator can be found in [272]. We will skip it here and concentrate instead on the better alternating step generator and shrinking generator.

7.2   The Alternating Step Generator

The alternating step generator [120] is an improvement of the stop-and-go generator. It consists of three LFSRs. The register $R1$ controls the clock and decides which of the registers $R2$ and $R3$ is clocked (see Algorithm 7.1 and Fig. 7.2).

The alternating step generator preserves the higher speed of the stop-and-go generator in comparison to the step-once-twice generator, but avoids the statistical weakness of the stop-and-go generator.

Algorithm 7.1 The alternating step generator

---

clock the register $R1$
if the output of $R1$ is 1 **then**
    clock the register $R2$ and repeat the previous output of $R3$
else
    clock the register $R3$ and repeat the previous output of $R2$
end if
add the output of $R2$ and $R3$ and return it as the result

Fig. 7.2 The alternating step
generator

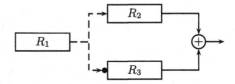

The alternating step generator is normally described as a sum of two stop-and-go generators with complementary control sequences. This has the advantage that many properties follow directly from the stop-and-go generator. For example, Theorem 7.1 immediately translates to the alternating step generator (see also [120]).

For some applications the following, which describes the alternating step generator as two interleaved LFSR sequences, is a better approach:

Let $(w_i)_{i\in\mathbb{N}}$, $(x_i)_{i\in\mathbb{N}}$ and $(y_i)_{i\in\mathbb{N}}$ be the output of the LFSRs L_1, L_2 and L_3, respectively. Let $\tau_n = \sum_{i=1}^n w_i$ be the number of 1s that LFSR L_1 produces in the first n steps. Then the shrinking generator sequence is $z_n = x_{\tau_n} + y_{n-\tau_n}$.

Consider the linear transformation $\hat{z}_n = z_n + z_{n+1}$. For cryptographic purposes the sequence $(\hat{z}_n)_{n\in\mathbb{N}}$ is as good as the original sequence $(z_n)_{n\in\mathbb{N}}$.

Since either $\tau_{n+1} = \tau_n$ (if $w_{n+1} = 0$) or $\tau_{n+1} = \tau_n + 1$ (if $w_{n+1} = 1$) we get

$$\hat{z}_n = \begin{cases} x_{\tau_n} + x_{\tau_n+1} & \text{if } w_{n+1} = 1, \\ y_{n-\tau_n} + y_{n-\tau_n+1} & \text{if } w_{n+1} = 0. \end{cases}$$

If $(x_i)_{i\in\mathbb{N}}$ is an LFSR sequence then $\hat{x}_i = x_i + x_{i+1}$ is also an LFSR sequence with the same feedback polynomial.

Thus the sequence $(\hat{z}_i)$ is generated by Algorithm 7.2. The linear feedback shift registers L_1, L_2 and L_3 have the same feedback polynomial as the register used in Algorithm 7.1 but the initial values are changed.

7.3 The Shrinking Generator

The shrinking generator is a remarkably simple cipher which has now been unbroken for more than 15 years.

Algorithm 7.2 The alternating step generator (alternative form)

---

clock the register $R1$
if the output of $R1$ is 1 **then**
    clock the register $R2$ and output its result
else
    clock the register $R3$ and output its result
end if

Fig. 7.3 The shrinking
generator

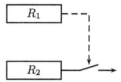

7.3.1 Description of the Cipher

The shrinking generator [58] is a variation of the step-once-twice generator. As for the step-once-twice generator, its output is a subsequence of an LFSR sequence. The difference is that the shrinking generator can delete several successive bits. The rule is if the ith bit of the control sequence is 1, then the ith bit of the generator sequence become a part of the output. If the ith bit of the control sequence is 0, then the corresponding bit of the generator sequence is deleted. Figure 7.3 illustrates the shrinking generator.

The advantage in comparison to the step-once-twice generator is that the deletion points are less regular, which make correlation attacks harder. A disadvantage is that the generation speed is variable, which makes it necessary to use a buffer to prevent side channel attacks.

7.3.2 Linear Complexity of the Shrinking Generator

The shrinking generator has a large period and a high linear complexity, as the following theorem shows. The method of proof is quite similar to Theorem 7.1.

Theorem 7.2 *Replace the LFSR sequence generated by R_1 by a de Bruijn sequence $(c_i)_{i\in\mathbb{N}}$ of period 2^c.*

Assume that R_2 has a primitive feedback polynomial p and the period of the sequence $X = (x_i)_{i\in\mathbb{N}}$ generated by R_2 is $2^n - 1$.

Then the shrinking generator has period $2^{c-1}(2^n - 1)$ and its minimal feedback polynomial is p^j for some j with $2^{c-2} < j < 2^{c-1}$.

Proof Let $Y = (y_i)_{i\in\mathbb{N}}$ be the output sequence of the shrinking generator.

Let τ_i denote the ith occurrence of a 1 in the control sequence $(c_i)_{i \in \mathbb{N}}$. Consider a full period of the de Bruijn sequence. In these 2^c steps the shrinking generator produces 2^{c-1} output bits, i.e. $\tau_{i+2^{c-1}} = \tau_i + 2^c$ or $y_{i+2^{c-1}} = x_{\tau_i+2^c}$.

Since the period of R_2 is $2^n - 1$ and $\gcd(2^c, 2^n - 1) = 1$ it follows that the cipher first repeats after $2^c(2^n - 1)$ time steps, in which it generates $2^{c-1}(2^n - 1)$ output bits.

Since the feedback polynomial p of R_2 is primitive, $(c_i)_{i \in \mathbb{N}}$ has the constancy on cyclotomic cosets property (see Theorem 2.4), i.e. p is not only the minimal polynomial of $(x_i)_{i \in \mathbb{N}}$ but also of $(x_{i2^c})_{i \in \mathbb{N}}$.

Next we prove that $p(x^{2^{c-1}}) = p(x)^{2^{c-1}}$ is a feedback polynomial of the sequence generated by the shrinking generator. Let $p = \sum_{i=0}^n p_i x^i$ then:

$$\sum_{i=0}^n p_i y_{j+2^{c-1}i} = \sum_{i=0}^m p_i x_{\tau_j+i2^c}$$

$$\left(\text{a full period of the de Bruijn sequence results in } 2^{c-1}\right.$$

$$\left.\text{output bits}\right),$$

$$= \sum_{i=0}^m p_i x_{\tau'_j}$$

$$\left(\tau'_j \text{ exists by the constancy on cyclotomic cosets property}\right),$$

$$= 0$$

$$(\text{since } p \text{ is the feedback polynomial of the LFSR } R_2).$$

Since p is irreducible, this proves that the minimal polynomial of the shrinking generator sequence is $p(x)^j$ for some $j \leq 2^{c-1}$.

Assume that $j \leq 2^{c-2}$. Then

$$p(x)^j \big| (x^{2^n-1} - 1)^j \big| (x^{2^n-1} - 1)^{2^{c-2}} \big| x^{(2^n-1)2^{c-2}} - 1,$$

i.e. the shrinking generator sequence has period $(2^n - 1)2^{c-2}$, a contradiction as we have already proved that the period is $(2^n - 1)2^{c-1}$.   □

Note that the proof for the period length does not make full use of the properties of the de Bruijn sequence. For the maximal period it is only necessary that the periods of R_1 and R_2 are relatively prime. Since the linear complexity is much lower than the period and an m-sequence differs from a de Bruijn sequence only if one looks at a full period, we have also proved a lower bound for the shrinking generator with a control sequence generated by an LFSR R_1 with primitive feedback polynomial.

7.3.3 Correlation Attacks Against the Shrinking Generator

As already mentioned at the beginning of this section, shrinking generators with reasonably large registers are still unbroken. One of the strongest known attacks against shrinking generators is a correlation attack found by Golic [113].

Let $X = (x_i)_{i\in\mathbb{N}}$ be the generator sequence, $C = (c_i)_{i\in\mathbb{N}}$ be the control sequence and $Y = (y_i)_{i\in\mathbb{N}}$ be the output sequence of a shrinking generator. By X^n we denote the subsequence of the first n bits. By X_m^n we denote the subsequence $x_m, \ldots, x_n$ of $n - m + 1$ bits.

For a correlation attack against the generator sequence the attacker must compute the *a posteriori* probability $\hat{p}_i = P(x_i = 1 \mid Y^n)$.

In our model we replace the control and generator sequence by iid. random variables with $p(x_i = 1) = p(c_i = 1) = \frac{1}{2}$. The output sequence is also a sequence of iid. random variables.

Note that the first i output bits of the generator sequence can at most influence the first i output bits of the shrinking generator. Thus

$$\hat{p}_i = P(x_i = 1 \mid Y^n) = P(x_i = 1 \mid Y^i).$$

By Bayes' theorem

$$\hat{p}_i = \frac{P(x_i = 1)P(Y^i \mid x_i = 1)}{P(Y^i)} = 2^{i-1} P(Y^i \mid x_i = 1). \tag{7.2}$$

The problem is to compute the probability $P(Y^i \mid x_i = 1)$. We start by distinguishing the cases $w(C^i) = j$ for $j = 0, \ldots, i$. This gives:

$$P(Y^i \mid x_i = 1) = \sum_{j=0}^{i} P(Y^i, w(C^i) = j \mid x_i = 1),$$

$$= \sum_{j=0}^{i} P(Y_{j+1}^i \mid Y^j, w(C^n) = j, x_i = 1),$$

$$P(Y^j, w(C^n) = j \mid x_i = 1).$$

If the weight of C^i is j, the output Y_{j+1}^i is computed from the tails $C_{i+1} = (c_k)_{k\geq i+1}$ and $Y_{i+1} = (y_k)_{k\geq i+1}$. Thus we can drop the condition in the first probability.

$$P(Y^i \mid x_i = 1) = \sum_{j=0}^{n} P(Y_{j+1}^i) P(Y^j, w(C^n) = j \mid x_i = 1)$$

$$= \sum_{j=0}^{i} 2^{i-j} P(Y^j, w(C^n) = j \mid x_i = 1).$$

When computing the probability $P(Y^j, w(C^i) = j \mid x_i = 1)$ we must distinguish two cases:

- In the first case, $c_i = 0$. In this case the ith output of the generator sequence is deleted. Thus it has no influence at all, whether it is 1 or 0. Thus

$$P(Y^j, w(C^i) = j, c_i = 0 \mid x_i = 1) = P(c_i = 0)P(Y^j)P(w(C^{i-1}) = j)$$

$$= \frac{1}{2} \cdot \frac{1}{2^j} \cdot \frac{\binom{i-1}{w}}{2^{i-1}}.$$

- The second case is $c_i = 1$. In that case the ith bit x_i of the generator sequence becomes a part of the output sequence. Since $w(C^i) = j$ we have $y_j = x_i = 1$. Thus

$$P(Y^j, w(C^i) = j, c_i = 1 \mid x_i = 1)$$

$$= P(c_i = 1)P(Y^j, y_j = 1)P(w(C^{i-1}) = j - 1)$$

$$= \frac{1}{2} \cdot \frac{y_j}{2^j} \cdot \frac{\binom{i-1}{w-1}}{2^{i-1}}.$$

Plugging everything into Eq. (7.2) we get

$$\hat{p}_i = 2^{1-i} \sum_{j=0}^{i} \binom{i-1}{j} + 2\binom{i-1}{j-1} y_j$$

$$= \frac{1}{4} + 2^{-i} \sum_{j=1}^{i} \binom{i-1}{j-1} y_j \qquad (7.3)$$

which is Theorem 2 in [113].

Next we want to estimate

$$\left| \hat{p}_i - \frac{1}{2} \right| = 2^{-i} \left| \sum_{j=1}^{i} \binom{i-1}{j-1} \left(y_j - \frac{1}{2} \right) \right|. \qquad (7.4)$$

The factor $\binom{i-1}{j-1}$ increases quickly for $j \leq (i + 1)/2$ and decreases quickly for $j \geq (i + 1)/2$. Thus only the middle terms of the sum make a significant contribution.

By Stirling's formula $n! \approx \sqrt{2\pi n}\, n^n e^{-n}$ we can estimate the size of the binomial coefficients. For $k \in [n/2 - \sqrt{n}, n/2 + \sqrt{n}]$ we get

$$\binom{n}{k} = \Theta\left(\frac{2^n}{\sqrt{n}} \right).$$

So we may continue from Eq. (7.4) by

$$\left| \hat{p}_i - \frac{1}{2} \right| \approx c2^{-i} \left| \sum_{j=i-\sqrt{i}}^{i/2+\sqrt{i}} \frac{2^i}{\sqrt{i}} \left(y_j - \frac{1}{2} \right) \right|$$

$$\approx \frac{c}{\sqrt{i}} \left| \sum_{j=i-\sqrt{i}}^{i/2+\sqrt{i}} y_j - \frac{1}{2} \right|.$$

Thus the problem is reduced to the average absolute difference of a binomial variable and its mean. By the central limit theorem we may replace the binomial distribution by a normal distribution and get

$$\left| \sum_{j=i-\sqrt{i}}^{i/2+\sqrt{i}} \left(y_j - \frac{1}{2} \right) \right| = \Theta \left(\sqrt[4]{n} E \left(|\mathcal{N}(0, 1)| \right) \right) = \Theta \left(\sqrt[4]{i} \right)$$

and thus

$$\left| \hat{p}_i - \frac{1}{2} \right| = \Theta \left(\frac{1}{\sqrt{i}} \right) \Theta \left(\sqrt[4]{i} \right) = \Theta \left(\frac{1}{\sqrt[4]{i}} \right), \tag{7.5}$$

i.e. the correlation between the shrinking generator sequence and its LFSR generator sequence decreases with $\sqrt[4]{i}$. The constant hidden in the Θ-term of Eq. (7.5) is about 0.15 (see [113]). We can expect to use 10000 to 100000 bits in a correlation attack.

The correlation is very weak and for reasonable large LFSRs (about 128 bits) the shrinking generator is still secure, but the analysis shows that attacking irregular clocked stream ciphers is not impossible. Advanced versions of this attack can use probabilities of bigrams and so on.

7.4 Side Channel Attacks

Even if a cipher is mathematically secure, the attacker can still hope to succeed. He can try to measure the time used for encryption or the energy consumed by the chip and use this measurement to draw conclusions on the performed calculations. Attacks of this kind are called *side channel attacks*. Of all stream ciphers, the irregular clocked shift registers are most vulnerable to side channel attacks, so it is appropriate to discuss this type of attack in this chapter.

A *Simple Power Analysis (SPA)* is possible when some operations are directly dependent on the value of the key. The classical example is the implementation of the square and multiply algorithm for computing $x^e \bmod N$. If $e = (e_{k-1} \ldots e_0)_2$, the i-th step requires a square operation and a multiplication if $e_i = 1$, but only one square operation if $e_i = 0$. If the implementation takes no precaution, one can determine e_i directly from the consumed energy.

Irregularly clocked stream ciphers are highly vulnerable against simple power analysis and timing attacks. For example, in a straightforward implementation of the shrinking generator, the time needed to generate a bit is proportional to the number of consecutive 0s in the control sequence. Every implementation must take countermeasures such as inserting dummy operations, using caching to hide the single steps, etc. Normally a simple power analysis can be prevented effectively by such countermeasures. The price for using such countermeasures is always extra gates. One should never forget this when comparing the relatively small size of irregular clocked ciphers with the larger regular clocked ciphers.

The advanced *Differential Power Analysis (DPA)* [162] exploits the fact that the power consumption of an operation is influenced by the value of the manipulated bits. This influence is too small to be measured directly, so differential power analysis needs to measure several thousand encryptions. The general form of the attack is as follows: Assume that we can measure for several encryptions the power used for the encryption. The attacker guesses a part of the key K'. Under the assumption that the guess is correct he can separate the encryptions into two sets S_0 and S_1. Each of the encryptions in S_0 must calculate $f(0)$ and each encryption in S_2 must calculate $f(1)$ for some simple function f. All other operations distribute equally over the sets S_0 and S_1. If the power used in the calculation of $f(0)$ is different from the power used for $f(1)$ then

$$\Delta = \frac{\text{power}(S_0)}{|S_0|} - \frac{\text{power}(S_1)}{|S_1|} \approx \text{power}\big(f(0)\big) - \text{power}\big(f(1)\big)$$

will differ significantly from 0. If the guess K' is incorrect, then all operations will distribute randomly over the sets S_0 and S_1 and we will get $\Delta \approx 0$. This allows us to deduce whether the guess K' was correct. The art of differential power analysis is to identify good functions f and determine the separation of the encryptions into S_0 and S_1.

The design of a synchronous stream cipher provides a high security against a differential power analysis. The key stream generation is separated from the encryption, which is only a simple XOR-operation. Thus it is nearly impossible to collect the several thousands of measurements needed for a differential power analysis.

With respect to differential power analysis, the weakest point is the initialization of the stream cipher. It is quite common for stream ciphers to frequently reseed (one reason for this is to make encryption and decryption parallelizable and to add synchronization points). In this case the key scheduling runs several times with the same key and different initialization vectors. Running an algorithm several times with the same key and different inputs is exactly what is needed for a differential power analysis. In his PhD thesis [167] J. Lano shows for the examples of E_0 (see Sect. 6.3.2) and $A5/1$ (see Sect. 8.3) how differential power analysis might be applied to the key scheduling of stream ciphers.

Power analysis is the most important side channel attack. Other side channel attacks use the cache and the branch prediction unit of modern processors (see, for example, [22]). Since stream ciphers are mostly used on dedicated hardware or on simple embedded processors, this class of side channel attack is less important for stream ciphers.

In fault analysis (see [129]) it is assumed that the attacker can insert faults in the execution of a cipher (e.g., always setting a bit to 0 or skipping an instruction). Although these attack are not impossible, they are less practical.

Part II
Some Special Ciphers

Chapter 8
The Security of Mobile Phones (GSM)

8.1 The GSM Protocol

GSM is the most widely used protocol for mobile phones. There are three types of cryptography used in the protocol: a stream cipher named A5 that is used to encrypt the data; an authentication code named A3; and a key agreement algorithm named A8. Figure 8.1 shows how the algorithms interact.

First the central computer chooses a random number. The SIM card of the mobile phone computes a MAC from its key and this random number using the Algorithm A3. The central computer, which also knows the key of the SIM card, checks if the received MAC is correct. This procedure ensures that the phone company knows that they are communicating with the right mobile phone.

The key generation algorithm A8 takes the same random vector to compute a key for the current session. The data itself is secured by the A5 stream cipher, which exists in two variants: the "strong" version A5/1, which has export limitations; and the weaker export version A5/2. The newer UMTS protocol uses the block cipher KASUMI, which is sometimes called A5/3. As indicated in Fig. 8.1 the protocol adds an error-correcting code to the message before it gets encrypted. This is a serious protocol flaw (compare with Algorithm 1.1). As we will see in the next sections, in the case of the GSM protocol, this flaw is fatal.

All ciphers were initially kept secret. It is never a good idea to violate Kerckhoffs' principle and try to ensure the security of a system by keeping the cipher secret. The GSM protocol shows once more that this leads only to a weak system.

The design of A3 and A8 is not specified by the GSM protocol and can be freely chosen by the telephone company. Most companies choose the example, called COMP128, given in the GSM specification. The specification was never officially published, but in 1988 it was leaked to the public [36]. Immediately afterwards the ciphers were broken, allowing the cloning of a GSM smart card [37].

Both variants of the A5 algorithm were reverse engineered in 1999 by M. Briceno [38]. The first attacks were published immediately afterwards. In the following section we will describe the ciphers and the attacks against them.

A. Klein, *Stream Ciphers*, DOI 10.1007/978-1-4471-5079-4_8,
© Springer-Verlag London 2013

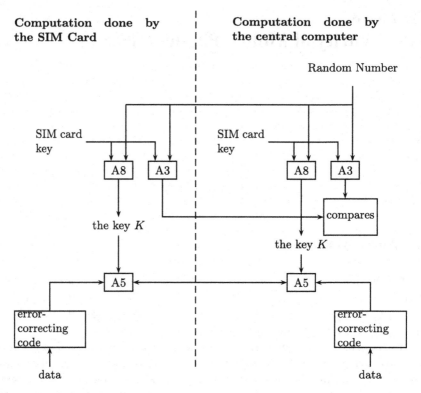

Fig. 8.1 Outline of the GSM protocol

8.2 A5/2

8.2.1 Description of A5/2

The export version $A5/2$ of $A5$ consists of four linear feedback shift registers of length 19, 22, 23 and 17. The registers R_1, R_2 and R_3 are used with a non-linear filter generator and the register R_4 controls the clock for the first three registers.

Figure 8.2 illustrates $A5/2$. As you can see in the figure, the registers have feedback polynomials $f_1 = z^{19} + z^5 + z^2 + z + 1$, $f_2 = z^{22} + z + 1$, $f_3 = z^{23} + z^{15} + z^2 + z + 1$ and $f_4 = z^{17} + z^5 + 1$, respectively.

In each time step the following happens:

- The bits $R_4[6]$, $R_4[13]$ and $R_4[9]$ form the input of the clocking unit, which computes the majority function of the three bits. The register R_1 is clocked if $R_4[6]$ agrees with the majority. Similarly R_2 and R_3 are clocked if $R_4[13]$ or $R_4[9]$ agrees with the majority, respectively.
- A majority function is applied as a non-linear filter to each of the first three registers. For register R_1 we compute $\mathrm{maj}(R_1[3], \overline{R_1[4]}, R_1[6])$, for register R_2 the filter is $\mathrm{maj}(\overline{R_2[5]}, R_2[8], R_2[12])$ and for R_3 the filter is $\mathrm{maj}(R_3[4], R_3[6], \overline{R_3[9]})$.

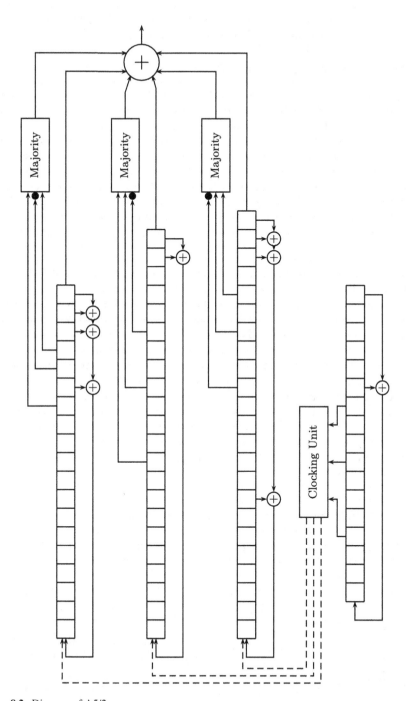

Fig. 8.2 Diagram of A5/2

- The XOR of the outputs of the three non-linear filters and the outputs of the three registers R_1, R_2 and R_3 is the output bit of $A5/2$.
- The register R_4 is clocked.

  $A5/2$ has numerous design flaws:

- First the total state consists of just 81 bits, which is hardly enough to prevent a brute force attack. In any modern application we should use at least 128 bit security to be on the safe side.
- The register $R4$ which controls the clocking unit is, with just 17 bits, very small. An attacker can simply iterate over all possible states of R_4 and attack just the weak non-linear combiners.
- The non-linear combiners of three registers R_1, R_2 and R_3 are of degree 2. This gives an algebraic complexity of just $\binom{19}{2} + 19 + \binom{22}{2} + 22 + \binom{23}{2} + 23 = 702$. (Note that if you put the $19 + 22 + 23 = 64$ bits of the registers $R1$, $R2$ and $R3$ into a single LFSR, a quadratic filter would achieve a higher algebraic complexity ($\binom{64}{2} + 64 = 2080$)!)

Before we can describe the attack on A5/2 we must describe the initialization process and the operation mode.

The operation mode is of particular interest. It has two points relevant for the attack. The first point is that after 228 output bits, A5/2 is reseeded with a new initialization vector. This is a good idea, it makes it impossible for the attacker to gather enough material for an algebraic attack before A5/2 is reseeded. As a consequence, we must analyze data from different frames in one attack, which introduces some complications.

The second important point is that GSM uses error-correcting codes in addition to the encryption, and that the error-correcting code is applied before the encryption. This is a terribly bad idea. It adds redundancy to the plaintext, which converts a known plaintext attack into a ciphertext-only attack.

To complete the description of A5/2 we show how the registers are initialized (see Algorithm 8.2).

8.2.2 An Instance of a Ciphertext-Only Attack

The best attack against A5/2 is due to E. Barkan, E. Biham and N. Keller [15] and breaks A5/2 secured communication in real time. We will develop their attack step by step.

For the moment we ignore the operation mode and the initialization process of A5/2 used in the GSM protocol. We simply assume that we can observe enough output bits to reconstruct the initial state.

As we have seen in Chap. 7, the simplest attack against any LFSR-based cipher with irregular clock control is to guess the state of the clock register and then attack the resulting (weak) cipher.

In this special case the clock register has only 19 bits (18 bits if we take into account that $R_4[6]$ is set to 1 in the GSM protocol), which makes this kind of attack fast enough in practice.

After guessing the register $R4$ we are left with a linear combination of three non-linear combiners. Each combiner has algebraic degree 2, so linearization gives us a system of linear equations with $\binom{19}{2} + 19 + \binom{22}{2} + 22 + \binom{23}{2} + 23 = 702$ variables. (This number reduces to 656 if we take into account that one bit of each shift register is set to 1 in the initialization phase of the GSM protocol, i.e. it is known to the attacker.) We can solve the linear system either with the normal Gaussian algorithm or with the XL-Algorithm (see Sect. 6.1.2.3). In the estimation of the complexity of the attack we will use the Gaussian algorithm. To solve a system of n variables with the Gaussian algorithm we need about $\frac{1}{2}n^3$ operations. So the attack needs $2^{19}(\frac{702^3}{2}) \approx 2^{46}$ bit operations, if we take into account that a word operation performs many bit operations in parallel and there are about 2^{41} word operations on a standard computer. This is already fast, but the attack still takes several minutes. It needs only 702^2 bits of memory (less than 2 kB) to store the system of linear equations and we need to observe less than 800 bits to obtain enough equations. Note that in the GSM protocol the cipher is reseeded after only 228 bits, so obtaining the data is still a problem. We solve this problem later in this section.

Our next goal is to speed up the attack. The idea is that we should be able to identify a wrong guess of the register R_4 without solving a system of linear equations. More precisely we can move the time needed to solve the system of linear equations into a pre-processing step.

The trick is that can we express for each possible initialization value of R_4 the output sequence of A5/2 by quadratic functions in the initial values of the registers R_1, R_2 and R_3. By linearization we transform this into linear expressions in 702 variables and with the Gaussian algorithm we can find linear relations satisfied by the output sequence. For each possible initialization value of R_4 we store about 30 of these relations. This is feasible since each relation connects only 1000 bits, so we must store only $30 \cdot 1000 \cdot 2^{19}$ bits (≈ 1.8 GB).

In the attack we can loop over all initial values of the register R_4 and check if the observed sequence satisfies the stored relations. If we guess the correct initial value of the register R_4 all pre-computed relations are satisfied. For a wrong guess each relation is satisfied only with probability $\frac{1}{2}$. Thus we must check on average $\sum i 2^{-i} = 2$ relations to identify a wrong guess. To check a relation we must just evaluate the bitwise-and-function for bit fields of less than 1000 bits and then perform a sideway addition mod 2 (see Algorithm 2.5). The checks are very fast.

By this method we identify the right initial value for the register R_4. The initial values for the remaining registers are found by solving a system of 702 linear equations as in the basic version of the attack. Solving the system of linear equations is approximately as expensive as the search over all 2^{19} initial states of R_4. The new attack needs less than a second on a standard computer.

Now we go back to the GSM protocol. The problem with the attack we have so far is that we need about 800 bits from the output sequence. In the GSM protocol the cipher is reseeded after only 228 output bits, so we need data from at least 4 frames for the attack.

First note that for an algebraic attack it does not matter that the first 99 output bits of the cipher are discarded (step 5 in Algorithm 8.1). Only the number of observed

Algorithm 8.1 A5/2 initialization

1. Set all registers to zero. Now clock every register 64 times and in each step add one bit of the key (from the least significant to the most significant bit) to the most significant bit of the registers.
2. Clock every register for 22 additional cycles and this time add the bits of the initialization vector to the most significant bit of the register. The initialization vector is increased by one in the next frame.
3. Set the bits $R_1[3]$, $R_2[5]$, $R_3[4]$ and $R_4[6]$ to 1.
4. Run the cipher for 99 steps and discard the output.

bits is important. Discarding the first output bits does not defend against algebraic attacks, but against attacks utilizing a weak key scheduling. RC4 (see Chap. 9) is a famous example of a cipher that is strong against algebraic attacks, but which has a weak key scheduling. In such cases discarding the first output bits is essential for the security of the system.

Note also that if we obtain the initial state of the shift registers (the state after step 4 of Algorithm 8.1), we can guess the old values of the bits that were set to 1 in step 3. Since we know the initialization vector (the frame number) we can invert step 2.

The states of the registers after step 1 of Algorithm 8.1 depend only on the main key. If we want, we can interpret the state after step 1 of Algorithm 8.1 as the main key. This would be enough to predict the cipher in the future, but retrieving the original key from this state is also no problem. All operations in step 1 are linear, so we need only solve a system of 64 linear equations.

Now let us deal with the problem of using four frames in one algebraic attack. In the simplest version we just guess the state of the register R_4 after the first step of Algorithm 8.1. Then we can simulate steps 2 to 4 of Algorithm 8.1 for all frames in question and generate the equations for the algebraic attack.

For the advanced form of the attack with the pre-computed relations note that, if $R_4(f)$ and $R_4(f')$ are the states of the register R_4 after step 2 of Algorithm 8.1, the value $R_4(f) \oplus R_4(f')$ depends only on $f \oplus f'$ and not on the unknown key.

If f is divisible by 4, i.e. the two least significant bits of f are zero, we have $f + 1 = f \oplus 1$, $f + 2 = f \oplus 2$ and $f + 3 = f \oplus 3$.

This allows us to transfer the time-memory trade-off to the GSM protocol. What we do is the following: For all possible values $R_4(f)$ of the register R_4 after step 2 of Algorithm 8.1 we can compute $R_4(f + 1)$, $R_4(f + 2)$ and $R_4(f + 3)$ under the condition that f is divisible by 4. Once we know the values $R_4(f)$, $R_4(f + 1)$, $R_4(f + 2)$ and $R_4(f + 3)$ we can simulate the four frames and search, using the Gaussian algorithm, for relations in the output bits. This can be done in the pre-processing phase.

In the attack we have to wait until we get the output from four frames f, $f + 1$, $f + 2$ and $f + 3$ with f is divisible by 4. Then we can loop over all possible values $R_4(f)$ and check if the pre-computed relations are satisfied. The test is fast and once we know $R_4(f)$ we need only solve a system of 656 linear equations to reconstruct the initial states of the registers R_1, R_2 and R_3. All computations can be done in less than one second.

There are some variations of the attack. Numbers divisible by 4 are not the only numbers for which we can compute the XOR of f, $f + 1$, $f + 2$ and $f + 3$. For example if $f \equiv 1 \mod 8$ we have $f + 1 = f \oplus 3$, $f + 2 = f \oplus 2$, $f + 3 = f \oplus 5$. If we want we can also pre-compute relations for these cases. This increases the time needed in the pre-computation phase and the disk space needed to store the data. The small benefit is that we don't have to wait for a frame number divisible by 4.

Another variation is that we can assume that f is divisible by 8. Then we can compute $R_4(f + 4)$ from $R_4(f)$ and use the relation between the output in the frames $f + 4$, $f + 5$, $f + 6$ and $f + 7$ in addition to the relations we have the basic version. This doubles the number of available relations, which allows us to store fewer relations per possible value of $R_4(f)$. However, disk space is cheap, so this variant is of less interest.

8.2.3 Other Attacks Against A5/2

The attack presented in the previous subsection is so good that there is no need to look at other attacks. We will describe only one other attack. A5/2 was originally kept secret and the reverse engineered algorithm was presented at Crypto 1999. At the same conference, Ian Goldberg, David Wagner and Lucky Green [107] presented an attack against the A5/2. It took them less than five hours to find this attack. In the following we will sketch this first attack.

The key observation is that since the initialization (Algorithm 8.1) sets the bit $R_4[6]$ to 1, the initial value of R_4 is the same for two initialization vectors which differ only at the 11th bit. (Remember that the initial value is added to the most significant bit and $R_4[6]$ is the 11th bit of R_4 if you start counting at the most significant bit.)

So two frames at distance 2^{11} will have the same R_4 register and the registers R_1, R_2 and R_3 will differ by a known δ. We guess the value of R_4. Then the output of A5/2 is a known quadratic function in R_1, R_2 and R_3.

Since for a quadratic function f and fixed δ the difference $g(x) = f(x + \delta) - f(x)$ is a linear function in x, the difference of the two frames at distance 2^{11} will give us a linear system of equations for the internal values of R_1, R_2 and R_3.

Note that, as in the attack described in the previous section, we can use the redundancy of the output to rapidly check if our guess for R_4 is correct or not. Thus we must solve the linear system only for a small number of possible values of R_4, i.e. the attack needs only about 2^{16} operations.

In comparison to the advanced attack described in the previous section this first attack has some drawbacks: It is a known plaintext attack and it needs two frames at distance 2^{11} (which lie about six seconds apart). In contrast the advanced attack needs only a few milliseconds before it can start and it is a ciphertext-only attack.

Algorithm 8.2 A5/1 initialization

---

1. Set all registers to zero. Now clock every register 64 times and in each step add one bit of the key (from the least significant to the most significant bit) to the most significant bit of the registers.
2. Clock every register for 22 additional cycles and this time add the bits of the initialization vector to the most significant bit of the register. The initialization vector is increased by one in the next frame.
3. Run the cipher for 100 steps and discard the output.

Nevertheless, the attack of Goldberg, Wagner and Green is very good and it becomes even more impressive if we remember that it was published only a few days after the description of A5/2.

8.3 A5/1

8.3.1 Description of A5/1

A5/1 uses the same registers R_1, R_2 and R_3 as A5/2, but a different clocking mechanism. In each step the clocking unit reads the bits $R_1[10]$, $R_2[11]$ and $R_3[12]$ (marked in gray in Fig. 8.3) and computes the majority function of these three bits. The register R_1 is clocked if $R_1[10]$ agrees with the majority. Similarly R_2 and R_3 are clocked if $R_2[11]$ or $R_3[12]$ respectively agrees with the majority.

The protocol is also nearly identical to the protocol used for A5/2. Again the error-correcting code is applied before the cipher and the cipher is reseeded after 228 bits. The initialization of A5/1 (Algorithm 8.2) is very similar to Algorithm 8.1.

8.3.2 Time-Memory Trade-off Attacks

The main flaw of A5/1 is the very small internal state of only 64 bits (equal to the key size). As mentioned in Sect. 3.3.1, this allows the attacker to search for the internal state instead of the key. Since the internal state is always changing, there are several possible time-memory trade-offs that make this attack practicable. The following attack was developed by A. Biryukov, A. Shamir and D. Wagner [28] and later improved in [14].

Let us start with the basic form of the attack. We must create a table which lists 2^k internal states and the corresponding output. Now observe 2^{64-k} time steps of the output of A5/1. By the birthday paradox we can expect that one of the 2^{64-k} internal states belonging to the output will be in the pre-computed table of 2^k states. Choosing $k \approx 32$ we already get an attack of quite reasonable time and space complexity. The space needed to store the pre-computed table is some gigabytes, which is very acceptable, but the number of bits we need to observe is bit higher than we would like.

Fig. 8.3 Diagram of A5/1

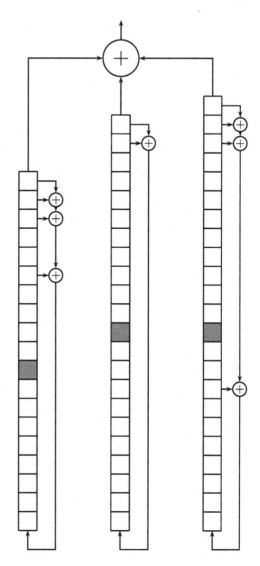

The attack recovers an internal state, but A5/1 can be effectively inverted, i.e. for an internal state one can compute the list of all predecessors. The majority rule used for the clock control makes it possible for a state to have up to four predecessors. The average number of predecessors is one and hence moving even several steps backward in time will result only in a small list of predecessors. As A5/1 is reseeded quite often we can reconstruct the initial state of a frame.

The initialization of A5/1 (see Algorithm 8.2) is linear in the key. Thus an attack that recovers an internal state is also a key recovering attack.

Now let us see how to improve the attack. The main problem is that one needs a large list of internal states and for each fraction of the output sequence we must

---

Algorithm 8.3 Enumerating special states of A5/1

Input: A 16 bit pattern **Output:** The list of internal states that generate the 16 bit pattern as output

1. Choose any value for the shortest register R_1 (2^{19} possibilities).
2. Choose any value for the leftmost 11 bits in the registers R_2 and R_3.
3. This gives a list of 2^{41} partial states. We extend each of the partial states to a special state that generates the 16 bit pattern.
4. Given a partial state, the clocking tabs are known, so we can determine the clock control for the next few time steps. We can determine the clock control until either R_2 or R_3 clocks 12 times. Each register is clocked with probability $3/4$, so we expect that we can predict the clock for about 16 time steps.

   (Of course there are some "wild" partial states in which R_2 (or R_3) is always clocked and we can predict the clock control for only 11 time steps. The practical way to deal with these wild partial states is to throw them away. We do not want to enumerate all 2^{48} special states, because it would take too much space and time, so it doesn't hurt if we drop some wild partial states.)
5. Due to the majority clocking rule either R_2 or R_3 must be clocked at every time step. This moves a fresh bit (unspecified by the partial state) to the right. By assigning the suitable value to this fresh bit we can force the output to the given pattern.

---

probe the disk to see if we have stored the corresponding internal state. Disk probing is very slow in comparison to computation in the processor, hence we want to avoid it. A simple idea that always works is to store only special states which correspond to a given output form in the table of pre-computed states. The advantage is that we can check the observed output and must only probe the disk if it has a special form. The generic way to create such a table of special states is to choose a random state, compute the output and check if it is special. So using a table of special states reduces the time needed in attack but increases the pre-processing time dramatically. In the case of A5/1 we are lucky. We can generate all internal states with a given output directly (see Algorithm 8.3).

Algorithm 8.3 generates only special states and no special state is missed. So one can generate any number of $c < 2^{48}$ random special states in $O(c)$ time steps.

The next trick to reduce the complexity of the attack uses the fact that A5/1 is not uniquely invertible. On average, every initial state has one predecessor, but the number of predecessor can vary between zero and four. For every state we can compute the tree of all its predecessors at level $1, 2, \dots$.

Since the A5/1 protocol (Algorithm 8.2 Step 3) discards the first 100 bits of the output we know that there is no need to store a state which has no predecessor at level 100.

Experiments (see [28]) show that this removes about 85 % of all states because their tree dies out before reaching level 100. For the remaining states the size of the tree can reach more than 26000, but the average size over all states is still 1. The probability that a state occurs in a collision is proportional to the size of the tree of predecessors. So in order to improve the success probability one should store only special states with large trees of predecessors.

In [28] the authors chose to use a table 2^{35} special states. This means they selected only one state out of 2^{19} possible states. This small fraction means that the average size of the tree of predecessors of the selected states is about 12500. The probability of hitting one of these states is $\approx 60\%$ (given that one can observe 71 frames which corresponds to about 2 minutes transmission time).

For an attack based on the birthday paradox this is high. Pushing the probability towards 1 becomes more and more expensive. Remember that the only way to get the probability to exactly 1 is to store all possible states.

There is one more optimization we must mention. All the calculations described above must simulate the three linear shift registers R_1, R_2 and R_3. In Sect. 2.6.2 we have seen several algorithms for this problem. Which of these algorithms should we use in this special application? The answer is none of them. The largest register R_3 has size 23 and period $2^{23} - 1$. This is small enough to pre-compute the whole period of the linear shift register sequence and replace all computations by a table look-up. This makes the simulation very efficient on standard PCs.

In the form as described above the time-memory trade-off attack is already practical. In [28] even more trade-offs between pre-processing time, computational power needed for the attack and the number of observed bits are described.

8.3.3 Correlation Attacks

The attack presented in the last section is sufficient to break A5/1, but we could imagine a generalized A5/1 in which the three registers are bigger. Then time-memory trade-off attacks rapidly become impractical. The following attack due to P. Ekdahl and T. Johansson [84] utilizes the ideas of correlation attacks to avoid this problem. This was later improved in [181]. For the small parameters used in the real A5/1 the time-memory trade-off attacks are better, but the correlation attack will also work with bigger shift registers.

The basic observation is the following. Let us approximate the clocking procedure by a random process. In each time step the clocking unit selects one of four possibilities, clock R_1 and R_2, clock R_1 and R_3, clock R_2 and R_3 or clock all three with probability $1/4$. Thus we can compute for each time step t the a priori probability $p_{a,b,c}^{(t)}$ for the register R_1 to be clocked exactly a times, the register R_2 to be clocked exactly b times and the register R_3 to be clocked exactly c times.

The values $p_{a,b,c}^{(t)}$ satisfy the following recurrence

$$p_{0,0,0}^{(0)} = 1, \tag{8.1}$$

$$p_{a,b,c}^{(0)} = 0 \quad \text{for } (a, b, c) \neq 0, \tag{8.2}$$

$$p_{a,b,c}^{(t+1)} = \frac{1}{4}\left(p_{a-1,b-1,c}^{(t)} + p_{a-1,b,c-1}^{(t)} + p_{a,b-1,c-1}^{(t)} + p_{a-1,b-1,c-1}^{(t)}\right). \tag{8.3}$$

For example, for time step 101 (the first observable output) we get $p_{76,76,76}^{(101)} = 9.74 \cdot 10^{-4}$. That means the probability that the first output z_1 of A5/1 is the sum of the output values $x_{76}^{(1)}, x_{76}^{(2)}, x_{76}^{(3)}$ of the three LFSRs after 76 clockings is $p_{76,76,76}^{(101)} + (1 - p_{76,76,76}^{(101)})\frac{1}{2} \approx 0.5 + 4.87 \cdot 10^{-4}$. Since the initial values of the LFSRs are linear combinations of the key and the frame number this gives us a correlation between the observable output and the key. It may seem that this is a very weak correlation, but one should not forget that the attacker has access to a large number of these correlations (one for each a, b, c, t and each frame number).

Before we proceed with the attack we will modify the basic relation a bit. Let us look at the sum $z_t + z_{t+1}$ of two successive output bits. We assume that the clocking unit advances only the registers R_1 and R_2 in that time step (with probability $1/4$). Then the output of register R_3 is the same in both time steps and cancels. Thus the attacker need only guess the current state of the registers R_1 and R_2. The probability that at time step t the register R_1 is clocked exactly a times and the register R_2 is clocked exactly b times is $p_{a,b,\star}^{(t)} = \sum_c p_{a,b,c}^{(t)}$. Since $1/4 p_{a,b,\star}^{(t)} > p_{a,b,c}^{\star}$ the correlation with two consecutive bits is stronger than simple correlation with one output bit. Similarly one also gets correlations for the case in which the registers R_1 and R_3 or R_2 and R_3 are clocked.

What the attacker has at this point is a lot of probabilities of the form

$$P(f_{a,b,\star}(K) = 0) = p \tag{8.4}$$

where $f_{a,b,\star}$ is some linear function in the key bits and p is close but different from $\frac{1}{2}$. The important thing is that the linear function $f_{a,b,\star}$ does not depend on the time step t and the frame number. This means that if the attacker has access to several thousand frames the attack will have many observations of the form (8.4) with the same linear function $f_{a,b,\star}$. Under the (simplified) assumption that all these observations are independent, one can compute the joint probability. Assuming we have N independent observations of the form $P(f_{a,b,\star}(K) = 0) = p_k$ then

$$P(f_{a,b,\star}(K) = 0) = \frac{\prod_{k=1}^{N} p_k}{\prod_{k=1}^{N} p_k + \prod_{k=1}^{N}(1 - p_k)}.$$

Normally the joint probability will be far from $1/2$.

At this point the attacker has a standard decoding problem to solve. He has a number of linear equations which are satisfied with a high probability and must find the solution. For the parameters of A5/1 the code size is small and standard decoding algorithms from the theory of error-correcting codes will do the job. Simulations [84] show that 70000 frames are enough for a successful attack.

The above analysis was under the simplified assumption that all events are independent, which is in the real world not the case. In [181] the authors show how to take the dependent events into account, which improves the attack. This lowers the number of frames needed for a successful attack to 10000.

A important lesson we can learn from the correlation attack against A5/1-like ciphers is not to use a linear function in the frame counter and the key as initialization. If in the above attack the attacker cannot combine many different frames, he would obtain only equations which are satisfied with a probability of $\frac{1}{2} \pm 10^{-3}$. The decoding of the error-correcting code would be much more complicated in this case (most likely the number of necessary operations would be impractical). This is a quite typical behavior. We will see in the next chapter that RC4 is also quite weak if one uses session keys of the form initialization vector‖main key. One way to avoid these problems is to use a hash function to compute the session key from the main key and the initialization vector. Even a weak hash function like MD5 will be sufficient. This prohibits related key attacks which use several frames effectively and is easy to implement.

Chapter 9
RC4 and Related Ciphers

9.1 Description of RC4

In 1987 Ron Rivest designed for RSA data security Inc. the stream cipher RC4, which is optimized for use in 8-bit processors. The design was kept secret until 1994, when the secret leaked out in a usenet posting [258]. The cipher is extremely fast and exceptionally simple. It is used in several network protocols, of which wireless LAN is the best known. This makes it, in the words of its author, "the most widely-used stream cipher in the world" [221].

There is a book [207] devoted completely to RC4 which, of course, covers more details than this chapter. [207] is focused on identifying potential weaknesses (like Theorem 9.12), but is less concerned with turning them into real attacks. Section 9.4 and Sect. 9.5, which cover attacks that utilize weak key scheduling, are more detailed than the corresponding sections in [207].

The internal state of RC4 consists of two pointers i and j, and a permutation of the values $0, \ldots, 255$. The value of pointer i is easy to predict, so i is called the *public pointer*, and j is the *private pointer*.

The key scheduling (Algorithm 9.1) will generate the initial permutation from a (random) key of length l bytes. Typically, l will be in the range between 5 and 32. (Due to some export limitations, the key length of RC4 was originally restricted to 5 bytes. Some commercial implementations still have this restriction. Of course, 5 bytes or 40 bits are not enough to secure the cipher against brute force attacks.) The maximal key length is $l = 256$ bytes. The main part of the algorithm is a pseudo-random generator (Algorithm 9.2) that produces one byte of output in each step. As usual for stream ciphers, the encryption will be an XOR of the pseudo-random sequence with the message.

For the analysis of RC4 it is convenient to replace the original algorithm that works on bytes ($\mathbb{Z}/256\mathbb{Z}$) by a generalization that works on $\mathbb{Z}/n\mathbb{Z}$ for some $n \in \mathbb{N}$. For $n = 256$ we obtain the original algorithm.

RC4 does not specify how to construct the session key from the main key and the initialization vector. The following three variants are in use.

A. Klein, *Stream Ciphers*, DOI 10.1007/978-1-4471-5079-4_9,
© Springer-Verlag London 2013

Algorithm 9.1 RC4 key scheduling

 1: {initialization}
 2: **for** i **from** 0 **to** $n - 1$ **do**
 3:     $S[i] \leftarrow i$
 4: **end for**
 5: $j \leftarrow 0$
 6: {generate a random permutation}
 7: **for** i **from** 0 **to** $n - 1$ **do**
 8:     $j \leftarrow (j + S[i] + K[i \bmod l]) \mod n$ {K is the session key}
 9:     Swap $S[i]$ and $S[j]$
10: **end for**

Algorithm 9.2 RC4 pseudo-random generator

 1: {initialization}
 2: $i \leftarrow 0$
 3: $j \leftarrow 0$
 4: {generate pseudo-random sequence}
 5: **loop**
 6:     $i \leftarrow (i + 1) \mod n$
 7:     $j \leftarrow (j + S[i]) \mod n$
 8:     Swap $S[i]$ and $S[j]$
 9:     $k \leftarrow (S[i] + S[j]) \mod n$
10:     **output** $S[k]$
11: **end loop**

---

1. Use session keys of the form IV‖main key. This version is used in the WEP protocol.
2. Use session keys of the form main key‖IV. All published attacks that break the first variant can also break this variant.
3. The RC4 key scheduling algorithm is not very strong. Both of the first two variants fail two key recovering attacks. To protect the main key it is wise to compute the session key as H(IV, main key), where H is some hash function. Any hash function seems to be feasible. The WPA protocol uses a newly constructed hash function. None of the known attacks are able to break RC4 using this variant of key scheduling.

9.2 Application of RC4 in WLAN Security

9.2.1 The WEP Protocol

In 1997 the IEEE released the first version of the 802.11 standard for wireless networking. Part of the standard is the encryption protocol WEP (Wireless Equivalent Privacy).

The session key in WEP has the form IV‖main key. The initialization vector is public and 3 bytes long. The length of the main key can vary but the default is 5 bytes. The messages are generated with a standard network protocol, i.e. they contain fixed headers and use a CRC-code to detect errors.

WEP contains many flaws that have nothing to do with RC4:

1. A 5 byte (40 bit) key is much too short. Any modern application should use 128 bit keys.
2. An initialization vector of only 3 bytes is much too short. If we use random numbers as IV the birthday paradox says that we can expect a collision after only $2^{12} = 4096$ sessions.

    Of course we can use a counter as IV instead of a random number, but this leads to other problems. Some implementations of WEP use a counter that starts at 0 after each reboot. Since (private) PCs are rebooted quite often, the attacker can easily observe different sessions with the same IV.
3. If one wants to use error-correcting codes and cryptography together, the correct order is to first encrypt the message and then add parity check symbols (see Algorithm 1.1). If you add parity check symbols first and encrypt later, as in the case of WEP, you add extra redundancy to the message that can be used by an attacker. We will see in Sect. 9.2.3 how to explore this error.
4. Publicly known header information should not be encrypted. First of all it is a waste of time to encrypt something which is not secret and secondly it allows known plaintext attacks. Every successful attack against WEP uses this weakness.

The WEP protocol is now fully broken. There are publicly available programs like `aircrack-ng` [4, 216] which break WEP secured networks in less than one minute. These programs use some variant of the attack described in Sect. 9.5. For a recent summary of all attacks against WEP (including hacking tricks not related to cryptography), see [262, 263].

9.2.2 The WPA Protocol

After the weaknesses of the WEP protocol were published, the IEEE released a new protocol WPA (Wi-Fi Protocol Access). It differs from WEP in the following points:

- WPA includes a key hash function to prevent simple key recovering attacks. The designers did not use a standard hash function, but invented a new one.
- A message integrity code (MIC) was added to prevent manipulations of the encoded message.

WPA retains some of the weaknesses of WEP.

- It still encrypts publicly known protocol information, which turns every known plaintext attack into a ciphertext-only attack.

Algorithm 9.3 Temporal Key Hash

1: {initialization of $P1K$}
2: $P1K[0] \leftarrow Lo16(IV32)$
3: $P1K[1] \leftarrow Hi16(IV32)$
4: $P1K[2] \leftarrow Mk16(TA[1], TA[0])$
5: $P1K[3] \leftarrow Mk16(TA[3], TA[2])$
6: $P1K[4] \leftarrow Mk16(TA[5], TA[4])$
7: {shuffle $P1K$}
8: **for** i from 0 to 7 **do**
9:    $j \leftarrow 2 \cdot (i\&1)$ {$j = 0$ if i is even and $j = 2$ for i odd}
10:    $P1K[0] \leftarrow P1K[0] + S[P1K[4] \oplus Mk16[TK[1 + j], TK[0 + j]]]$
11:    $P1K[1] \leftarrow P1K[1] + S[P1K[0] \oplus Mk16[TK[5 + j], TK[4 + j]]]$
12:    $P1K[2] \leftarrow P1K[2] + S[P1K[1] \oplus Mk16[TK[9 + j], TK[8 + j]]]$
13:    $P1K[3] \leftarrow P1K[3] + S[P1K[2] \oplus Mk16[TK[13 + j], TK[12 + j]]]$
14:    $P1K[4] \leftarrow P1K[4] + S[P1K[3] \oplus Mk16[TK[1 + j], TK[0 + j]]] + i$
15: **end for**
16: {initialization of PPK}
17: $PPK[0] \leftarrow P1K[0]$
18: $PPK[1] \leftarrow P1K[1]$
19: $PPK[2] \leftarrow P1K[2]$
20: $PPK[3] \leftarrow P1K[3]$
21: $PPK[4] \leftarrow P1K[4]$
22: $PPK[5] \leftarrow P1K[5] + IV16$
23: {shuffle PPK}
24: $PPK[0] \leftarrow PPK[0] + S[PPK[5] \oplus Mk16(TK[1], TK[0])]$
25: $PPK[1] \leftarrow PPK[1] + S[PPK[0] \oplus Mk16(TK[3], TK[2])]$
26: $PPK[2] \leftarrow PPK[2] + S[PPK[1] \oplus Mk16(TK[5], TK[4])]$
27: $PPK[3] \leftarrow PPK[3] + S[PPK[2] \oplus Mk16(TK[7], TK[6])]$
28: $PPK[4] \leftarrow PPK[4] + S[PPK[3] \oplus Mk16(TK[9], TK[8])]$
29: $PPK[5] \leftarrow PPK[5] + S[PPK[4] \oplus Mk16(TK[11], TK[10])]$
30: $PPK[0] \leftarrow PPK[0] + RotR[PPK[5] \oplus Mk16(TK[13], TK[12])]$
31: $PPK[1] \leftarrow PPK[1] + RotR[PPK[0] \oplus Mk16(TK[15], TK[14])]$
32: $PPK[2] \leftarrow PPK[2] + RotR[PPK[1]]$
33: $PPK[3] \leftarrow PPK[3] + RotR[PPK[2]]$
34: $PPK[4] \leftarrow PPK[4] + RotR[PPK[3]]$
35: $PPK[5] \leftarrow PPK[5] + RotR[PPK[4]]$
36: {compute the RC4 key K}
37: $K[0] \leftarrow Hi8(IV16)$
38: $K[1] \leftarrow (Hi8(IV16)|0x20)\&0x7F$
39: $K[2] \leftarrow Lo8(IV16)$
40: $K[3] \leftarrow Lo8((PKK[5] \oplus Mk16(Tk[1], Tk[0])) \ggg 1)$
41: **for** i from 0 to 6 **do**
42:    $K[4 + 2 \cdot i] = Lo8(PPK[i])$
43:    $K[5 + 2 \cdot i] = Hi8(PPK[i])$
44: **end for**

- It uses an error-correcting code in the same wrong fashion as WEP. In contrast to WEP, the design adds some ad hoc rules to avoid the worst consequences of this mistake.

The most interesting part of WPA is the temporal key hash (Algorithm 9.3). The algorithm uses a 48 bit initialization vector, which is split into 16 least significant bits $IV16$ and the most significant 32 bits $IV32$. Further, it takes a 6 byte transmitter address TA and a 16 byte key TK as input.

The notation $Lo\,i(X)$ denotes the lower i bits of X, $Hi\,i(X)$ denotes the higher i bits of X, $Rot\,R$ is a rotation to the right by one bit, $\oplus$ is the XOR and all additions are done modulo 2^{16}. By & we denote the bitwise AND (as in the language C), i.e. $i\,\&\,1 = 0$ if i is even and $i\,\&\,1 = 1$ for odd i.

As non-linear part it uses an S-box that is based on the AES S-box. The S-box is described in Algorithm 9.4. The non-linear step interprets a byte as an element of $\mathbb{F}_{256} = F[x]/\langle x^8 + x^4 + x^3 + x + 1\rangle$ and computes the multiplicative inverse. The other steps are linear and are added just to increase the algebraic complexity of the operation.

Algorithm 9.4 Temporal Key Hash S-box

1: **for** j from 0 to 1 **do**
2:     {The next two steps are the AES S-box and are performed by table look up.}
3:     Invert $X[j]$ as an element of $\mathbb{F}_{256}$; 0 goes to 0.
4:     Apply the $\mathbb{F}_2^8$ affine transformation

$$b_i' \leftarrow b_i + b_{(i+4)\bmod 8} + b_{(i+5)\bmod 8} + b_{(i+6)\bmod 8} + b_{(i+7)\bmod 8} + c_i,$$

   where the b_i denote the bits of $X[j]$ and $c = 01100011$.
5:     Compute $Y[i] = x \cdot X[j]$ in $\mathbb{F}_{256}$ {by table look up.}
6: **end for**
7: **return** $Mk16(Y[0] \oplus X[1] \oplus Y[1], X[0] \oplus Y[0] \oplus Y[1])$

The temporal key hash is rather complicated, so we include some figures that visualize the structure of the Algorithm. The S-Box (Algorithm 9.4) is displayed in Fig. 9.1. Figure 9.2 visualizes lines 9–14 of Algorithm 9.3 and Fig. 9.3 shows the computation done in lines 24–35.

Algorithm 9.3 is not a good hash function. As Moen, Raddum and Hole [194] show, it is possible to invert the temporal key hash with just 2^{32} operations given two values with the same $IV32$. At the moment, it is unknown if one can turn their result into a practical attack against the WPA protocol.

9.2.3 A Weakness Common to Both Protocols

It is not clear why the protocols require that the error-correcting code has to be applied before the cryptographic code. Most likely it was simply the easiest way

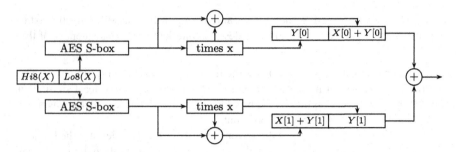

Fig. 9.1   The S-box of the Temporal Key Hash (Part 1)

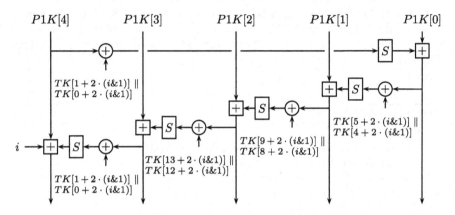

Fig. 9.2   Temporal Key Hash (Part 1)

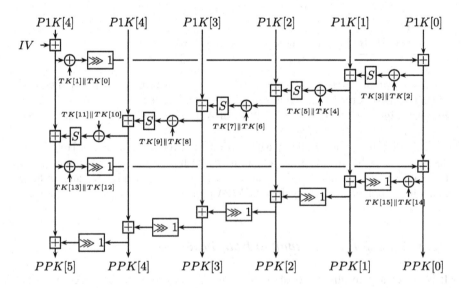

Fig. 9.3   Temporal Key Hash (Part 2)

to add the cryptography to the older network protocol. We have already seen in the previous chapter that applying an error-correcting code before the cryptographic code is a serious mistake that weakens the whole system. In the case of the WEP protocol one can use this error to break the system without breaking RC4 itself.

This was first observed by W.A. Arbaugh [7]. In the following we describe a variant of this attack that is better suited for a real implementation [164]. The problem is that the used CRC-code is of variable length. This allows the attacker to obtain many relations.

A *cyclic redundancy check code* (CRC-code) is described by a generator polynomial $g(x) \in \mathbb{F}_2$ of degree k. The encoding and decoding is done by Algorithms 9.5 and 9.6.

Algorithm 9.5 CRC encoding

1: Get message $m(x) = m_0 + \cdots + m_{n-k} x^{n-k} \in \mathbb{F}_2[x]$
2: $r(x) = x^k m(x) \mod g(x)$
3: Send $w(x) = x^k m(x) + r(x)$

Algorithm 9.6 CRC decoding

1: Get received codeword $w(x) = w_0 + \cdots + w_n x^n \in \mathbb{F}_2[x]$
2: $r(x) = w(x) \mod g(x)$
3: **if** $r(x) \neq 0$ **then**
4:     Report an error and require that the message is transmitted again
5: **else**
6:     Decode as $m(x) = w_k + w_{k+1} x + \cdots + w_n x^{n-k}$
7: **end if**

Now suppose that we receive a cryptographically secured CRC-codeword, i.e. we receive $c_i = w_i + x_i$ where w_i are the coefficients of a codeword computed by Algorithm 9.5 and $x_0, \ldots, x_n$ is some (pseudo-)random sequence. Normally the leading coefficient will be sent first, i.e. the word will be sent in the order $c_n, \ldots, c_0$ and the pseudo-random sequence x will also be computed by the stream cipher in the order $x_n, \ldots, x_0$.

Assume that the receiver accepts not only the codewords of length $n + 1$ but also codewords of length $n + 1 - m$. Since the CRC-code does not specify the length of the codewords this is a reasonable assumption. Most real applications will accept any number of bytes as a codeword, i.e. we have $m = 8$. The attacker tries to alter the last m bits of the sequence $c_n, \ldots, c_m$ in such a way that it will be accepted as a CRC-code word. To that end he simply enumerates all 2^m possible words $t_0, \ldots, t_{m-1}$ and sends $c_n, \ldots, c_{2m}, c_{2m-1} + t_m, \ldots, c_m + t_0$. The receiver will decrypt it as $w_n, \ldots, w_{2m}, w_{2m-1} + t_{m-1}, \ldots, w_m + t_0$ and use this word as input for the CRC-decoding algorithm. By the protocol (line 4 of Algorithm 9.6) he has to report an error if $(w_n x^{n-m} + \cdots + w_m) + (t_{m-1} x^{m-1} + \cdots + t_0)$ is not a multiple of $g(x)$. In this way the attacker learns the polynomial $t(x) = t_{m-1} x^{m-1} + \cdots + t_0$

for which $w'(x) + t(x)$ with $w'(x) = w_n x^{n-m} + \cdots + w_m$ is a multiple of $g(x)$. In addition he knows that $w(x) = w_n x^n + \cdots + w_0$ is a multiple of $g(x)$ and hence

$$0 \equiv w(x) + x^m \big(w'(x) + t(x) \big) \quad \mathrm{mod}\ g(x),$$

$$0 \equiv \big(w_{m-1} x^{m-1} + \cdots + w_0 \big) + x^m t(x) \quad \mathrm{mod}\ g(x),$$

$$x^m t(m) \equiv \big(w_{m-1} x^{m-1} + \cdots + w_0 \big) \quad \mathrm{mod}\ g(x).$$

This congruence has a unique solution $w_{m-1}, \ldots, w_0$. So the attacker has learned with only 2^m operations the last m bits of the message. Continuing in this fashion, he can recover all but the first m bits of the message.

As we have seen, this attack works even if the cipher is a true one-time pad. The main problem is that, besides the mistake of applying the error-correcting code before the cryptographic code, the receiver accepts different messages encrypted with the same (pseudo-)random sequence. The correct response to an error for the receiver would be to reject the pseudo-random sequence $x_0, \ldots, x_n$ and require that the whole message be encrypted again with a new unused pseudo-random sequence. The WPA protocol has some elements that go in this direction.

The attack is therefore a variation of the old theme, that the security of a one-time pad vanishes immediately if you use it twice. If the protocol designers had chosen the correct order, applying the error-correcting code after the cryptographic code, then the receiver would have avoided this trap automatically.

This it is an important point that many people miss. Applying an error-correction code before the cryptographic code is a serious mistake. It helps the attacker to mount a known plaintext attack. In the best case the cipher is still good enough to ensure the security of the system, but even then there is no reason to help the attacker. In the worst case it is a protocol failure that kills the security of the whole system. If you choose the correct order and apply the error-correcting code after the cryptographic code, there are no problems at all.

9.3   Analysis of the RC4 Key Scheduling

The key scheduling algorithm is the weakest part of RC4. At the moment all successful attacks use at least some weakness of the key scheduling. The first published analysis of the RC4 key scheduling dates back to the year 1981 [225] and is even older than RC4 itself.

For the analysis we assume that the key has full length. In this case we can write line 8 of Algorithm 9.1 as $j \leftarrow f(K_0, \ldots, K_{i-1}) + K_i \quad \mathrm{mod}\ n$. We prove that j will be a uniformly distributed random variable.

Lemma 9.1 *Assume that $K_0, \ldots, K_i$ are iid. random variables uniformly distributed in the set $\{0, \ldots, n-1\}$. Then for any function f the variable*

$$X_i = \big(f(K_0, \ldots, K_{i-1}) + K_i \big) \quad \mathrm{mod}\ n$$

is uniformly distributed in $\{0, \ldots, n-1\}$ and independent from $K_0, \ldots, K_{i-1}$.

Proof We claim that

$$p\big(K_0 = a_0, \ldots, K_{i-1} = a_{i-1}, \big(f(K_0, \ldots, K_{i-1}) + K_i\big) \mod n = a_i\big) = \frac{1}{n^{i+1}}$$

for all possible values $(a_0, \ldots, a_i)$. But this is clear, since

$$p(K_0 = a_0, \ldots, K_{i-1} = a_{i-1}, X_i = a_i)$$
$$= p\big(K_0 = a_0, \ldots, K_{i-1} = a_{i-1}, K_i = \big(a_i - f(a_0, \ldots, a_{i-1})\big) \mod n\big).$$

By summing over the elementary events we find

$$p\big(\big(f(K_0, \ldots, K_{i-1}) + K_i\big) \mod n = a\big) = \frac{1}{n}$$

for $0 \le a < n$ and

$$p\big((K_0, \ldots, K_{i-1}) \in \mathcal{A}, X_i = a\big) = \frac{|\mathcal{A}|}{n^{i+1}} = p\big((K_0, \ldots, K_{i-1}) \in \mathcal{A}\big) p(X_i = a),$$

i.e. $(f(K_0, \ldots, K_{i-1}) + K_i) \mod n$ is uniformly distributed in $\{0, \ldots, n-1\}$ and is independent from $K_0, \ldots, K_{i-1}$.   □

So we reach the idealized RC4 key scheduling given by Algorithm 9.7.

Algorithm 9.7 Idealized RC4 key scheduling

1: {initialization}
2: **for** i **from** 0 **to** $n - 1$ **do**
3:     $S[i] \leftarrow i$
4: **end for**
5: {generate a random permutation}
6: **for** i **from** 0 **to** $n - 1$ **do**
7:     Choose j random in $\{0, \ldots, n-1\}$
8:     Swap $S[i]$ and $S[j]$
9: **end for**

A well-known algorithm for generating true random permutations (see for example [152]) replaces line 7 of Algorithm 9.7 by "Choose j random in $\{i, \ldots, n-1\}$".

9.3.1 The Most Likely and Least Likely RC4 Permutation

As we can see, Algorithm 9.7 produces n^n results of equal probability. Since $n!$ is not a divisor of n^n, some permutations must be more likely than others. In fact there is an asymptotically exponential gap between the probability of the most likely and least likely permutation.

We begin our study with the determination of the least likely permutation.

Theorem 9.2 (Robbins and Bolker [225]) *Algorithm 9.7 produces the right cycle, i.e. the permutation $S[0] = n$, $S[i] = i - 1$ for $i > 0$, with probability $\frac{2^{n-1}}{n^n}$.*

Proof After the i-th loop in Algorithm 9.7, we say the elements $S[0], \ldots, S[i - 1]$ lie in the past. From the definition of the algorithm, we obtain:

1. A value that lies in the past will remain in the past in all future iterations.
2. Values that lie in the past can move only to the right (from position i to a position $i' > i$).
3. If a value moves to the left, it will be in the past afterwards.
4. Every value can move at most once to the left.

Now let us count the number of shuffles which produce the right cycle as a result. We prove by induction that there are 2^{n-1} such shuffles. For $n = 1$ there is nothing to prove. Now let $n > 1$.

Since n can move at most once to the left, there are only two ways to get n at position 0.

One possibility is that n moves to position 0 in the first step and stays there in the remaining steps. Then after the first step we have the permutation $n, 1, \ldots, n - 1, 0$ and the remaining $n - 1$ steps must cycle the permutation $1, \ldots, n - 1, 0$ to the right. By induction there are 2^{n-2} such shuffles.

The second possibility is that n moves only in the last step. So the first $n - 1$ steps must produce the permutation $n - 1, 0, \ldots, n - 2, n$ and the last step exchanges $n - 1$ and n. By induction there are 2^{n-2} shuffles which produce the right cycle $n - 1, 0, \ldots, n - 2$.

In total we have 2^{n-1} shuffles which produce the right cycle.   □

The probability $\frac{2^{n-1}}{n^n}$ is very low in comparison to the average probability $\frac{1}{n!} \approx \frac{1}{\sqrt{2\pi n}} \frac{e^n}{n^n}$. The right cycle is indeed the least likely permutation (although for small n there are other permutations that are equally likely).

Theorem 9.3 (Robbins and Bolker [225]) *Every permutation is generated with at least probability $\frac{2^{n-1}}{n^n}$.*

Proof We will compute the exact probability for the identity permutation later (Theorem 9.7). Now we prove for all other permutations S that there are at least 2^n shuffles that produce S.

Since S is not the identity, we find a position i with $j = S[i] < i$.

Consider the permutation $S' = (i\ j)S$. It fixes i and by induction there are at least 2^{n-2} ways to write S' in the form

$$S' = (n - 1\ r_{n-1}) \ldots (i + 1\ r_{i+1})(i\ r_i) \ldots (1\ r_1)(0\ r_0)$$

Fig. 9.4 The graph
representation of
$S = (0\ 1)(2\ 3) =$
$(3\ 1)(2\ 3)(1\ 2)(0\ 1)$

where all r_k differ from i. Multiply by $(i\ j)$ on the right. By conjugating we can
move $(i\ j)$ from the right to the middle and get

$$S = \left(n - 1\ r'_{n-1}\right) \ldots \left(i + 1\ r'_{i+1}\right)(i\ j)(i - 1\ r_{i-1}) \ldots (1\ r_1)(0\ r_0),$$

where $(k\ r'_k) = (k\ r_k)^{(i\ j)} = (i\ j)(k\ r_k)(i\ j)$ for $k > i$. At this point we use $j < i$ so
that k is fixed by $(i\ j)$.

Similarly we can start with the permutation $S'' = S(i\ j)$ which fixes j to find
2^{n-2} solutions of

$$S'' = (n - 1\ r_{n-1}) \ldots (j + 1\ r_{j+1})(j - 1\ r_{j-1}) \ldots (0\ r_0)$$

or after multiplication by $(i\ j)$ on the left

$$S = (n - 1\ r_{n-1}) \ldots (j + 1\ r_{j+1})(j\ i)\left(j - 1\ r'_{j-1}\right) \ldots \left(0\ r'_0\right)$$

where $r_k \neq j$. Again we use $j < i$.

Thus we have altogether 2^{n-1} shuffles that produce S. These shuffles are all
different since in the first 2^{k-1} shuffles the j-th transposition is $(j\ r_j)$ with $r_j \neq i$
and in the last 2^{n-1} shuffles the j-th transposition is $(j\ i)$. □

The determination of the most likely permutation is more complicated. We start
with a graph representation of our problem. Let S be a permutation which is obtained by the shuffle $S = (n - 1\ r_{n-1}) \ldots (0\ r_0)$. Then we represent the shuffle by a
directed graph with vertex set $\{0, \ldots, n - 1\}$ and edge set $\{(0, r_0), \ldots, (n - 1, r_{n-1})\}$.
We also say that the graph represents S.

Every vertex has out degree 1. So each connected component of the graph must
contain a unique directed cycle. To avoid confusion with the cycles of the permutation we call the graph cycle a ring. In addition to the ring the connected component
can contain trees with root on the ring and edges pointing towards the ring. In the
example of Fig. 9.4 we have the ring $1 \to 2 \to 3 \to 1$ and one tree with root 1.

Permutations from different connected components do not interact. So the shuffle
induces permutations on connected components of the graph.

The next lemma will help us to count the number of shuffles producing a given
permutation.

Lemma 9.4 (Schmidt and Simion [234]) *If the ring has length 1, the permutation*
S' induced on the connected component with m vertices is an m-cycle. If the ring
has length at least 2 the induced permutation S' consists of two cycles whose lengths
add up to m.

Proof If the ring has length 1, then we have a representation of S' as rooted tree with a loop at the root. The loop has no effect on the generated permutation, so we ignore it. This leaves us with a permutation of the form

$$S' = (a_{m-1} \; b_{m-1}) \cdots (a_1 \; b_1) \tag{9.1}$$

whose corresponding graph has edge set $E = \{(a_i, b_i) \mid i = 1, \ldots, m-1\}$ and vertex set V.

We prove by induction on m that S' is an m-cycle. For $m = 1$ this is trivial. For $m > 1$ choose an edge (a_i, b_i) of the graph with a_i as a leaf vertex. Deleting this edge gives us a tree on $V \setminus \{a_i\}$. By induction

$$S'' = (a_{i-1} \; b_{i-1}) \cdots (a_1 \; b_1)(a_m \; b_m) \cdots (a_{i+1} \; b_{i+1})$$

is a cycle, i.e.

$$S'' = (c_i \; h_1 \; \ldots \; h_{m-2}).$$

Then the permutation S' is conjugate to

$$S''(a_i \; b_i) = (c_i \; b_i \; h_1 \; \ldots \; h_{m-2}).$$

The conjugate of an m-cycle is an m-cycle, which completes the proof.

Now suppose that the ring has length at least 2. Let

$$S' = (d_m \; e_m) \cdots (d_1 \; e_1).$$

Choose an edge $(d_i \; e_i)$ from the ring. Then

$$S'' = (d_{i-1} \; e_{i-1}) \cdots (d_1 \; e_1)(d_m \; e_m) \cdots (d_{i+1} \; e_{i+1})$$

is of the form (9.1) and we have just proved that S'' is an m-cycle, i.e.

$$S'' = (d_i \; g_1 \; \ldots \; g_j \; e_i \; g_{j+1} \; \ldots \; g_{m-2}).$$

So S' is conjugate to

$$S''(d_i \; e_i) = (e_i \; g_1 \; \ldots \; g_j)(d_i \; g_{j+1} \; \ldots \; g_{m-2}),$$

which is the product of two cycles whose lengths add up to m. Conjugation does not change the cycle structure, so the proof is complete.   □

We define $N(S)$ as the number of shuffles that generate the permutation S. By $N^{tree}(S')$ we denote the number of shuffles whose graph has a ring of length 1 that generates S' and by $N^{ring}(S')$ we denote the number of shuffles whose graph has a ring of length at least two and that generates S. With this notation we obtain:

Lemma 9.5 *For any permutation S which is the product* $\pi_q \cdots \pi_1$ *of q disjoint cycles, we have*

$$N(S) = \sum_{\substack{\chi \text{ is involution } \chi(i)=i \\ \text{of } \{1,\ldots,q\}}} \prod N^{tree}(\pi_i) \prod_{\chi(i)=j<i} N^{ring}(\pi_i, \pi_j). \qquad (9.2)$$

Proof Any shuffle that generates S defines a partition of the set $\{\pi_1, \ldots, \pi_q\}$ into sets of size one and two. The one element blocks are those cycles that are generated by a tree and the two element blocks are pairs of cycles that are generated by a graph with ring length at least 2. For each partition

$$\chi = \left\{ \{\pi_{i_1}\}, \ldots, \{\pi_{i_k}\}, \{\pi_{j_1}, \pi_{j'_1}\}, \ldots, \{\pi_{j_{k'}}, \pi_{j'_{k'}}\} \right\}$$

we get the number of corresponding graphs by multiplying the numbers $N^{tree}(\pi_{i_l})$ and $N(\pi_{j_{l'}}, \pi_{j'_{l'}})$. The number $N(S)$ is obtained by summing over all partitions. $\square$

As an immediate consequence we get that the RC4-shuffle (Algorithm 9.7) produces the n-cycles as often as it should.

Corollary 9.6 (Robbins and Bolker [225]) *Algorithm 9.7 produces an n-cycle with probability* n^{-1}.

Proof By Lemma 9.5 the frequency of an n-cycle π is $N(\pi) = N^{tree}(\pi)$. To get the frequency of all n-cycles we just count the directed trees with n vertices.

Cayley's theorem (see Sect. 16.3) says that there are n^{n-2} labeled trees on n vertices and thus n^{n-1} directed trees.

Thus the probability of an n-cycle is $\frac{n^{n-1}}{n^n} = \frac{1}{n}$. $\square$

With Lemma 9.5 we directly obtain the probability for the identity permutation.

Corollary 9.7 (Robbins and Bolker [225]) *Algorithm 9.7 produces the identity permutation with probability* $\frac{t_n}{n^n}$, *where* t_n *denotes the number of involutions in* S_n.

Proof The identity permutation consists of n cycles $\pi_1 \cdots \pi_n$ of length 1. For 1-cycles π_i we have $N^{tree}(\pi_i) = 1$ and $N^{ring}(\pi_i, \pi_j) = 1$, so (9.2) simplifies to $N(id) = \sum_{\chi \text{ is involution of } \{1,\ldots,q\}} 1 = t_n$. $\square$

Since $t_n = \frac{1}{\sqrt{2}} n^{n/2} e^{-n/2 + \sqrt{n} - 1/4} (1 + O(n^{-1/2}))$ (Sect. 16.2) the identity permutation occurs much more often than it should. For $n > 18$ it is the most likely shuffle.

Theorem 9.8 (Goldstein and Moews [108]) *For* $n \geq 18$ *the most like permutation generated by Algorithm 9.7 is the identity. For* $4 \leq n \leq 17$ *the most frequent permutation is* $(n-1 \cdots m)(m-1 \cdots 0)$ *where* m *is* $n/2$ *for* n *even and either* $\lfloor n/2 \rfloor$ *or* $\lceil n/2 \rceil$ *for* n *odd.*

Proof See [108]. $\square$

9.3.2 Discarding the First RC4 Bytes

The results in the previous subsection show that the permutations produced by Algorithm 9.7 are far from random. This has serious consequences for the security of RC4. The first output bytes of RC4 are particularly highly biased. For example, the second output byte of RC4 is zero with a probability twice as high as it should be [93].

As a consequence, many researchers have suggested discarding the first output bytes of RC4 or to run the key scheduling several times. The natural question that arises at this point is how many bytes should we discard?

The best answer to this question is due to I. Mironov [192]. He found the following distinguisher for RC4 permutations.

Theorem 9.9 *Assume that the permutation S is generated by t shuffling steps of the form $(a_i\ b_i)$ where the variables b_i are iid. and uniformly distributed in $\{0, n\}$. Then*

$$P\big(\mathrm{sign}(S) = (-1)^t\big) = \sum_{i=0}^{\lfloor n/2 \rfloor} \binom{t}{2i}\left(1 - \frac{1}{n}\right)^{t-2i} \frac{1}{n^{2i}}.$$

For $n, t \to \infty$ and $n/t \to \lambda$, we have

$$P\big(\mathrm{sign}(S) = (-1)^t\big) \xrightarrow[\substack{n,t\to\infty \\ n/t\to\lambda}]{} \frac{1}{2}\big(1 + e^{-2\lambda}\big).$$

Proof The swap $(a_i\ b_i)$ changes the sign of the permutation S if $a_i \neq b_i$ and leaves the sign unchanged for $a_i = b_j$. Since b_i is uniformly distributed the probability of $a_i \neq b_i$ is $1 - \frac{1}{n}$ and the probability of $a_i = b_i$ is $\frac{1}{n}$. So the probability of having exactly k passes with $a_i = a_i$ is $\binom{t}{k}(1 - \frac{1}{n})^{t-k}\frac{1}{n^k}$.

The sign of S is $(-1)^t$ if we have an even number of steps i with $a_i = b_i$, i.e.

$$P\big(\mathrm{sign}(S) = (-1)^n\big) = \sum_{i=0}^{\lfloor t/2 \rfloor} \binom{t}{2i}\left(1 - \frac{1}{n}\right)^{t-2i} \frac{1}{n^{2i}}.$$

To compute the limit for $n, t \to \infty$, $t/n \to \lambda$, we factor:

$$= \left(1 - \frac{1}{n}\right)^t \left(\sum_{i=0}^{\lfloor n/2 \rfloor} \frac{\prod_{j=0}^{2i-1}(t - j)}{n^{2i}}\left(1 - \frac{1}{n}\right)^{-2i}\frac{1}{(2i)!}\right).$$

The first factor tends to $e^{-\lambda}$. For fixed i the summand $\frac{\prod_{j=0}^{2i-1}(t-j)}{n^{2i}}(1 - \frac{1}{n})^{-2i}\frac{1}{(2i)!}$ of the second factor tends to $\frac{\lambda^{2i}}{(2i)!}$.

So the whole sum tends to

$$\sum_{i=0}^{\infty} \frac{\lambda^{2i}}{(2i)!} = \frac{1}{2} \sum_{i=0}^{\infty} \frac{\lambda^i}{i!} + \frac{1}{2} \sum_{i=0}^{\infty} \frac{(-\lambda)^i}{i!} = \frac{1}{2} e^{\lambda} + \frac{1}{2} e^{-\lambda}.$$

The order of the limits is not an issue since all terms are positive.

So

$$\lim_{\substack{n,t\to\infty \\ t/n\to\lambda}} P\left(\text{sign}(S) = (-1)^t\right) = e^{-\lambda} \left(\frac{1}{2} e^{\lambda} + \frac{1}{2} e^{-\lambda} \right) = \frac{1}{2}\left(1 + e^{-2\lambda}\right). \qquad \square$$

If we observe a permutation S and want to guess if it was generated by λn iterations of Algorithm 9.7 or uniformly at random, we compute $\text{sign}(S)$ and conclude that S comes from λn iterations of Algorithm 9.7 if $\text{sign}(S) = (-1)^{\lambda n}$. Our chance of guessing correctly is about $\frac{1}{2} e^{-2\lambda}$.

If we want to lower the quality of the distinguisher below ϵ we should choose $\lambda > \frac{1}{2} \ln \frac{1}{2\epsilon}$. For example $\epsilon = \frac{1}{256}$ would give a bound of $\lambda \approx 2.4$. Under these circumstances, discarding the first $2n - 3n$ outputs seems to be a reasonable precaution. As we will see in Sect. 9.5.3, there are practical attacks against RC4 if we discard only the first n outputs.

Now we ask the natural question: Can we find an upper bound for the number of outputs we have to discard to prevent any attack that is based on the weak key scheduling? The answer is yes, as we will show in the remaining part of this section.

The success probability of a distinguisher is bounded by the variation distance (see Sect. 15.2.1). The appropriate tool to obtain a bound for the variation distance in our case is the concept of strongly uniform stopping times.

Definition 9.1 (Diaconis [79]) Suppose a process generates a sequence Q_i, $i \in \mathbb{N}$, of random permutations of $\{0, \ldots, n-1\}$.

A stopping time τ is said to be *strongly uniform* if Q_τ is uniformly distributed under the condition $\tau < \infty$, i.e. if

$$P(Q_\tau = \pi \mid \tau < \infty) = \frac{1}{n!}$$

for any permutation π of $\{0, \ldots, n-1\}$.

Theorem 9.10 (Diaconis [79]) *Let Q be a shuffling process and let τ be a strongly uniform stopping time. Then for all times t, the variation distance between Q_t and the uniform shuffle U satisfies*

$$\|Q_t - U\| \le P(\tau > t).$$

Proof See [79]. $\qquad \square$

So we have to find a uniform stopping time τ and compute the probabilities $P(\tau > t)$.

We define τ by the following procedure.

1. At the beginning the elements $S[0], \ldots, S[n-2]$ are unmarked and $S[n-1]$ is marked.
2. Every time we swap $S[i]$ with $S[j]$, $S[i]$ becomes marked if either $j = i$ or $S[j]$ is already marked. Note this is not symmetric in i and j, we will not mark $S[j]$ if $S[i]$ is already marked.
3. τ is the first time step in which all elements of the permutation are marked.

Theorem 9.11 (Mironov [192]) *The stopping time defined by the above procedure is strongly uniform.*

Proof We claim that at any time step the permutation of all marked symbols is uniformly chosen.

At $t = 1$ this is trivial, since only one symbol is marked.

Now suppose this is true for the time $t - 1$. If at time t two different unmarked symbols are exchanged, then the permutation of marked symbols does not change, so it stays uniformly distributed.

If $S[i]$ is marked and $S[j]$ is unmarked, the swap of $S[i]$ and $S[j]$ corresponds to a permutation π of the marked symbols. If π' is a uniformly distributed permutation then $\pi \circ \pi'$ is uniformly distributed, which proves this case.

Now suppose that $S[i]$ becomes marked at time step t. In this case, j must be either i or point to a marked position. All alternative values of j have the same probability. This means we insert $S[i]$ at a random position of the permutation of all marked symbols. This operation again yields a uniform distribution of marked symbols. □

The computation of the exact distribution of τ is rather complicated. To demonstrate the method we will compute the expected value for the simplified case in which i is also chosen at random.

If i is chosen at random, the probability that a new symbol becomes marked in one step if x symbols are already marked is $p_x = \frac{n-x}{n} \cdot \frac{x+1}{n}$.

Thus the expected number of steps we need to mark one more symbol if x symbols are already marked is $1/p_x$. Summing over x gives us the expected value

$$E(\tau) = \sum_{i=1}^{n-1} 1/p_x = n^2 \sum_{i=1}^{n-1} \frac{1}{(n-x)(x+1)} = \frac{n^2}{n+1}\left(\frac{1}{n-x} + \frac{1}{x+1}\right).$$

Using Euler's approximation for the harmonic numbers $H_n = \sum_{i=1}^{n} \frac{1}{i} = \ln(n) + \gamma + O(\frac{1}{n})$ (see Theorem 16.2 in Sect. 16.1) we obtain $E(\tau) = \frac{n^2}{n+1}(2\ln(n) + 2\gamma - 1 + O(\frac{1}{n}))$.

For $n = 256$ this means that we should discard the first $12 \cdot 256$ elements of the pseudo-random sequence to be on the safe side.

An interesting modification of the RC4 algorithm would be to run the shuffle until we reach the stopping time τ. For this modification, we find some interesting research problems.

- The real key has fixed length and is cyclically repeated, so we cannot apply Theorem 9.11 to conclude that S_τ is uniformly distributed. Indeed, S_τ cannot be uniformly distributed since $n!$ is not a divisor of n^l. How large is the difference between S_τ and the uniform distribution? Experiments with small n suggest that the new algorithm produces a better distribution than the original RC4 key scheduling.
- Prove that τ must be finite for fixed key length.
- The running time of the modified key scheduling depends on the key. This allows side channel attacks. How can we prevent them?

9.4 Chosen IV Attacks

The first really practical attack against RC4 is due to S. Fluhrer, I. Mantin and A. Shamir [93]. It is a chosen IV attack or related key attack that explores the weakness of the key scheduling algorithm.

The attack assumes that the attacker has control over some bytes of the session key. This is the case if the session key is of the form IV∥main key or main key∥IV.

9.4.1 Initialization Vector Precedes the Main Key

Assume that the attacker knows the first A bytes of the session key, either because they are part of the public initialization vector or because he has recovered them in previous steps of the attack.

Assume further that the attacker can choose the first two bytes of the IV to be $(A, n-1)$. The following happens in the first two steps of the key scheduling (Algorithm 9.1). In the first step, j is increased by $S[0] + K[0] = A$ and we swap $S[0] = 0$ and $S[A] = A$. In the next step, i is 1 and j is increased by $S[1] + K[1] \equiv 0 \mod n$, so we swap $S[1]$ and $S[A]$. After the first two swaps, we have $S[0] = A$ and $S[1] = 0$ (see Fig. 9.5).

What happens in the next $A - 2$ steps of the key scheduling is known to the attacker, since we assume the first A bytes of the session key to be known. So we can express the value that is swapped to $S[A]$ in the $(A + 1)$th step as a function of $K[A]$.

The probability that $S[0]$, $S[1]$ and $S[A]$ remain unchanged during the remaining $n - A - 1$ steps of the key scheduling phase is approximately $(1 - \frac{3}{n})^{n-1-A} \approx \frac{1}{e^3} \approx 0.05$.

If this happens, the first step of RC4 will set i to 1 and j will stay 0, since $S[1] = 0$, so the output will be $S[A] = f(K[A])$.

The attacker will compute $f^{-1}(X)$ for every session, where X is the first output byte. As we have seen above, $f^{-1}(X)$ is with probability ≈ 0.05 equal to $K[A]$. In the other cases $f^{-1}(X)$ will approximately be equally distributed in $\{0, \ldots, n-1\}$.

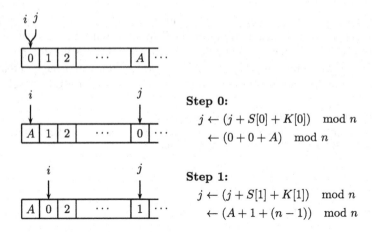

Fig. 9.5 The FMS-attack key scheduling

So the attacker gets the correct value of $K[A]$ with probability $\approx 0.05 + 0.95\frac{1}{n}$ and a wrong value with probability $\approx 0.95\frac{1}{n}$. For $n = 256$, about 60 sessions with a chosen IV are enough to recover $K[A]$ almost certainly.

It is a remarkable property of the FMS-attack that it becomes stronger as n increases. In the limit $n \to \infty$, we can stop the FMS-attack at the first time we get the same value $f^{-1}(X)$ for two distinct sessions. With a probability of 50 %, this is the case after just 33 sessions.

Since in most applications the attacker cannot manipulate the IV directly, he has to wait until by pure chance he observes enough sessions which have an IV of the required value. Since we must prescribe two IV bytes, this will slow down the attack by a factor of n^2 provided that the IV is chosen at random.

If the initialization vector is generated by a counter, the number of sessions will depend on whether the counter is little or big endian [93]. A little endian counter ($K[0]$ increments quickly) causes no problem; the attack will succeed after about 4,000,000 observed sessions. The case of a big endian counter is much more complicated. For an IV of only 3 bytes, the attack will be even faster (1,000,000 sessions), but we can run into some problems if the counter does not start with zero (see [93] Sect. A.2). For larger IVs the attacker will have to wait for more than n^3 sessions before the counter starts generating IVs of the required form. However, there are ways to adapt the attack to these cases (see also Sect. 9.5.3).

9.4.2 Variants of the Attack

After the publication of the FMS-attack, people tried to secure WEP encrypted networks by avoiding initialization vectors of the form $(*, n - 1, *)$. This only helps against inexperienced attackers, since the FMS-attack is very variable and can be mounted with a lot of different initialization vectors. Perhaps the most complete list of FMS-like attacks is [165].

Some of these variations do not use the first but the second byte of the output, other variants do not find a hint for a value, but exclude some possible values for the key with a high probability. We give an example for both variants.

Suppose the attacker knows the first A bytes of the session key and wants to get information about the $(A + 1)$th byte. The attack computes the first A steps of the RC4 key scheduling (Algorithm 9.1) and uses only sessions which satisfy the following conditions:

- After A steps of the key scheduling, we have $S[1] = B$ and $S[B] = 0$ for some $B < A$.
- The first byte of the RC4 pseudo-random sequence is B.

If the remaining scheduling steps do not change $S[1]$ and $S[B]$, we always get B as first byte of the pseudo-random sequence. This event occurs with probability e^{-2}.

If at least $S[0]$ or $S[B]$ is changed in the key scheduling, we still have a $\frac{1}{n}$ chance to get B as the first pseudo-random byte.

So the conditional probability that $S[1]$ and $S[B]$ remain unchanged given the observation of B as the first pseudo-random byte is

$$\frac{e^{-2}}{e^{-2} + (1 - e^2)\frac{1}{n}} = \frac{n}{n + e^2 - 1}.$$

This means with high probability ($\approx 97.5\,\%$ for $n = 256$) we can conclude from the observation of B as the first byte of the pseudo-random sequence that $S[1]$ and $S[B]$ remain unchanged in the last $n - A$ steps of the key scheduling. This excludes some possible values for the key.

The oldest variant of a chosen IV attack that uses the second instead of the first pseudo-random byte is due to Hulton [133].

The structure of the attack is quite similar to the FMS-attack. The attacker chooses an A byte initialization vector such that after the first A steps of the key scheduling the following holds: $S_A[1] < A$, $S_A[2] < A$, $S_A[1] + S_A[2] = S < A$, $S_A[2] + S_A[S] = A$, $S_A[1] \neq 2$, $S_A[2] \neq 0$. In the next step, $S[A]$ is a value $X = f_{IV}(K[0])$ that depends only on the IV and the first unknown key byte.

With probability $(1 - \frac{4}{n})^{n-A} \approx e^{-4}$, the four S-box entries at the positions 1, 2, S and A are not changed in the remaining part of the key scheduling. Then the first step of the RC4 pseudo-random generator advances j to $S_A[1]$. At the second step it sets j to $S = S_A[1] + S_A[2]$. Since $S_A[2] + S_A[S] = A$, the second pseudo-random byte will be $f_{IV}(K[0])$.

So the attacker learns $f_{IV}(K[0])$ with probability e^{-4}. This is lower than the success probability of the FMS-attack, but still acceptable. The real drawback is that the attacker has to prescribe 3 S-box values which reduces the number of usable initialization vectors.

9.4.3 Initialization Vector Follows the Main Key

If the initialization vector follows the main key, the FMS-attack becomes more complicated. Instead of directly recovering the Ath key byte the attacker tries to reconstruct the values $S_A[0], \ldots, S_A[A-1]$ of the internal state after the first A steps of the key scheduling.

We explain the basic version of the attack which works only for weak keys with $S_A[1] = X < A$ and $X + S_A[X] = A$. The fraction of weak keys is very small ($\approx 0.6\,\%_0$ for $A = 13$), but the class of weak keys will be extended later.

The attacker must guess the value j_A of the pointer j after the first A steps, i.e. the attacker will try all possible values for j_A. If j_A is assumed to be known, the attacker can compute the $(A+1)$th step of the key scheduling. He will investigate only those initialization vectors which will swap $S[A]$ with $S[Y]$ for some $Y \in \{0, 2, \ldots, X-1, X+1, \ldots, A-1\}$.

With probability $(1 - \frac{3}{n})^{n-A-1} \approx e^{-3}$ the next $n - a - 1$ shuffles of the key scheduling will not touch $S[1]$, $S[X]$ and $S[A]$. In the first step of the RC4 algorithm we will set $i = 1$, $j = S[i] = X$ and $k = S[i] + S[j] = X + S[X] = A$. So the output will be $S[A]$, which is identical to the byte $S_A[Y]$.

By observing about 60 sessions (for $n = 256$) with such an IV, the attacker learns $S_A[Y]$. There is still a technical problem, since the value of j after the first $A + 1$ steps will always be the same. Different observations from different sessions are not independent. This will cause the attack to fail for some values of Y, but we learn enough values $S_A[Y]$ to find the rest by brute force (see [93] for the details).

By this procedure, the attacker learns the values $S_A[0], \ldots, S_A[A-1]$ of the S-box after the first A steps of the key scheduling. From these values, he has to derive $K[0], \ldots, K[A-1]$. Since there are n^A possible keys, but only $n \cdot (n-1) \cdots (n-A+1) = \frac{n!}{(n-A)!}$ different partial S-boxes, the attacker must test different possible keys. One way to do this is described by Algorithm 9.8 (see [93]).

The algorithm will need $A^{\lambda+1}$ time steps where λ is the number of values $0 \le i < A$ with $S_A[i] < A$. Typically this bound is quite small, so the algorithm is efficient. We will need it again in Sect. 9.5.1. An alternative algorithm which is a little faster for large values of A can be found in [148].

If the initialization vector has length at least 3, we can extend the class of weak keys. If the length is at least 4, we can attack almost all keys, but the success probability of the attack will decrease. A detailed description can be found in [93].

As we can see, the FMS-attack does not perform well in this case. It is now outdated by the attack described in Sect. 9.5.1.

9.5 Attacks Based on Golić's Correlation

In 2000 Golić mentioned in the paper [112] a correlation between the output of RC4 and parts of the internal state. Namely

$$p\big(S[k] + S[j] \equiv i \mod n\big) \approx \frac{2}{n}. \tag{9.3}$$

Algorithm 9.8 Computing the key from an early permutation state

---

1: {initialization}
2: **for** i **from** 0 **to** $n-1$ **do**
3:   $S[i] \leftarrow i$
4: **end for**
5: **for** i **from** 0 **to** A **do**
6:   $X \leftarrow S^{-1}[S_A[i]]$
7:   **if** $i < X < A$ **then**
8:     Choose X non-deterministic in $\{0, \ldots, A-1\}$ and run the remaining part of the algorithm for all possible choices.
9:   **end if**
10:   $K[i] \leftarrow X - j - S[i]$
11:   $j \leftarrow X$
12:   Swap $S[i]$ and $S[j]$
13: **end for**
14: Verify that $S[i] = S_A[i]$ for $i = 0, \ldots, A$

He gave no proof for this observation and found no useful application. Later it was rediscovered by several authors. Recently, Mantin [175] and I [148] found independent attacks that are based on Golić's correlation. Attacks from this class are currently the best known attacks against RC4.

We will derive Golić's correlation in a way that makes it easy to generalize to other RC4 like ciphers. For the theorem, we need only steps 9 and 10 (computation of k and output of $S[k]$) of the RC4 pseudo-random generator (Algorithm 9.2).

Theorem 9.12 (Klein [148]) *Assume that the internal states are uniformly distributed. Then for a fixed public pointer i, we have:*

$$P\big(S[j] + S[k] \equiv i \mod n\big) = \frac{2}{n}, \tag{9.4}$$

$$P\big(S[j] + S[k] \equiv i \mod n \mid S[j] = x\big) = \frac{2}{n} \text{ for all } x \in \{0, \ldots, n-1\}. \tag{9.5}$$

For $c \not\equiv i \mod n$, we have:

$$P\big(S[j] + S[k] \equiv c \mod n\big) = \frac{n-2}{n(n-1)}, \tag{9.6}$$

$$P\big(S[j] + S[k] \equiv c \mod n \mid S[j] = x\big) = \frac{n-2}{n(n-1)} \text{ for all } x \in \{0, \ldots, n-1\}. \tag{9.7}$$

Proof First note that (9.4) follows from (9.5), and (9.6) follows from (9.7).

To prove (9.5), we count all internal states with $S[j] + S[k] \equiv i \mod n$ and $S[j] = x$. First we use $k \equiv S[j] + S[i] \mod n$ (line 9 of Algorithm 9.2) to write $S[j] + S[k] \equiv i \mod n$ as $k + S[k] \equiv i + S[i] \mod n$.

We have to distinguish between two cases:

1. $i = k$.

   Then $S[i] = S[k]$ and the equivalence is satisfied trivially. In this case, $S[i] = S[k] \equiv i - S[j] \equiv i - x \mod n$ and there are $(n-1)!$ ways to choose the remaining $n-1$ entries of the S-box.

2. $i \neq k$.

   Then we have to set $S[k] \equiv i - x \mod n$ and $S[i] = k + S[k] - i \mod n$. We may still choose k ($n-1$ possibilities) and the remaining $n-2$ entries of the S-box ($(n-2)!$ possibilities).

Hence there are altogether $(n-1)! + (n-1)(n-2)! = 2(n-1)!$ ways in which $S[j] + S[k] \equiv i \mod n$ and $S[j] = x$, but there exist $n!$ possible internal states where $S[j] = x$, which proves (9.5).

The proof of (9.7) is quite similar.

Here we have to distinguish between three cases:

1. $i = k$.

   In this case, $S[i] = S[k]$ and $k + S[k] \equiv c + S[i] \mod n$ cannot be satisfied, since $c \neq i = k$.

2. $c = k$.

   In this case, $k + S[k] \equiv c + S[i] \mod n$ implies $S[k] \equiv S[i] \mod n$. But this is impossible, since $k \neq i$ and therefore $S[i] \neq S[k]$.

3. $i \neq k$ and $c \neq k$.

   In this case, we have to set $S[k] \equiv i - x \mod n$ and $S[i] = k + S[k] - c \mod n$. We may still choose k ($n-2$ possibilities) and the remaining $n-2$ entries of the S-box ($(n-2)!$ possibilities).

Therefore only the third case yields a contribution to the number of possible internal states. Thus

$$P\big(S[j] + S[k] \equiv c \mod n \mid S[j] = x\big) = \frac{(n-2)(n-2)!}{n!} = \frac{n-2}{n(n-1)}.$$

This proves the theorem. $\square$

9.5.1 Initialization Vector Follows the Main Key

Attacks based on Golić's correlation use the observed output to recover a part of the internal space and then use the weak key scheduling algorithm to conclude the main key from the internal space. In the case that the initialization vector follows the main key, the nature of the attack as an internal space recovering attack can be seen in its purest form.

Let A be the length of the main key, then the first A steps of the key scheduling depend only on the main key. Let $S_A[0], \ldots, S_A[A-1]$ be the first A bytes of the

S-box after the first A steps of the key scheduling algorithm. As in Sect. 9.4.3 the attacker aims at recovering these bytes.

For $1 \leq i < A$, we find that the probability that $S[i]$ remains unchanged in the last $n - A$ steps of the key scheduling algorithm and the first $i - 1$ steps of the pseudo-random generator is $(1 - \frac{1}{n})^{(n-A)+(i-1)} \approx \frac{1}{e}$.

If this happens, the ith step of Algorithm 9.2 will do the following: j will be set to some unknown value and $S[i] = S_A[i]$ is swapped with $S[j]$. Now we have $S[j] = S_A[i]$. Then the algorithm outputs $S[k]$ as the ith pseudo-random byte, which can be observed by the attacker.

By Theorem 9.12, we have

$$p\big(S[j] + S[k] \equiv i \bmod n\big) = \frac{2}{n}.$$

The values $S[k] = X[i]$ and i are known to the attacker so he can conclude that $S[j] = S_A[i]$.

If $S[i]$ is changed during the last $n - A$ steps of the key scheduling algorithm and the first $i - 1$ steps of the pseudo-random generator, the attacker still has the chance that $S_A[i] + X[i] \equiv i \bmod n$ by accident.

This leads us to the probability

$$p\big(S_A[i] + X[i] \equiv i \bmod n\big) \approx \frac{1}{e} \cdot \frac{2}{n} + \left(1 - \frac{1}{e}\right) \frac{n-2}{n(n-1)} \approx \frac{1.36}{n},$$

which is significantly larger than the probability $\frac{1}{n}$ we would expect from a true random sequence.

So what the attacker has to do is compute $i - X[i] \bmod n$ for every observed session and select the most frequent result as the value for $S_A[i]$. After 25000 sessions he will have the correct value $S_A[i]$ with high probability. Only $S_A[0]$ cannot be obtained by this method, but we can find that value by brute force.

By this method the attacker recovers the early state $S_A[0], \ldots, S_A[A - 1]$. We have already seen in Sect. 9.4.3 how to reconstruct the key from this information (Algorithm 9.8).

9.5.2 Initialization Vector Precedes the Main Key

The case where the initialization vector precedes the main key has drawn more attention, since this variant is used in the WEP protocol.

The basic idea is the following (see also [148]).

Since the initialization vector (A bytes) is known to the attacker, he can simulate the first A steps of the key scheduling. Thus he can express the value $S[A]$ after the $(A + 1)$-th step as a function $f_{IV}(K[0])$. With high probability $\approx e^{-1}$, the value of $S[A]$ is not changed during the remaining key scheduling and the first $A - 1$ steps of the RC4 pseudo-random generator.

If this happens, the A-th step of the RC4 pseudo-random generator will swap $f_{IV}(K[0])$ to $S[j]$ and by Theorem 9.12, we get the correlation

$$P\big(f_{IV}(K[0]) + X_A \equiv i \bmod n\big) = \frac{1}{e} \cdot \frac{2}{n} + \left(1 - \frac{1}{e}\right)\frac{n-2}{(n-1)n} \approx \frac{1.36}{n}. \qquad (9.8)$$

This allows us to derive $K[0]$. Interpreting $K[0]$ as part of the IV we can repeat the attack to obtain the other key bytes.

We can speed up the attack if we use additional correlations to determine the first key byte. If possible, we can also use the special IVs of the chosen IV attacks in Sect. 9.4, but these attacks use only one of n^2 IVs. For $n = 256$, we can expect to see at most one IV suitable for the FMS-attack before the attack based on Eq. (9.8) is finished. So the FMS-attack is of no help.

The following variant of Golić's correlation is better. It is possible that after the A-th step of the key scheduling $S[1] = A$. If this happens there is a high probability that neither $S[1] = A$ nor $S[A + 1] = f_{IV}(K[0])$ are changed in the remaining part of the key scheduling. This has two important consequences for our attack.

First the probability that $S[A + 1] = f_{IV}(K[0])$ remains unchanged in the key scheduling and the first $(A - 1)$ steps of the RC4 algorithm is just $e^{-1}(1 - e^{-1})$. (If the value $S[1]$ remains unchanged then $S[A+1]$ will be changed in the first RC4 key stream generation. So $S[1]$ must be changed, which explains the additional factor $1 - e$.) Thus we must replace Eq. (9.8) by

$$P\big(f_{IV}(K[0]) + X_A \equiv i \bmod n\big) = \frac{1}{e}\left(1 - \frac{1}{e}\right)\frac{2}{n} + \left(1 - \frac{1}{e}\left(1 - \frac{1}{e}\right)\right)\frac{n-2}{(n-1)n}$$

$$\approx \frac{1.23}{n}, \qquad (9.9)$$

i.e. we should trust the correlation less.

The second case is that $S[1] = A$ and $S[A + 1] = f_{IV}(K[0])$ remain unchanged. This happens with probability e^{-2}. In this case, the first step of the RC4 pseudo-random generator will swap $f_{IV}(K[0])$ to $S[1]$ and the output X_A of the A-th step will not be correlated to $f_{IV}(K[0])$. However, we can use a variant of Golić's correlation

$$P\big(S[i] + S[k] \equiv j \bmod n\big) = \frac{2}{n}.$$

The proof is identical to the proof of Theorem 9.12, we need only exchange i and j.

Normally this correlation is not helpful since it contains two unknowns $S[i]$ and j, but in this special case, j is known to be A. Thus we obtain the correlation

$$P\big(f_{IV}(K[0]) + X_0 \equiv A \bmod n\big) = \frac{1}{e^2} \cdot \frac{2}{n} + \left(1 - \frac{1}{e^2}\right)\frac{n-2}{(n-1)n} \approx \frac{1.14}{n}. \qquad (9.10)$$

The correlation of Eq. (9.10) is weaker than the correlation of Eq. (9.8) and it uses only $\frac{1}{n}$ of all possible IVs, but it still contributes enough information to significantly speed up the attack.

If we measure the complexity of the attack by counting the number of used sessions, this is the best we can do, but if we take into account that the computation of f_{IV} costs time, things are different. In the case of the WEP protocol, the attacker can easily observe many sessions by ARP-spoofing. The fact that we have to recover the key bytes sequentially rather than in parallel is particularly unsuitable for a fast implementation. In this case we can use the following variant from [216], which is implemented in the aircrack tool suite [4].

The basic idea is that in the normal case $f_{IV} = S[j_{IV} + K[0] + S[A]]$. It is very likely that in the next steps of key scheduling the swaps will play no role, so we can guess $\sum_{j=0}^{i} K[j]$ by calculating $S_{IV}^{-1}[A - 1 + i - X_{A-1+i}] - j_{IV} - \sum_{j=A}^{A+i} S_{IV}[j]$. The advantage of the new method is that it saves some computation steps and recovers the key bytes in parallel. The disadvantage is that there are strong keys for which the swaps matter. In these cases the attack will fail, but it is still faster to try this method first and switch to the slower, but exact, method only for strong keys.

9.5.3 Attacking RC4 with the First n Bytes Discarded

Combining the ideas of the chosen IV attacks and the attacks described in the previous subsections, one gets an attack which can break RC4 even if the first 256 bytes of the pseudo-random sequence are unobservable.

We assume that the initialization vector precedes the main key. The first byte of the main key gets the number b. The attacker knows the IV, so he can compute the first b steps of the key scheduling. The attacker uses only the sessions in which $S[1]$ is set to b. For random initialization vectors this will happen with probability $\frac{1}{n}$. Since the initialization vector is known to the attacker, he can compute the first b steps of the key scheduling phase and check if $S[1] = b$.

The next step of the key scheduling assigns the value $f(K[b])$ to $S[b]$. The attacker knows the function f and he wants to find the unknown key value $K[b]$. In the remaining steps of the key scheduling phase each of the values of $S[1] = b$ and $S[k] = f(K[b])$ will remain unchanged with a high probability of $\frac{1}{e^2}$.

The first step of the pseudo-random generator will set j to $S[1] = b$ and interchange $S[1]$ with $S[j] = S[b] = f(K[b])$. Thus we know that after the first step of the pseudo-random generator $S[1] = f(K[b])$ with probability $\frac{1}{e^2}$. In the next $n - 1$ steps of the pseudo-random generator the pointer j will be different from 1 with probability $\approx \frac{1}{e}$, i.e. at the beginning of the second round $S[1]$ will have the value $f(K[b])$ with probability $\approx \frac{1}{e^3}$.

Now we analyze the output of the first step of the second round. The pointer i will be equal to 1 and j has a value that is not known to us. After we have swapped $S[i]$ and $S[j]$, we know that $S[j] = f(K[b])$ with probability $\frac{1}{e^3}$. Theorem 9.12

says that $S[j] = 1 - S[k]$ with probability $\frac{2}{n}$. As in Sect. 9.5.2, we obtain

$$P\big(f(K[b]) = 1 - S[k]\big) \approx \frac{1}{e^3} \cdot \frac{2}{n} + \left(1 - \frac{1}{e^3}\right) \cdot \frac{n-2}{n(n-1)} \approx \frac{1.05}{n}. \qquad (9.11)$$

The difference with the uniform distribution is smaller than before, but still significant.

We obtain an estimator for $f(K[b])$ by observing $S[k]$. Inverting the known function f, we obtain $K[b]$. Since the success probability is smaller than the one used for the 1-round attack we have to study more sessions (approximately $813n \ln(n)$ sessions that satisfy $S[1] = b$).

After we have obtained $K[b]$, we can treat it as part of the initialization vector and apply the algorithm described above to obtain $K[b+1]$, $K[b+2]$, and so on.

Since we cannot choose the initialization vectors, we have to wait long enough to get enough initialization vectors (with $S[1] = b$, $S[1] = b+1$, etc.) for our attacks. For random initialization vectors, the estimated number of required sessions is therefore $n(813n \ln(n))$.

If the initialization vector is not random but generated by a counter, we must modify the attack a little.

If the initialization vector is generated by a little endian counter, the assumption $S[1] = b$ is satisfied in every nth session. Thus in this case we need $n(813n \ln(n))$ sessions for a successful attack.

If the initialization vector is generated by a big endian counter, things are a bit different. If $S[1] = b$ is satisfied once, it will remain equal to b for many sessions, but it will take a long time until the assumption $S[1] = b$ is first satisfied.

If the initialization is b bytes long and we want to attack the first byte $K[b]$ of the main key, we have to wait until $K[0] + K[1] = b - 1$ to obtain $S[1] = b$. This happens first for $K[0] = 0$ and $K[1] = b - 1$. The counter value is at this time $(b-1)n^{b-2}$. Thus we have to wait a long time before we can start our attack. For typical values like $n = 256$ and $b = 16$, this will be unacceptable.

The following modification of the attack will help in this case.

Instead of the investigation of initialization vectors that will result in $S[1] = f(K[b])$ after the first steps of the pseudo-random generator, we will look at initialization vectors that will result in $S[j] = f(K[b])$ after the jth step of the pseudo-random generator. This is possible, but now we must assume that the first j bytes of the S-box are not changed during the last $n - j$ steps of the key scheduling phase. The probability of this is $(1 - \frac{j}{n})^{n-j} \approx \frac{1}{e^j}$. This means that the success probability will be smaller, i.e. we have to observe more sessions, but if we choose j near to b, we can use every nth session for the attack. For certain values of b, j and n, this can be faster than the basic variant of the attack.

We can adapt the attack to the case that the initialization vector follows the main key, but then the attack becomes more difficult and less practical. The key length l should be small in comparison with n (e.g., $l = 16$ and $n = 256$). In the first l steps of the key scheduling phase, j will be set to a value j_l which depends only on the main key. The permutation S will be similar to the identity permutation.

If the key scheduling started with $i = 1$ and $j = j_l$ (S initialized with the identity), we would be able to apply the attack as in the case IV‖main key. Only minor modifications are necessary for the new starting values of i and j.

We do not know the value j_l, i.e. we have to guess it. Since after l steps the permutation S differs from the identity, we must further assume that the difference does not matter, i.e. the success probability for the attack becomes smaller.

9.5.4 A Ciphertext-Only Attack

There are many ways to vary attacks based on Golić's correlation [148]. Perhaps one of the most interesting variants is that we can mount a ciphertext-only attack.

Suppose that we have a small set of frequent symbols $\mathcal{F}$. The probability that a frequent symbol occurs is p. In many applications the set of printable ASCII characters could form this set.

For simplicity we assume that all frequent symbols have the probability $p/|\mathcal{F}|$ and that each of the rare symbols have the same probability $(1 - p)/(n - |\mathcal{F}|)$.

Let us take the case that the initialization vector follows the main key. We want to recover $S_A[1]$.

By Golić's correlation (see also Sect. 9.5.1) we have that the first output byte X_1 takes the value $1 - S_A[1]$ with probability $\frac{1}{e} \cdot \frac{2}{n} + (1 - \frac{1}{e})\frac{n-2}{(n-1)n} \approx \frac{1.36}{n}$ and it takes all other values with probability $\approx \frac{1}{n} - \frac{1.36}{n(n-1)}$.

We cannot observe X_1 directly, but we can see the ciphertext $C_1 = X_1 + M_1$. The known *a priori* distribution of the message bytes allows us to mount a likelihood attack.

The likelihood we find is

$$
P(S_A[1] = k) = \begin{cases}
\frac{1.36}{n}\frac{p}{|\mathcal{F}|} + \frac{1}{n} - \frac{1.36}{n(n-1)}(p\frac{1-|\mathcal{F}|}{|\mathcal{F}|} + (1-p)) \\
\quad \text{if } C_1 + k - 1 \in \mathcal{F}, \\
\frac{1.36}{n}\frac{1-p}{n-|\mathcal{F}|} + \frac{1}{n} - \frac{1.36}{n(n-1)}(p + (1-p)\frac{n-|\mathcal{F}|-1}{n-|\mathcal{F}|}) \\
\quad \text{if } C_1 + k - 1 \notin \mathcal{F}.
\end{cases}
$$

We will observe many sessions and select the value for $S_A[1]$ with the largest likelihood.

The attack can be improved by taking a better model for the *a priori* distribution of the message. For many real cases, we can achieve a ciphertext-only attack that works as fast as the FMS-attack with known plaintext.

9.6 State Recovering Attacks

As stated at the beginning of this chapter, the key scheduling is the weakest part of RC4. We have seen in Sect. 9.4 attacks that are based solely on the weakness of the key scheduling. Golić's correlation (Theorem 9.12) is a weakness of the RC4 pseudo-random generator, but the attacks that are based on this weakness still need the weak key scheduling to succeed. In this section we study attacks that aim at recovering the internal state of RC4 directly without using the key scheduling.

Since the internal state of RC4 is huge ($n \cdot n!$, so for $n = 256$, it has size 1692 bits), state recovering attacks are very difficult. Even a big speed-up in comparison to the exhaustive search would not be helpful. At the moment the fastest internal state recovering is due to A. Maximov and D. Khovratovich [182]. They estimate the complexity of their attack to 2^{241} operations for $n = 256$, which limits the effective key length to about 30 bytes. However, the attack needs to examine at least 2^{120} bytes of the output sequence.

The basic idea of all state recovering attacks is the following (see [149]).

At any time step, there are four unknown variables $S_t[i_t]$, j_t, $S_t[j_t]$ and k_t. We can simply simulate the RC4 pseudo-random generator (Algorithm 9.2) and guess an unknown value as soon as we need it. If we reach a contradiction we go back to the last guess and try the next value. This leads to Algorithm 9.9.

In Algorithm 9.9 "**Choose**" means a non-deterministic choice. Every time the algorithm reaches a "**Fail**" the corresponding computation path ends. In most programming languages one has to program a simulation of non-determinism explicitly, but some languages like Lisp allow a non-deterministic operator, which allows us to transform the pseudo-code of Algorithm 9.9 directly into real code.

It is remarkable that even the direct implementation of this idea leads to an algorithm that is much faster than the complete search. We want to calculate the expected running time. As in our previous analysis, we approximate the pointers j and k by true random variables.

If the algorithm is in line 5, we have two possibilities: Either $S[t + t']$ already has a value or it doesn't. Since we already have s S-box entries with an assigned value and t' of these values were assigned to $S[t], \ldots, S[t + t' - 1]$, the probability that $S[t + t']$ has an assigned value is $\frac{s-t'}{n-t'}$. This gives us the recursion:

$$c_1(t', s) = \frac{s - t'}{n - t'} c_2(t', s) + \left(1 - \frac{s - t'}{n - t'}\right)(n - s)c_2(t', s + 1). \qquad (9.12)$$

If the algorithm is in line 10, we have an $\frac{s}{n}$ chance that $S[j_t]$ already has an assigned value. If this is the case, there is a $\frac{1}{n}$ chance that $S[k]$ already has the correct value and a $(1 - \frac{s}{n})^2$ chance that $S[k]$ has no assigned value and that $X_{t+t'}$ was not assigned to some other S-box element. In the remaining case, we have found a contradiction and the computation path ends here (cost 1). If $S[j_t]$ has no assigned value, we go directly to line 23. This gives us the recursion:

$$c_2(t', s) = \frac{s}{n}\left[\left(\frac{2s}{n} - \frac{1}{n} - \frac{s^2}{n}\right) + \left(1 - \frac{s}{n}\right)^2 c_1(t' + 1, s + 1) + \frac{1}{n}c_1(t' + 1, s)\right]$$
$$+ \left(1 - \frac{s}{n}\right)c_3(t', s). \qquad (9.13)$$

Algorithm 9.9 A simple internal state recovering attack (see [149])

1: $t' = 0$,
2: $s = 0$ {s denotes the number of assigned S-box values}
3: **Choose** j_{-1} in $\{0, \ldots, n-1\}$
4: **loop**
5:     {From here the algorithm needs $c_1(t', s)$ steps}
6:     **if** $S[t + t']$ has no assigned value **then**
7:         **Choose** $S[t + t']$ different from all other S-box entries.
8:         $s \leftarrow s + 1$
9:     **end if**
10:     {From here the algorithm needs $c_2(t', s)$ steps}
11:     $j_{t'} \leftarrow j_{t'-1} + S[t + t']$
12:     Swap $S[t + t']$ and $S[j_{t'}]$
13:     **if** $S[t + t']$ has an assigned value **then**
14:         $k \leftarrow S[t + t'] + S[j_{t'}]$
15:         **if** $S[k]$ has an assigned value **then**
16:             Check if $S[k] = X_{t+t'}$. If Yes **goto** line 33, otherwise **Fail**
17:         **else**
18:             Assign $X_{t+t'}$ to $S[k]$ if possible, otherwise **Fail**
19:             $s \leftarrow s + 1$
20:             **goto** line 33
21:         **end if**
22:     **end if**
23:     {From here the algorithm needs $c_3(t', s)$ steps}
24:     **if** $X_{t+t'}$ was assigned to some S-box position **then**
25:         $k \leftarrow S^{-1}[X_{t+t'}]$
26:         Assign $k - S[t + t']$ to $S[j_{t'}]$ if possible, otherwise **Fail**
27:         $s \leftarrow s + 1$
28:     **else**
29:         **Choose** a value for $S[j_{t'}]$ such that no value is assigned to $S[k]$ with $k \leftarrow S[t + t'] + S[j_{t'}]$.
30:         Assign $X_{t+t'}$ to $S[k]$ is possible, otherwise **Fail**
31:         $s \leftarrow s + 2$
32:     **end if**
33:     {Test if we reached the end}
34:     **if** $s = n$ **then**
35:         Test if the reconstructed state produces the observed output. If Yes stop, otherwise **Fail**
36:     **else**
37:         $t' \leftarrow t' + 1$
38:     **end if**
39: **end loop**

Now we are in line 23. We have an $\frac{s}{n}$ probability that $X_{t+t'}$ was already assigned to some S-box element. In this case, we reach a contradiction if $k - S[t + t']$ was already assigned to some S-box element different from $S[j_{t'}]$. If neither $S[j_{t'}]$ has an assigned value nor $X_{t+t'}$ is assigned to some S-box element, we have $(n - s)$ choices for $S[j_{t'}]$, but only $(n - s)(1 - \frac{s}{n})$ do not lead to an immediate contradiction. Thus we have the recursion:

$$c_3(t', s) = \frac{s}{n}\left[\frac{s}{n} + \left(1 - \frac{s}{n}\right)c_1(t' + 1, s + 1)\right]$$

$$+ \left(1 - \frac{s}{n}\right)\left[(n - s)\left(1 - \frac{s}{n}\right)c_1(t' + 1, s + 2)\right]. \qquad (9.14)$$

The starting values for the recursion are $c_*(t', s) = 0$ for $s > n$ or $t' > s$ and $c_*(t', n) = 1$ for $t' \leq n$. Solving the recurrence, we find that the attack of the original RC4 ($n = 256$) needs 2^{780} steps. Simulations for small values of n support our analysis.

An improvement of the simple state recovering algorithm is the following idea (see [182]): If we know $S[i]$ we can compute the secret pointer j and simulate all swaps. Assume we know the value $S[i]$ for x consecutive steps. We call these x consecutive steps the search window. Now we can guess an $S[j]$ or k to proceed. Instead of always choosing the first unknown value as in the simple state recovering attack (Algorithm 9.9), we choose the value that implies as many other values as possible. This will reduce the number of necessary non-deterministic steps and speed up the search. In addition, we could be lucky and the search window might grow during the search. For details we refer to [182].

For a good start we should use a large search window. In [182] the authors suggest using d-order, g-generative patterns for this purpose.

Definition 9.2 A d-order pattern is a quadruple (i, j, S) where S is a partial function $\mathbb{Z}/n\mathbb{Z} \rightarrow \mathbb{Z}/n\mathbb{Z}$ which gives the value of d S-box entries.

The pattern is g-generative if we can compute the secret pointer j for the next g steps of the RC4 algorithm out of the pattern.

Table 9.1 illustrates a 7-generative pattern of order 4.

The largest pattern found in [182] is a 14-order, 76-generative pattern. If we were able to start with such a pattern, the attack would be quite fast (about 2^{240} operations), but such a pattern occurs only every $256 \cdot 256!/(256 - 14)! \approx 2^{120}$ output bytes, so it is near to impossible to find this special pattern. For a practical attack, we would need a huge number of good patterns, but at this time no one knows how to find or even store such a pattern set.

9.7 Other Attacks on RC4

An alternative class of attacks against RC4 searches for distinguishers between the RC4 pseudo-random sequence and a true random sequence.

Table 9.1 A 4-order, 7-generative pattern									
i	j	S[0]	S[1]	S[2]	S[3]	S[4]	S[5]	S[6]
-1	-2	6	-1	2	$*$	$*$	$*$	-2
0	4	$*$	-1	2	$*$	6	$*$	-2
1	3	$*$	$*$	2	-1	6	$*$	-2
2	5	$*$	$*$	$*$	-1	6	2	-2
3	4	$*$	$*$	$*$	6	-1	2	-2
4	3	$*$	$*$	$*$	-1	6	2	-2
5	5	$*$	$*$	$*$	-1	6	2	-2
6	3	$*$	$*$	$*$	-2	6	2	-1

The first published result of this kind is due to J.Dj. Golić [110, 111]. He proves that the sum of the last bits at time step t and $t+2$ is correlated to 1. The correlation coefficient is $\approx 15 \cdot 2^{-24}$. He then concludes that 2^{40} bytes of the RC4 pseudo-random sequence are distinguishable from a true random sequence.

9.7.1 Digraph Probabilities

Better distinguishers are based on the computation of digraph probabilities.

In [95] we find an analysis of digraph probabilities. Under the assumption that the internal state is random, we can compute the exact digraph probabilities by the following state counting algorithm.

Algorithm 9.10 Computing the digraph probabilities

1: **for** i **from** 0 **to** $n-1$ **do**
2:    **for** j **from** 0 **to** $n-1$ **do**
3:       {Loop over all possible values for i and j in line 9 of Algorithm 9.2.}
4:       Choose non-deterministic $S[i]$
5:       Choose non-deterministic $S[i+1]$
6:       Choose non-deterministic $S[j]$ if $j \notin \{i, i+1\}$
7:       Compute $k = S[i] + S[j]$, $i+1$ and $j' = j + S[i]$
8:       Choose non-deterministic $S[j']$ if $j' \notin \{i, i+1, j\}$
9:       Compute $k' = S[i+1] + S[j']$
10:      **for all** possible values of $S[k]$ and $S[k']$ **do**
11:         Increase $P[i, S[k], S[k']]$ by $\prod_{i=1}^{6 - |\{i,i+1,j,j',k,k'\}|}(n - 6 + i)$
12:      **end for**
13:   **end for**
14: **end for**

A naive implementation of Algorithm 9.10 would require a loop for every "Choose" which results in a run time of $O(n^8)$. For $n = 256$, this is still in the range of modern computers, but takes too long. Therefore the authors of [95] did the calculation only for $n = 8, 16, 32$ and extrapolated the results to $n = 256$.

However, it is possible to transform Algorithm 9.10 into a much faster program. To that end we have to successively apply the following transformation steps.

- If A and B are two independent program fragments we can replace

$$\textbf{for } i \in \mathcal{A} \textbf{ do } \text{A, B } \textbf{end for}$$

by

$$\textbf{for } i \in \mathcal{A} \textbf{ do } \text{A } \textbf{end for}, \textbf{ for } i \in \mathcal{A} \textbf{ do } \text{B } \textbf{end for}.$$

- We can replace

$$\textbf{for } i \in \mathcal{A} \textbf{ do if } i \notin \mathcal{B} \textbf{ then } \text{A } \textbf{end if } \textbf{end for}$$

by

$$\textbf{for } i \in \mathcal{A} \backslash \mathcal{B} \textbf{ do } \text{A } \textbf{end for}.$$

Similarly we can remove all if-statements after a for-statement.

We use these two rules to separate in the computation of Algorithm 9.10 the general case where $i, i + 1, j, j', k$ and k' are all different from the special cases in which some values coincide. The algorithm then looks like Algorithm 9.11.

As we can see, the 8 nested loops give a running time of $O(n^8)$. The special cases have at most 7 nested loops and contribute only an $O(n^7)$ to the total running time.

Now we use the following transform to reduce the number of nested loops.

- We can transform

```
1: for i ∈ 𝒜 do
2:    for j ∈ ℬ\{f(i)} do
3:       A
4:    end for
5: end for
```

to

```
1: for j ∈ ℬ do
2:    for i ∈ 𝒜\{f⁻¹(j)} do
3:       A
4:    end for
5: end for
```

This transform allows us to change the order of the for-loops.
- We can replace the loop

$$\textbf{for } i \in \mathcal{A} \textbf{ do } \text{Increase counter by } k \textbf{ end for}$$

by the single instruction

$$\text{Increase counter by } k \cdot |\mathcal{A}|.$$

Algorithm 9.11 Computing the digraph probabilities (1. Transformation)

1: {Do the general case first}
2: **for** i **from** 0 **to** $n-1$ **do**
3:   **for** $j \in \{0, \ldots, n-1\} \backslash \{i, i+1\}$ **do**
4:     **for** $S[i] \in \{0, \ldots, n-1\} \backslash \{i - S[i], i+1-S[i]\}$ **do**
5:       **for** $S[i+1] \in \{0, \ldots, n-1\} \backslash \{S[i]\}$ **do**
6:         **for** $S[j] \in \{0, \ldots, n-1\} \backslash \{S[i], S[i+1], i-S[i], i+1-S[i], j-S[i]\}$ **do**
7:           **for** $S[j'] \in \{0, \ldots, n-1\} \backslash \{S[i], S[i+1], S[j], i-S[i+1], i+1-S[i+1], j-S[i+1], (i+S[i])-S[i+1], (S[i]+S[j])-S[i+1]\}$ **do**
8:             **for** $S[k] \in \{0, \ldots, n-1\} \backslash \{S[i], S[i+1], S[j], S[j']\}$ **do**
9:               **for** $S[k'] \in \{0, \ldots, n-1\} \backslash \{S[i], S[i+1], S[j], S[j'], S[k']\}$ **do**
10:                 Increase $P[i, S[k], S[k']]$ by 1
11:               **end for**
12:             **end for**
13:           **end for**
14:         **end for**
15:       **end for**
16:     **end for**
17:   **end for**
18: **end for**
19: {Now do the special cases}
20: {This part is skipped}

We use the first of these two rules to move the loop over $S[j']$ (line 7 of Algorithm 9.11) into the loops over $S[k]$ and $S[k']$ (line 8 and 9 of Algorithm 9.11). Then we can use the second rule to replace the loop over $S[j']$ by a simple multiplication.

In our case lines 7 to 13 of Algorithm 9.11 are transformed to Algorithm 9.12.

Algorithm 9.12 Computing the digraph probabilities (2. Transformation, inner loops)

1: **for** $S[k] \in \{0, \ldots, n-1\} \backslash \{S[i], S[i+1], S[j]\}$ **do**
2:   **for** $S[k'] \in \{0, \ldots, n-1\} \backslash \{S[i], S[i+1], S[j], S[k']\}$ **do**
3:     Increase counter $P[i, S[k], S[k']]$ by
$|\{0, \ldots, n-1\} \backslash \{S[i], S[i+1], S[j], S[k], S[k'], i-S[i+1], i+1-S[i+1], (i+S[i])-S[i+1], (S[i]+S[j])-S[i+1]\}|$
4:   **end for**
5: **end for**

This saves us one loop and reduces the running time to $O(n^7)$.

Table 9.2 Digraph probabilities of RC4

Digraph	Value(s) of i	Difference from the uniform distribution
$(0, 0)$	$i = 1$	$+7.81\ ‰$
$(0, 0)$	$i \neq 1, 255$	$+3.91\ ‰$
$(0, 1)$	$i \neq 0, 1$	$+3.91\ ‰$
$(i + 1, 255)$	$i \neq 254$	$+3.91\ ‰$
$(255, i + 1)$	$i \neq 1, 254$	$+3.91\ ‰$
$(255, i + 2)$	$i \neq 0, 253, 254, 255$	$+3.91\ ‰$
$(255, 0)$	$i = 254$	$+3.91\ ‰$
$(255, 1)$	$i = 255$	$+3.91\ ‰$
$(255, 2)$	$i = 0, 1$	$+3.91\ ‰$
$(129, 129)$	$i = 2$	$+3.91\ ‰$
$(255, 255)$	$i \neq 254$	$-3.91\ ‰$
$(0, i + 1)$	$i \neq 0, 255$	$-3.91\ ‰$

This is not the end. We can again split the loop into the general part in which $S[i]$, $S[i + 1]$, $S[j]$, $S[k]$, $S[k']$, $i - S[i + 1]$, $i + 1 - S[i + 1]$ are all different and into the special part in which some of the values coincide. Then we apply the above transformation again to reduce the running time to $O(n^6)$.

Repeating these steps again and again we finally reach an $O(n^3)$ implementation of Algorithm 9.10, but the constant will be extremely large. Taking this into account, and that the transformation itself costs time, we will stop the transformation as soon as we reach an $O(n^5)$ algorithm. This is fast enough to compute the digraph probabilities for $n = 256$. Table 9.2 shows the largest differences from the uniform distribution.

A very interesting observation on digraph probabilities is the following.

Theorem 9.13 (Mantin [176]) *For small g, the probability of the pattern $ABSAB$ where S is a word of length g is*

$$\approx \left(1 + \frac{e^{(4-8g)/n}}{n}\right) \frac{1}{n^2}.$$

Proof Suppose that the following happens: Suppose further that in the $g - 2$ time steps from $t + 1$ to $t + g - 2$, none of the 8 S-box entries displayed in Fig. 9.6 are changed. Then the computation goes as shown in Fig. 9.6 and produces the pattern $ABSAB$.

The probability that all these conditions are satisfied is

$$\frac{1}{n^3} \left(1 - \frac{g + 4}{n}\right)^2 \left(1 - \frac{6}{n}\right)^{g-2}.$$

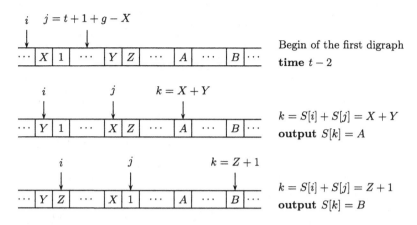

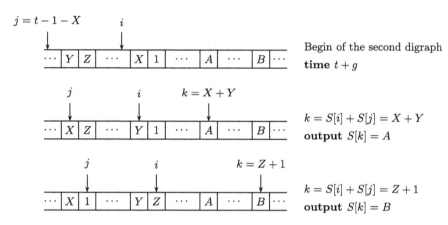

Fig. 9.6 Digraph repetition

($\frac{1}{n^3}$ is the probability that $j_{t-1} = t + g - 1$, $j_{t+g-1} = t - 1$ and $S_{t-1}[t] = 1$. Neither $X + Y$ nor $Z + 1$ may lie in the interval $[t - 1, t + g + 2]$, which gives us the probability $(1 - \frac{g+4}{n})^2$. The probability that j is never $X + Y$ or $Z + 1$ is $(1 - \frac{6}{n})^{g-2}$.)

Using the approximation $(1 - \frac{6}{n})^{g-2} \approx e^{6(g-2)/N}$ and $(1 - \frac{g+4}{n}) \approx e^{(g+4)/N}$ we find that our special event has probability $\frac{e^{(4-8g)/n}}{n} n^{-3}$.

If the special event does not happen, the pattern can still occur by chance (with probability $\frac{1}{n^2}$), which proves the theorem. □

Theorem 9.13 has a very high bias. This gives a good distinguishing attack.

I want to close this section with a research problem. From simulations with small values of n, I found the following correlation.

Fig. 9.7 Example of a
3-fortuitous state

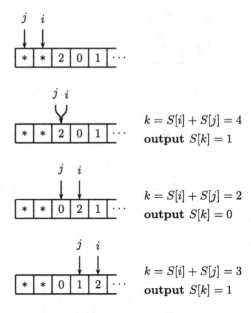

$$k = S[i] + S[j] = 4$$
output $S[k] = 1$

$$k = S[i] + S[j] = 2$$
output $S[k] = 0$

$$k = S[i] + S[j] = 3$$
output $S[k] = 1$

- If we observe the output $t - 1$ at time step t, then we may predict with probability $\frac{1}{n} + \frac{1}{n^2}$ the output 1 at time step $t + n$.

However, I have no proof for it and I have not found an application in an attack.

9.7.2 Fortuitous States

Another very interesting idea is the notion of fortuitous states introduced in [95].

Definition 9.3 A state i, j, S of RC4 is called k-*fortuitous* if the next k steps of the RC4 algorithm use only the values $S[i + 1], \ldots, S[i + k]$.

The name 'fortuitous state' refers to the fact that such a state is a fortune for the cryptanalyst. Figure 9.7 illustrates a 3-fortuitous state.

We can find fortuitous states by Algorithm 9.13. Every time the computation reaches a "Choose non-deterministic" (line 8 and 15), the computation branches to cover all possible choices.

Every time we must choose a non-deterministic value for $S[i]$ or $S[j]$ respectively, we have at most k choices which do not lead to an immediate contradiction. Thus the worst case complexity of the search is $O(n^2 k^k)$. However, most often the search will reach a dead end before we have assigned values to all k active S-box elements, so the computation is much faster than the worst case bound suggests. Algorithm 9.13 is good enough to enumerate k-fortuitous states for $k \le 8$. For larger k it becomes very slow.

Algorithm 9.13 Searching fortuitous states

1: **for** i **from** 0 **to** $n - 1$ **do**
2:    **for** j **from** 0 **to** $n - 1$ **do**
3:      {Search fortuitous states with pointers i, j}
4:      {The active part of the S-box is $S[i + 1], \ldots, S[i + k]$}
5:      **for** k times **do**
6:        advance i by 1
7:        **if** $S[i]$ has no assigned value **then**
8:          Choose non-deterministic S[i]
9:        **end if**
10:       $j \leftarrow j + S[i]$
11:       **if** $S[j]$ is not an active part of S-box **then**
12:         Stop the search {Not a k-fortuitous state}
13:       **end if**
14:       **if** $S[j]$ has no assigned value **then**
15:         Choose non-deterministic S[i]
16:       **end if**
17:       **if** $S[k]$ with $k \leftarrow S[i] + S[j]$ is not an active part of S-box **then**
18:         Stop the search {Not a k-fortuitous state}
19:       **end if**
20:       Swap $S[i]$ and $S[j]$
21:     **end for**
22:     Output i, j and S {k-fortuitous state found}
23:   **end for**
24: **end for**

The search can be improved significantly. There are two ideas in the new algorithm:

1. In the first pass, we do not choose a value for $S[j]$. We only store restrictions.
2. In the first pass, we have to choose values for $S[i]$ and calculate the pointers j. These operations are translation invariant, so we can do it in parallel for all i.

These ideas lead to the new Algorithm 9.14.

Algorithm 9.14 is a bit complicated. We now describe it in detail.

The algorithm stores in the pair (m, M) the smallest and the largest value of k, i.e. all $S[i] + S[j]$ lie in the interval $[m, M]$. Since we are working in $\mathbb{Z}/n\mathbb{Z}$, the terms "smallest" and "largest" are a bit ambiguous. It is entirely possible that $m = n - 1$ and $M = 1$, which means that all k are either $n - 1$, 0 or 1. As long as $r < n/2$ we will have no difficulty detecting which of two given numbers is the "largest". At the start the values (m, M) are undefined. Since we are searching for an r-fortuitous state, the values k have to be in the range $[i, i + r - 1]$ for some i, i.e. $M - m$ can be at most r. In line 11, we use this to cut off branches in the search tree.

Similarly to (m, M), we keep for every position $S[i]$ the smallest value $m[i]$ and the largest value $M[i]$ that was added to $S[i]$.

Algorithm 9.14 Enumerating fortuitous states (fast)

1: $i := n - 1$
2: **for** j **from** 0 **to** $n - 1$ **do**
3:      {Find possible $S[i]$}
4:      {The active part of the S-box is $S[1], \ldots, S[r]$}
5:      **for** r **times do**
6:          advance i by 1
7:          **if** $S[i]$ has no assigned value **then**
8:              Choose non-deterministic $S[i]$
9:              **if** $m[i]$ and $M[i]$ have values **then**
10:                 $(m, M) := (\min\{m, m[i] + S[i]\}, \max\{M, M[i] + S[i]\})$
11:                 Stop the search if $M - m \geq r$ {Not an r-fortuitous state}
12:             **end if**
13:         **end if**
14:         j := j +S[i]
15:         **if** $S[j]$ is not an active part of the S-box **then**
16:             Stop the search {Not an r-fortuitous state}
17:         **end if**
18:         **if** $S[j]$ has an assigned value **then**
19:             $(m, M) := (\min\{m, S[i] + S[j]\}, \max\{M, S[i] + S[j]\})$
20:             Stop the search if $M - m > r$ {Not an r-fortuitous state}
21:         **else**
22:             $(m[j], M[j]) := (\min\{m, S[i]\}, \max\{M, S[i]\})$
23:             Stop the search if $M[j] - m[j] \geq r$ {Not an r-fortuitous state}
24:         **end if**
25:         Swap((S[i],m[i],M[i]) ; (S[j],m[j],M[j]))
26:     **end for**
27:     {We have found suitable values for $S[i]$.}
28:     {Now loop over all i and translate the S-box.}
29:     **for** i **from** $M - r$ **to** m **do**
30:         {loop from 0 to $n - 1$ if (m, M) has no value}
31:         **for** x **from** 0 **to** $r - 1$ **do**
32:             **if** $S[x]$ has no assigned value **then**
33:                 Choose non-deterministic $S[x] \in \{i - m[x], \ldots, i + r - 1 - M[x]\}$
34:                 **if** No value for $S[x]$ left **then**
35:                     Stop the search {Not an r-fortuitous state}
36:                 **end if**
37:             **end if**
38:         **end for**
39:         Output fortuitous state found $i - 1, j + i - 1, S_{\to i}$
40:         {$S_{\to i}$ denotes the S-box shifted by i places to the right}
41:     **end for**
42: **end for**

When we reach line 27, we have found a partial state of RC4 with the following properties: We start with $i = n - 1$. In the next r steps the values j will be in the range $[0, r - 1]$. We can translate this partial state by a, just by adding a to all pointers and shifting the S-box to the right.

The loop in line 29 runs over all possible shifts for which all values k are in the range $[i, i + r - 1]$, i.e. for which $[m, M] \subseteq [i, i + r - 1]$.

Up to now we have assigned values only to the S-box elements $S[i]$ which we needed for computing the pointer j. Now we have to assign values to all other S-box elements $S[x]$; this is done in lines 32–38. Since we have collected all necessary information to decide which values are allowed in the previous steps, this part is fast.

Algorithm 9.14 still visits all fortuitous states. If our only goal is just to count them, we can improve the running time by replacing the enumeration in lines 29–42. When we reach line 29, we have to solve the following counting problem:

Given finite sets $A_1, \ldots, A_r$, what is the number of r-tuples $(x_1, \ldots, x_r)$ with $x_i \in A_i$ and $x_i \neq x_j$ for $i \neq j$. In Algorithm 9.14, we enumerate these r-tuples. There is no closed formula for the number $N(A_1, \ldots, A_r)$ of such r-tuples, but we can count them by using a special case of the Möbius inversion

$$N(A_1, \ldots, A_r) = \sum_{p \text{ is a partition of } \{1, \ldots, r\}} c(p) \prod_{\pi \text{ is a part of } p} \left| \bigcap_{i \in \pi} A_i \right|. \qquad (9.15)$$

The coefficients $c(p)$ satisfy the recurrence

$$c(\{\{1\}, \ldots, \{r\}\}) = 1 \quad \text{and} \quad \sum_{p' \text{ is a refinement of } p} c(p') = 0$$

for all partitions p different from $\{\{1\}, \ldots, \{r\}\}$.

What happens in Eq. (9.15) becomes clear if we look at the special case $r = 3$.

$$N(A_1, A_2, A_3) = |A_1| \cdot |A_2| \cdot |A_3| - |A_1 \cap A_2| \cdot |A_3| - |A_1 \cap A_3| \cdot |A_2|$$
$$- |A_2 \cap A_3| \cdot |A_1| + 2|A_1 \cap A_2 \cap A_3|.$$

The first term just counts all triples, the next three terms subtract the number of triples in which two of the three values are even. The last term corrects the error we made when we subtracted the triples of the form (x, x, x) three times.

The coefficients $c(p)$ are known as Möbius functions of the lattice of all partitions with respect to refinement. For this lattice one can prove $c((\pi_1, \ldots, \pi_k)) = \prod_{j=1}^{k} (-1)^{|\pi_k|} (|\pi_k| - 1)!$ (see, for example, [256] Sect. 3.10).

What makes the fortuitous state a fortune for the cryptanalyst is the following:

Let us take the 3-fortuitous state of Fig. 9.7. Under the assumption that every internal state has the same probability we know that the initial condition holds with probability $(256 \cdot 256 \cdot 255 \cdot 254)^{-1}$. Every time the initial condition is satisfied, the next three output bytes will be 1, 0, 1. The probability of generating this specific output sequence is about $\frac{1}{256^3}$. This means that the probability that the sequence 1, 0, 1 is caused by the fortuitous state is $\approx 256^3 / (256 \cdot 256 \cdot 255 \cdot 254) \approx \frac{1}{253}$. For

Table 9.3 The number of fortuitous states and their expected occurrence

Length	Number	Frequency	Expected False Hits
2	516	22.9	255
3	290	31.8	253
4	6540	35.2	250
5	25, 419	41.3	246
6	101, 819	47.2	241
7	1, 134, 639	51.8	236
8	14, 495, 278	56.1	229
9	169, 822, 607	60.5	222
10	1, 626, 661, 109	65.1	214
11	18, 523, 664, 501	69.6	205
12	233, 003, 340, 709	73.9	197
13	3, 250, 027, 739, 130	78.0	187
14	46, 027, 946, 087, 463	82.1	178
15	734, 406, 859, 761, 986	86.0	169
16	12, 652, 826, 949, 516, 042	93.1	159

the attacker this means that if he observes the output sequence $1, 0, 1$ beginning at a time with $i = 2$, he can conclude $S[2] = 2$, $S[3] = 0$, $S[1] = 1$ with probability $\frac{1}{253}$, which is much more than the probability $\frac{1}{256^3}$ we would expect for a random guess.

In Table 9.3 we list the number of k-fortuitous states together with the logarithm of the expected frequency (base 2) and the expected number of false hits. By a false hit, we mean that we observe an output pattern, which looks like the output of a fortuitous state, but which is not created by a fortuitous state.

As one can see, the occurrence rate of fortuitous states is not high. Even when one occurs and the attacker has guessed correctly a part of the internal state it is not clear how to proceed. The internal state of RC4 is big enough that even the knowledge of 8 bytes of the S-box does not seem very helpful.

The idea of fortuitous states can be extended to non-successive outputs or to states in which a known S-box values allow the prediction of b outputs (see [177, 209]).

9.8 RC4 Variants

9.8.1 An RC4 Variant for 32-Bit Processors

RC4 is designed for 8-bit processors. This is good for embedded devices but on a modern computer we would like to have a cipher that uses 32-bit or even 64-bit words. An interesting RC4 variant that has this property is described in [116].

The reason why we cannot extend RC4 directly to 32 bits is that for $n = 2^{32}$ the S-box would be far too large. Algorithm 9.15 and Algorithm 9.16 solve this problem by introducing a new parameter. The S-box still has only n elements, but all elements are taken in $\mathbb{Z}/m\mathbb{Z}$ for some multiple of n. The authors suggest taking $n = 256$ and m either 2^{32} or 2^{64}.

Algorithm 9.15 RC4(n,m) key scheduling

 1: {initialization}
 2: **for** i from 0 to $N-1$ **do**
 3:     $S[i] \leftarrow a_i$
 4: **end for**
 5: $j \leftarrow 0, k \leftarrow 0$
 6: {Shuffle S}
 7: **for** counter from 1 to r **do**
 8:     **for** i from 0 to $N-1$ **do**
 9:         $j \leftarrow (j + S[i] + K[i \bmod l]) \mod n$
10:         $Swap(S[i], S[j])$
11:         $S[i] \leftarrow S[i] + S[j] \mod m$
12:         $k \leftarrow k + S[i] \mod m$
13:     **end for**
14: **end for**

Algorithm 9.16 RC4(n,m) pseudo-random generator

 1: {initialization}
 2: $i \leftarrow 0, j \leftarrow 0$
 3: {Generate pseudo-random sequence}
 4: **loop**
 5:     $i \leftarrow i + 1 \mod n$
 6:     $j \leftarrow j + S[i] \mod n$
 7:     $k \leftarrow k + S[j] \mod m$
 8:     **output** $S[S[i] + S[j] \bmod n] + k \mod m$
 9:     $S[S[i] + S[j] \bmod n] = k + S[i] \mod m$
10: **end loop**

Note that the key scheduling is repeated r times, which should avoid attacks that explore the weak key scheduling. In [116] the authors use methods similar to that of Sect. 9.3.2 to determine good values for the parameter r in the key scheduling. They advise taking $r = 20$ for $m = 2^{32}$ and $r = 40$ for $m = 2^{64}$.

The biggest difference in comparison to the original RC4 is the counter k which is used during the update of the S-box and to obfuscate the output. Perhaps the best way to understand the role of k is to look at a preliminary version of Algorithm 9.16 which had no such counter (see Algorithm 9.17).

Algorithm 9.17 RC4(n,m) pseudo-random generator, old version from [121]

1: {initialization}
2: $i \leftarrow 0, j \leftarrow 0$
3: {Generate pseudo-random sequence}
4: **loop**
5:    $i \leftarrow i + 1 \mod n$
6:    $j \leftarrow j + S[i] \mod n$
7:    $Swap(S[i], S[j])$
8:    **output** $S[S[i] + S[j] \mod n]$
9:    $S[S[i] + S[j] \mod n] \leftarrow S[i] + S[j] \mod m$
10: **end loop**

Algorithm 9.17 has a serious flaw that allows strong distinguishers [284]. The problem is the update $S[S[i] + S[j] \mod n] = S[i] + S[j] \mod m$. It is possible that $S[i]$, $S[j]$ and $S[S[i] + S[j] \mod n]$ are selected as outputs. So we expect to find three values a, b, c with $c = a + b$ in the output sequence of Algorithm 9.17 much more frequently than we would expect in a true random sequence.

Another problem of Algorithm 9.17 is that it is susceptible to attacks based on Golić's correlation [148].

9.8.2 RC4A

Another interesting variant of the RC4 algorithm is the RC4A algorithm (Algorithm 9.18) described in [209]. The idea is to use two S-boxes in "parallel" to obtain a stronger algorithm.

In each step, RC4A produces two output bytes. It advances i only once per two output bytes, i.e. it needs one fewer instruction per two output bytes in comparison to RC4. In addition, RC4A can easily be parallelized, since most instructions touch only one of the two S-boxes.

RC4A is immune to most attacks that work against RC4. The most dangerous attack is based on a variant of Golić's correlation.

Theorem 9.14 *Assume that the permutations S_1 and S_2 are chosen independently and uniformly at random. Then:*

$$P\big(S_1[j_1] + S_1[k_1] + S_2[j_2] + S_2[k_2] \equiv 2i \mod n\big) = \frac{1}{n-1}. \qquad (9.16)$$

Proof The proof is very similar to the proof of Theorem 9.12.

We use $k_2 \equiv S_1[i] + S_1[j_1] \mod n$ and $k_1 \equiv S_2[i] + S_2[j_2] \mod n$ to write

$$S_1[j_1] + S_1[k_1] + S_2[j_2] + S_2[k_2] \equiv 2i \mod n$$

Algorithm 9.18 RC4A pseudo-random generator

1: {initialization}
2: $i \leftarrow 0$
3: $j_1 \leftarrow 0$, $j_2 \leftarrow 0$
4: {generate pseudo-random sequence}
5: **loop**
6:    $i \leftarrow (i + 1) \mod n$
7:    $j_1 \leftarrow (j_1 + S_1[i]) \mod n$
8:    Swap $S_1[i]$ and $S_1[j]$
9:    $k_2 \leftarrow (S_1[i] + S_1[j]) \mod n$
10:   **print** $S_2[k_2]$
11:   $j_2 \leftarrow (j_2 + S_2[i]) \mod n$
12:   Swap $S_2[i]$ and $S_2[j]$
13:   $k_1 \leftarrow (S_2[i] + S_2[j]) \mod n$
14:   **print** $S_1[k_1]$
15: **end loop**

---

as

$$\bigl(k_1 + S_1[k_1]\bigr) + \bigl(k_2 + S_2[k_2]\bigr) \equiv \bigl(i + S_1[i]\bigr) + \bigl(i + S_2[i]\bigr) \mod n.$$

Now we count the states that satisfy this condition. For this, we have to distinguish between several cases:

1. $k_1 = k_2 = i$
   In this case, we may choose S_1 and S_2 without any restriction. Thus we have $(n!)^2$ possible combinations.
2. $k_1 = i, k_2 \neq i$
   In this case, we may choose S_1 as we want ($n!$ possibilities). Next we may choose k_2 ($n - 1$ possibilities), then we have to choose $S_2[i]$ (n possibilities) which determines $S_2[k_2]$. Finally, we may choose the remaining part of S_2 ($(n - 2)!$ possibilities). Altogether we have $(n!)((n - 1)n(n - 2)!) = (n!)^2$ possibilities in this case. (Compare this with Theorem 9.12.)
3. $k_1 \neq i, k_2 = i$
   This is analogous to the previous case, with $(n!)^2$ possibilities.
4. $k_1 \neq i, k_2 \neq i$
   We distinguish between two subcases:
   (a) $k_1 + S[k_1] \equiv i + S_1[i] \mod n$
       This is equivalent to $k_2 + S[k_2] \equiv i + S_2[i] \mod n$.
       We may count the number of possibilities for both S-boxes separately. As in Theorem 9.12 or in the two cases above, we find $(n!)^2$ possibilities in this case.
   (b) $k_1 + S[k_1] \not\equiv i + S_1[i] \mod n$
       In this case, we may first choose k_1 (n possibilities) and $S_1[i]$ (n possibilities). Now we choose $S_1[k_1]$ ($n - 2$ possibilities, since $S_1[k_1] \neq S_1[i]$ and

$k_1 + S[k_1] \not\equiv i + S_1[i] \mod n$ by assumption). There are $(n-2)!$ possibilities left to choose the remaining part of S_1.

By assumption $k_2 \neq i$. Furthermore, $(k_1 + S[k_1]) + k_2 \not\equiv (i + S_1[i]) + i \mod n$, since $S_2[k_2] \neq S_2[i]$. Thus we have $n - 2$ possibilities to choose k_2. Now we may choose $S_2[i]$ (n possibilities), which determines $S_2[k_2]$. For the remaining part of S_2 we have $(n-2)!$ possibilities.

This makes $[(n-1)n(n-2)(n-2)!][(n-2)n(n-2)!]$ possibilities in this case.

Altogether there are $n \cdot n!(n-2)!$ internal states with

$$\left(k_1 + S_1[k_1]\right) + \left(k_2 + S_2[k_2]\right) \equiv \left(i + S_1[i]\right) + \left(i + S_2[i]\right) \mod n.$$

Since the total number of possible internal states is $(n \cdot n!)^2$, this proves the theorem. □

There are also analogs to Eqs. (9.5), (9.6) and (9.7) of Theorem 9.12.

The correlation of Theorem 9.14 is weaker than Golić's correlation, but it is still strong enough to be exploited in an attack.

Theorem 9.14 allows us to estimate $S_1[j_1] + S_2[j_2] \mod n$. Whether we are able to use this information to obtain the main key depends on the key scheduling algorithm.

Suppose we have two independent subkeys K_1 and K_2 and use these subkeys with the normal RC4 key scheduling algorithm (Algorithm 9.1) to initialize S_1 and S_2 respectively. Then we can use Theorem 9.14 and the technique of our 1-round attack to compute a relation between K_1 and K_2. This reduces the effective key length by $\frac{1}{2}$.

If one generates K_2 from K_1 by a simple RC4 algorithm as suggested in [209], we can mount the following attack. Assume that the initialization vector precedes the main key. With a chosen initialization vector, we may enforce that the first two bytes of K_2 depend in a simple way on the first two bytes of the main key. (Use initialization vectors similar to the one used in the FMS-attack.) Now we analyze the first two output bytes of the RC4A pseudo-random generator with Theorem 9.12. $S_1[j]$ depends only on the first two bytes of K_1. These bytes are part of the initialization vector which are known to the attacker. Thus the attacker may compute $S_2[j_2]$ from the observed values. This byte depends only on the first two bytes of the main key. The attack now follows the pattern of attack described in Sect. 9.5.1.

For the attack described above, more information about the initialization vector is needed than for the FMS-attack, furthermore the correlation is rather weak. This means that it needs an unrealistically high number of observed sessions.

As a conclusion of this subsection, we may say: RC4A has weaknesses that are very similar to the weaknesses of RC4, but the consequences are not so dramatic.

9.8.3 Modifications to Avoid Known Attacks

A common design principle in cryptography is to study successful attacks and only make small changes to the algorithm to avoid attacks of these classes. In this section, we collect some ideas for improvements of RC4 that are based on this idea.

The chosen IV attacks use the fact that the first bytes of the S-box depend with high probability only on the first bytes of the session key. Besides discarding the first output bytes or iterating the key scheduling, we can simply let i run from $n-1$ down to 0 in the key scheduling, i.e. we use Algorithm 9.19.

Algorithm 9.19 RC4 key scheduling

1: {initialization}
2: **for** i **from** 0 **to** $n-1$ **do**
3:     $S[i] \leftarrow i$
4: **end for**
5: $j \leftarrow 0$
6: {generate a random permutation}
7: **for** i **from** $n-1$ **down to** 0 **do**
8:     $j \leftarrow (j + S[i] + K[i \mod l]) \mod n$
9:     Swap $S[i]$ and $S[j]$
10: **end for**

Besides preventing chosen IV attacks, Algorithm 9.19 also slows down attacks that are based on Golić's correlation (see [148]).

Golić's correlation says that $P(i + S[i] \equiv k + S[k] \bmod n) = \frac{2}{n}$. We cannot change this probability but we can modify RC4 in such a way that it is less useful to the attacker. For example:

- We move in Algorithm 9.2 the swap of $S[i]$ and $S[j]$ behind the output of $S[k]$. Now the attacker still has the correlation $P(S[j] + S[k] \equiv i \bmod n) = \frac{2}{n}$, but he has less control over $S[j]$ (see [148]).
- We compute k as $a(S[i] + S[k])$ for some $a \neq 1$ which is coprime to n. Now the attacker cannot remove $S[i]$ from

$$P\big(i + S[i] \equiv a\big(S[i] + S[j]\big) + S[k] \bmod n\big) = \frac{2}{n}.$$

Having all three values $S[i]$, $S[j]$ and $S[k]$ in the equation makes it much less useful.

Another suggestion, made by G.K. Paul in his PhD thesis [208], is to use Algorithm 9.20 to obtain the initial permutation. He gives several simulations and theoretical results to support his suggestion.

The algorithm also uses random swaps, but changes the order in which the cards are swapped. Experiments indicate that Algorithm 9.20 removes most irregularities from the RC4 key shuffle.

Algorithm 9.20 Paul's suggestion for key scheduling

1: {initialization}
2: **for** i **from** 0 **to** $n - 1$ **do**
3:     $S[i] \leftarrow i$
4: **end for**
5: $j \leftarrow 0$
6: **for** i **from** 0 **to** $n - 1$ **do**
7:     $j \leftarrow (j + S[i] + K[i])$ {K is the main key}
8:     Swap $S[i]$ and $S[j]$
9: **end for**
10: **for** i **down from** $n/2 - 1$ **to** 0 **do**
11:     $j \leftarrow (j + S[i]) \oplus (K[i] + IV[i])$ {IV initialization vector (session id)}
12:     Swap $S[i]$ and $S[j]$
13: **end for**
14: **for** i **from** $n/2$ **to** $n - 1$ **do**
15:     $j \leftarrow (j + S[i]) \oplus (K[i] + IV[i])$ {IV initialization vector (session id)}
16:     Swap $S[i]$ and $S[j]$
17: **end for**
18: **for** y **from** 0 **to** $n - 1$ **do**
19:     **if** y is even **then**
20:         $i \leftarrow y/2$
21:     **else**
22:         $i \leftarrow n - (y + 1)/2$
23:     **end if**
24:     $j \leftarrow (j + S[i] + K[i])$
25:     Swap $S[i]$ and $S[j]$
26: **end for**

Chapter 10
The eStream Project

The ECRYPT Stream Cipher Project [85] was run from 2004 to 2008 to identify a portfolio of promising new stream ciphers. Currently none of these ciphers has proved itself in a widespread application, but all of them show state of the art developments in stream cipher design. Since there is already a whole book [226] devoted to the eStream finalists, we keep this chapter short and describe only three examples.

10.1 Trivium

Trivium is a synchronous stream cipher designed by C. De Cannière and B. Preneel [75]. It is optimized for hardware implementations. One design goal was to achieve the security with minimal effort, so the cipher has no large security margins.

Trivium uses 80-bit keys and 80-bit initialization vectors. The internal state is has size 288 bits.

The key stream generation is described by Algorithm 10.1. Figure 10.1 shows a graphical representation of the key stream generation process.

The design is based on the following key ideas:

- Trivium should have a compact implementation to allow low power applications. Thus it is based on shift registers. Since non-linear combinations of LFSRs need larger internal states, Trivium uses non-linear feedback functions, even if this makes the cipher harder to analyze.
- To allow high speed implementations it is necessary that Trivium is parallelizable. After an internal state of Trivium is changed, it is not used for at least 64 steps. Therefore a hardware implementation of Trivium can generate 64 bits of output per clock cycle. Such an implementation needs only 288 flip-flops, 192 AND-gates and 704 XOR-gates. In comparison to the bit-oriented implementation with 288 flip-flops, 3 AND-gates and 11 XOR-gates this is quite cheap. The price is that Trivium uses sparse update functions, which is always risky.

A. Klein, *Stream Ciphers*, DOI 10.1007/978-1-4471-5079-4_10,
© Springer-Verlag London 2013

Algorithm 10.1 Trivium key stream generation

1: **loop**
2:     $t_1 \leftarrow s_{66} + s_{93}$
3:     $t_2 \leftarrow s_{162} + s_{177}$
4:     $t_3 \leftarrow s_{243} + s_{288}$
5:     **output** $t_1 + t_2 + t_3$
6:     $t_1 \leftarrow t_1 + s_{91}s_{92} + s_{171}$
7:     $t_2 \leftarrow t_2 + s_{175}s_{176} + s_{264}$
8:     $t_3 \leftarrow t_3 + s_{286}s_{287} + s_{69}$
9:     $(s_1, s_2, \ldots, s_{93}) \leftarrow (t_3, s_1, \ldots, s_{92})$
10:    $(s_{94}, s_{95}, \ldots, s_{177}) \leftarrow (t_1, s_{94}, \ldots, s_{176})$
11:    $(s_{178}, s_{179}, \ldots, s_{288}) \leftarrow (t_2, s_{178}, \ldots, s_{287})$
12: **end loop**

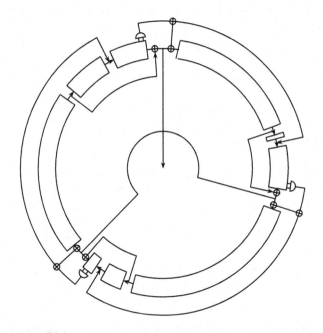

Fig. 10.1 The cipher Trivium

The key scheduling of Trivium is also quite simple (Algorithm 10.2). It writes the main key and the initialization vector into the internal memory. To avoid related key attacks the registers are clocked for four full cycles.

Even two full cycles are enough to guarantee that every bit of the internal state depends on every bit of the key. So four rounds seem to be sufficient to mask the main key.

The output of Trivium can be described by sparse equations, but the degree of these equations increases fast enough to make the standard linearization techniques

---

Algorithm 10.2 Trivium key scheduling

1: $(s_1, \ldots, s_{93}) \leftarrow (K_1, \ldots, K_{80}, 0, \ldots, 0)$
2: $(s_{94}, \ldots, s_{177}) \leftarrow (IV_1, \ldots, IV_{80}, 0, \ldots, 0)$
3: $(s_{178}, \ldots, s_{288}) \leftarrow (0, \ldots, 0, 1, 1, 1)$
4: **for** i **from** 1 **to** $4 \cdot 288$ **do**
5:     $t_1 \leftarrow s_{66} + s_{93} + s_{91} s_{92} + s_{171}$
6:     $t_2 \leftarrow s_{162} + s_{177} + s_{175} s_{176} + s_{264}$
7:     $t_3 \leftarrow s_{243} + s_{288} + s_{286} s_{287} + s_{69}$
8:     $(s_1, s_2, \ldots, s_{93}) \leftarrow (t_3, s_1, \ldots, s_{92})$
9:     $(s_{94}, s_{95}, \ldots, s_{177}) \leftarrow (t_1, s_{94}, \ldots, s_{176})$
10:    $(s_{178}, s_{179}, \ldots, s_{288}) \leftarrow (t_2, s_{178}, \ldots, s_{287})$
11: **end for**

---

impracticable. At the moment no practical algebraic attack is known. The best attack of this type is a so-called cube-attack [80].

The idea of a cube-attack is to shift computation from the attack into pre-processing by using chosen initialization vectors. Let v denote the initialization vector and let x be the key. Let $\{v_{i_1}, \ldots, v_{i_k}\}$ be a part of the initialization vector. Assume that you have an algebraic relation of the form

$$F(v, x) = v_{i_1} \cdots v_{i_k} f(v, x) + g(v, x) = 0 \tag{10.1}$$

where $f(v, x)$ does not depend on $v_{i_1}, \ldots, v_{i_k}$ and $g(v, x)$ contains no monomial divisible by $v_{i_1} \cdots v_{i_k}$.

Assume that the attacker can observe 2^k sessions in which the initialization vector takes all possible values in $\{v_{i_1}, \ldots, v_{i_k}\}$ and is fixed in all other positions, then summing over the 2^k possibilities gives the equation

$$\sum_{v=v^{(1)}}^{v=v^{(2^k)}} F(v, x) = f(v, x) = 0.$$

The new equation is of course of much lower degree than the old one.

The value of a cube-attack should not be overestimated, since a good cipher does not allow one to find many equations of the form (10.1) (see [23]). If you are able to find enough equations for a cube-attack it is quite probable that a straightforward algebraic attack will succeed anyway. The real advantage of the cube-attack is that the search for equations of the form (10.1) can be done in a pre-processing phase, i.e. it can afford a lot of computation without slowing down the actual attack.

Now consider a variation of Trivium that is clocked only 672 times instead of 1152 times during the key scheduling phase (see line 4 of Algorithm 10.2). Note that in the reduced variant every bit of the internal state already depends on every key bit, but with the cube-attack one can even find linear equations. For example, starting with the set $\{v_2, v_{13}, v_{20}, v_{24}, v_{43}, v_{46}, v_{53}, v_{55}, v_{57}, v_{67}\}$ we obtain the linear term

$f = 1 + x_0 + x_9 + x_{50}$, which corresponds to the 675 output bit (starting the count with the 672 bits that are discarded during the key scheduling).

In [80] the authors give a list of 63 linear equation that can be obtained in this way, which is enough to recover the 80 bit key by trying only 2^{17} possibilities. Thus the reduced Trivium variant is broken.

However, the non-linear elements in Trivium increase the degree of the equations quickly and there is no obvious way to extend the attack. Even for the reduced Trivium, the attack seems less practical, since the attacker must have access to an enormous number of sessions. Assume that the initialization vector is just a counter. Then you must wait for $2^{67} + 2^{57} + \cdots + 2^2 \approx 1.4 \cdot 10^{20}$ sessions before you observe a full cube of type $\{v_2, v_{13}, v_{20}, v_{24}, v_{43}, v_{46}, v_{53}, v_{55}, v_{57}, v_{67}\}$. If the initialization vector is pseudo-random, it is very unlikely that you can get such a cube in shorter time. The protocol will most likely force a key change earlier.

Even with all these problems, the cube-attack is probably the most dangerous algebraic attack against Trivium.

Perhaps the most likely attack against Trivium would be of the state recovering type. The straightforward attack guess $(s_{25}, \ldots, s_{93})$, $(s_{97}, \ldots, s_{177})$ and $(s_{244}, \ldots, s_{288})$ and the rest of the internal state can be immediately determined from the keystream. This reduces the effective internal state to 195 bits. It is quite possible that this can be reduced further, but it is still far from the 80 bit key size.

10.2   Rabbit

Rabbit is a synchronous stream cipher developed by M. Boesgaard, M. Vesterager, T. Peterson, J. Christiansen and O. Scavenius [34]. Rabbit works internally on 32-bit words which makes it suitable for software implementation.

The internal state consist of eight 32-bit variables $x_{j,t}$ ($0 \le j < 8$) and eight 32-bit counters $c_{j,t}$ ($0 \le j < 8$). Here t denotes the number of iterations. In addition, a carry bit $\phi_{7,t}$ has to be stored during the iterations. Altogether the internal state has size 513 bits.

The core of the Rabbit algorithm is the next-state function.

Let

$$g_{j,t} = \left((x_{j,t} + c_{j,t+1})^2 \oplus \left((x_{j,t} + c_{j,t+1})^2 \gg 32\right)\right) \quad \bmod 2^{32}$$

where $\oplus$ denotes the XOR operation and all additions are done modulo 2^{32}.

The internal states $x_{j,t}$ are updated according to

$$x_{0,t+1} = g_{0,t} + (g_{7,t} \lll 16) + (g_{6,t} \lll 16),$$

$$x_{1,t+1} = g_{1,t} + (g_{0,t} \lll 8) + g_{7,t},$$

$$x_{2,t+1} = g_{2,t} + (g_{1,t} \lll 16) + (g_{0,t} \lll 16),$$

$$x_{3,t+1} = g_{3,t} + (g_{2,t} \lll 8) + g_{1,t},$$

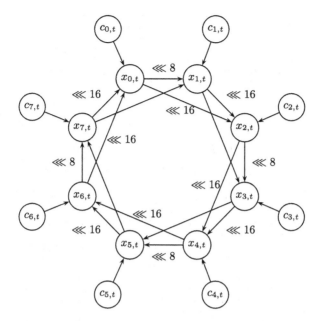

Fig. 10.2 The cipher Rabbit

$$x_{4,t+1} = g_{4,t} + (g_{3,t} \lll 16) + (g_{2,t} \lll 16),$$

$$x_{5,t+1} = g_{5,t} + (g_{4,t} \lll 8) + g_{3,t},$$

$$x_{6,t+1} = g_{6,t} + (g_{5,t} \lll 16) + (g_{4,t} \lll 16),$$

$$x_{7,t+1} = g_{7,t} + (g_{6,t} \lll 8) + g_{5,t}.$$

(Here $\lll$ denotes left rotation.)

Figure 10.2 gives a graphic impression of the next-state function.

The counters are updated according to:

$$c_{0,t+1} = c_{0,t} + a_0 + \phi_{7,t} \qquad \mathrm{mod}\ 2^{32},$$

$$c_{1,t+1} = c_{1,t} + a_1 + \phi_{0,t+1} \quad \mathrm{mod}\ 2^{32},$$

$$c_{2,t+1} = c_{2,t} + a_2 + \phi_{1,t+1} \quad \mathrm{mod}\ 2^{32},$$

$$c_{3,t+1} = c_{3,t} + a_3 + \phi_{2,t+1} \quad \mathrm{mod}\ 2^{32},$$

$$c_{4,t+1} = c_{4,t} + a_4 + \phi_{3,t+1} \quad \mathrm{mod}\ 2^{32},$$

$$c_{5,t+1} = c_{5,t} + a_5 + \phi_{4,t+1} \quad \mathrm{mod}\ 2^{32},$$

$$c_{6,t+1} = c_{6,t} + a_6 + \phi_{5,t+1} \quad \mathrm{mod}\ 2^{32},$$

$$c_{7,t+1} = c_{7,t} + a_7 + \phi_{6,t+1} \quad \mathrm{mod}\ 2^{32},$$

where $\phi_{j,t+1}$ is the carry bit when computing $c_{j,t+1}$, i.e.

$$\phi_{j,t+1} = \begin{cases} 1 & \text{if } j = 0 \text{ and } c_{0,t} + a_0 + \phi_{7,t} > 2^{32}, \\ 1 & \text{if } j \neq 0 \text{ and } c_{j,t} + a_j + \phi_{j-1,t+1} > 2^{32}, \\ 0 & \text{otherwise.} \end{cases}$$

The constants a_j are

$$a_0 = 0x4D34D34D, \qquad a_1 = 0xD34D34D3,$$
$$a_2 = 0x34D34D34, \qquad a_3 = 0x4D34D34D,$$
$$a_4 = 0xD34D34D3, \qquad a_5 = 0x34D34D34,$$
$$a_6 = 0x4D34D34D, \qquad a_7 = 0xD34D34D3.$$

At each time step 128 output bits are extracted from the internal state. The extraction is done according to

$$s_t^{[15\ldots0]} = x_{0,t}^{[15\ldots0]} \oplus x_{5,t}^{[31\ldots16]}, \qquad s_t^{[31\ldots16]} = x_{0,t}^{[31\ldots16]} \oplus x_{3,t}^{[15\ldots0]},$$
$$s_t^{[47\ldots32]} = x_{2,t}^{[15\ldots0]} \oplus x_{7,t}^{[31\ldots16]}, \qquad s_t^{[63\ldots48]} = x_{2,t}^{[31\ldots16]} \oplus x_{5,t}^{[15\ldots0]},$$
$$s_t^{[79\ldots64]} = x_{4,t}^{[15\ldots0]} \oplus x_{1,t}^{[31\ldots16]}, \qquad s_t^{[95\ldots80]} = x_{4,t}^{[31\ldots16]} \oplus x_{7,t}^{[15\ldots0]},$$
$$s_t^{[111\ldots96]} = x_{6,t}^{[15\ldots0]} \oplus x_{3,t}^{[31\ldots16]}, \qquad s_t^{[127\ldots112]} = x_{6,t}^{[31\ldots16]} \oplus x_{1,t}^{[15\ldots0]}.$$

Here $x_{j,t}^{[15\ldots0]}$ denotes the rightmost 16 bits of $x_{j,t}$ and similarly $x_{j,t}^{[31\ldots16]}$ denotes the leftmost 16 bits. The numbers in the brackets indicate the position of the extracted bits.

Rabbit borrows ideas from chaos theory, which explains the use the square function as non-linear part. The idea of using the square function for pseudo-random generation dates back to the (very weak) von-Neumann-generator, which takes the middle digits of the square as next state function. The combination of the highest and the lowest bits, as done in Rabbit, provides better results (smaller correlation coefficients).

The counter system is used to guarantee a large cipher period, based on a technique developed in [242].

The design of Rabbit is based on 32-bit registers, but it can be also be efficiently implemented on 8-bit processors. The only critical operation is computing the square of a 32-bit integer. In 8-bit arithmetic this requires 10 multiplications, which is still good.

The simple symmetric structure of Rabbit also helps in hardware implementations. In the design paper the authors mention a design with 4100 gates and a speed of 500 MBit/s (optimized for size) and a design with 100000 gates and a speed of 12.4 GBit/s (optimized for speed). This is much better than a modern block cipher, but lies far behind a stream cipher designed for hardware implementation such as Trivium.

There are no known attacks against Rabbit. What currently comes closest to an attack is a bias in the keystream of Rabbit found by J.-P. Aumasson [11]. It leads to a distinguisher which requires 2^{254} bits of the keystream.

10.3 Mosquito and Moustique

Self-synchronizing stream ciphers are very rare. Only two proposals in the eStream project were self-synchronizing stream ciphers. SSS [123] was a candidate in the first round which was broken by J. Daemen, J. Lano and B. Preneel [70].

The second candidate, Mosquito [68], was designed by J. Daemen and P. Kitsos. It was broken during round two of the eStream competition by A. Joux and F. Muller [139]. Based on this experience, the authors of Mosquito designed the updated variant Moustique [69], which removes some flaws in the cipher. Moustique was broken during the third and final round of the eStream project by E. Käsper, V. Rijmen, T.E. Biørstad, C. Rechberger, M. Robshaw and G. Sekar [145].

We describe Moustique and its attack as an example of the difficulty of designing self-synchronizing stream ciphers.

The internal state of Moustique consists of 8 bit fields $a^{(0)}, \ldots, a^{(7)}$. The size of $a^{(0)}$ is 128 bits and the bits are indexed from 1 to 128. The size of $a^{(1)}$, $a^{(2)}$, $a^{(3)}$, $a^{(4)}$ and $a^{(5)}$ is 53 bits and their bits are indexed from 0 to 52. The register $a^{(6)}$ has 12 bits indexed from 0 to 11 and $a^{(7)}$ has 3 bits indexed $a_0^{(7)}$, $a_1^{(7)}$ and $a_2^{(7)}$.

Figure 10.3 gives an overview of the cipher. The eight levels of bit fields make the delay of the self-synchronizing stream cipher equal to 9.

The update is done by the simple Boolean functions:

$$g_0(a, b, c, d) = a + b + c + d,$$

$$g_1(a, b, c, d) = a + b + c(d + 1) + 1,$$

$$g_1(a, b, c, d) = a(b + 1) + c(d + 1) + 1.$$

At any time step the output bit z of Moustique is computed as

$$z = a_0^{(7)} + a_1^{(7)} + a_2^{(7)}$$

and the message bit m is encrypted to the cipher bit c by

$$c = m + z = g_0\big(m, a_0^{(7)}, a_1^{(7)}, a_2^{(7)}\big).$$

The next state of the bit field $a^{(i)}$ ($1 \leq i \leq 7$) depends only on the bit field $a^{(i-1)}$. The update of the bit field $a^{(i)}$ ($1 \leq i \leq 7$) is given in the next equation.

$$a_i^{(7)} \leftarrow g_0\big(a_{4i}^{(6)}, a_{4i+1}^{(6)}, a_{4i+2}^{(6)}, a_{4i+3}^{(6)}\big) \qquad \text{for } 0 \leq i < 3,$$

$$a_i^{(6)} \leftarrow g_1\big(a_{4i}^{(5)}, a_{4i+3}^{(5)}, a_{4i+1}^{(5)}, a_{4i+2}^{(5)}\big) \qquad \text{for } 0 \leq i < 12,$$

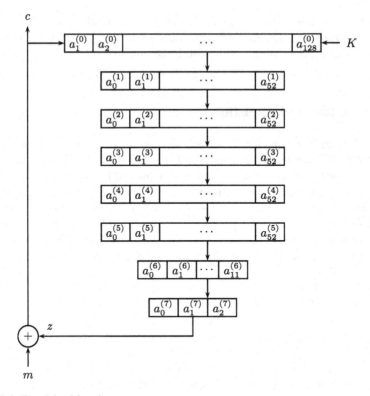

Fig. 10.3 The cipher Moustique

$$a_{4i \bmod 53}^{(5)} \leftarrow g_1\left(a_i^{(4)}, a_{i+3}^{(4)}, a_{i+1}^{(4)}, a_{i+2}^{(4)}\right) \qquad \text{for } 0 \le i < 53,$$

$$a_{4i \bmod 53}^{(4)} \leftarrow g_1\left(a_i^{(3)}, a_{i+3}^{(3)}, a_{i+1}^{(3)}, a_{i+2}^{(3)}\right) \qquad \text{for } 0 \le i < 53,$$

$$a_{4i \bmod 53}^{(3)} \leftarrow g_1\left(a_i^{(2)}, a_{i+3}^{(2)}, a_{i+1}^{(2)}, a_{i+2}^{(2)}\right) \qquad \text{for } 0 \le i < 53,$$

$$a_{4i \bmod 53}^{(2)} \leftarrow g_1\left(a_i^{(1)}, a_{i+3}^{(1)}, a_{i+1}^{(1)}, a_{i+2}^{(1)}\right) \qquad \text{for } 0 \le i < 53,$$

$$a_{4i \bmod 53}^{(1)} \leftarrow g_1\left(a_{128-i}^{(0)}, a_{i+18}^{(0)}, a_{113-i}^{(0)}, a_{i+1}^{(0)}\right) \quad \text{for } 0 \le i < 53.$$

The bit field $a^{(0)}$ is organized as what the authors call a conditional complementing shift register (CCSR). To describe the update of $a^{(0)}$ we rename the bits. The 128 bits of $a^{(0)}$ are partitioned into 96 cells q^i indexed from 1 to 96. The number of bits in cell i is denoted by n_i and given by Table 10.1.

The mapping between the labels q_j^i and $a_k^{(0)}$ is illustrated in Fig. 10.4. The index i is given on the right, the index j on the bottom and the index k stands in the cells. For example q_2^{95} is identical to $a_{107}^{(0)}$.

Table 10.1 Number of bits per cell in the CCSR of Moustique

Range of i	n_i
$1 \le i \le 88$	1
$89 \le i \le 92$	2
$93 \le i \le 94$	4
$i = 95$	8
$i = 96$	16

Fig. 10.4 Mapping between q_j^i and $a_k^{(0)}$

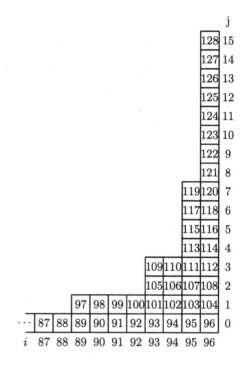

Formally we set $q_0^0 = c + 1$ and $n_0 = 1$ where c is the last cipher text bit. The update of the CCSR is then done by

$$q_j^{96} \leftarrow g_2\left(q_{j \bmod 8}^{95}, q_0^{95-j}, q_j^{94} {}_{\bmod 4}, q_{[j \le 5]}^{94-j}\right)$$

for $0 \le j \le 16$.

Here $[j \le 5]$ denotes the indicator function, i.e. $[j \le 5]$ is 1 for $j \le 5$ and 0 otherwise.

$$q_j^i \leftarrow g_x\left(q_{j \bmod n_{i-1}}^{i-1}, k_{j-1}, q_{j \bmod n_v}^v, q_{j \bmod n_w}^w\right)$$

for $0 \le j \le n_i$ and $i = 95, \dots, 3$.

Table 10.2 Bit updating in the CCSR of Moustique

Index		Function g_x	v	w
$(i - j) \equiv 1$	mod 3	g_0	$2(i - j - 1)/3$	$i - 2$
$(i - j) \equiv 2$	mod 3	g_1	$i - 4$	$i - 2$
$(i - j) \equiv 3$	mod 6	g_1	0	$i - 2$
$(i - j) \equiv 0$	mod 6	g_1	$i - 5$	0

Here k_{j-1} denotes the jth bit of the key. The values of x, v and w are given by Table 10.2

$$q_0^2 \leftarrow q_0^1 + k_1 + 1 = g_1(q_0^1, k_1, 0, 0),$$
$$q_0^1 \leftarrow c + k_1 + 1 = g_0(q_0^0, k_0, 0, 0).$$

Now we sketch the attack from [145] which removed Moustique from the list of eStream finalists. The starting point is the following observation:

Lemma 10.1 *Let $i \leq 77$ and $i \equiv 2 \bmod 3$ then q_0^i occurs in the update function of the CCSR only linearly.*

Proof The value q_0^i occurs only in the following update steps:

- $q_0^{i+1} = g_1(q_0^i, k_i, q_0^v, q_0^w)$ and g_1 is linear in its first two components.
- $q_0^{i+2} = g_0(q_0^{i+1}, k_{i+1}, q_0^{2(i+1)/3}, q_0^{i-2})$, but q_0 is linear.
- It cannot occur as $v = i' - 4$ with $i' = i + 4$ in the computation of q_0^{i+4} since $(i + 4) - 0 \equiv 0 \not\equiv 1 \bmod 3$.
- It cannot occur as $v = i' - 5$ with $i' = i + 5$ in the computation of q_0^{i+5} since $(i + 5) - 0 \equiv 1 \bmod 3$, i.e. $(i + 5) - 0 \equiv 0 \bmod 6$.
- It occurs as $v = 2(i' - j' - 1)/3$ in the computation of $q_0^{i'}$ with $i' = (3i + 1)/2$ and $j' = 0$, if $i' \equiv 1 \bmod 3$. But then the equation is $q_0^{i'} = g_0(q_0^{i'-1}, k_{i'-1}, q_0^i, q_0^{i'-2})$ and g_0 is linear. $\square$

The practical consequence of Lemma 10.1 is that it can be used to mount a related key attack.

Assume that the attacker can observe the output of two instances of Moustique, which encrypt the same plain text. The difference between these two instances is that the initial bit q_0^{71} is switched. In addition, the bits k_{70}, k_{71} and k_{72} of the key differ in both instances. By Lemma 10.1 the bit q_0^{71} occurs only linearly in the update function of the CCSR and then by switching k_{70}, k_{71} and k_{72} we assure that during the iteration of the CCSR the differences remain in the bit q_0^{71}.

The only way that this difference can affect the output is if it affects the register $a^{(1)}$. This is done via the equation

$$
\begin{aligned}
a_9^{(1)} &= a_{4\times 42 \bmod 53}^{(1)} \\
&= g_1\left(a_{86}^{(0)}, a_{60}^{(0)}, a_{71}^{(0)}, a_{43}^{(0)}\right) \\
&= g_1\left(q_0^{86}, q_0^{60}, q_0^{71}, q_0^{43}\right) \\
&= q_0^{86} + q_0^{60} + q_0^{71}\left(q_0^{43} + 1\right) + 1.
\end{aligned}
$$

If $q_0^{43} = 1$, the error does not propagate.

The attacker observes the two related cipher texts and searches for the first difference. This immediately gives one bit of information (the corresponding bit q_0^{43} must be 0). However, the attacker can also guess the key bits $k_0, \ldots, k_{42}$ and compute the first 43 bits $q_0^1, \ldots, q_0^{43}$ of the CCSR. In all previous steps the bit q_0^{43} must be 1. This allows the attacker to check if the guess was correct.

This is only the most basic version of the attack. There is still much work to do to effectively test the keys and to exclude false positives. See the original paper [145] for the details.

Related key attacks like this may be a bit exotic and it is not immediately clear how the attack should be used in practice (see also the discussion in [24]). However the attack is a clear sign that something is wrong in Moustique and it was absolutely correct to remove it from the eStream portfolio.

Moustique provides a good example of the problems that occur when designing a self-synchronizing stream ciphers. Everyone who is interested in self-synchronizing stream ciphers should study the examples of Moustique and its predecessors Mosquito, $\Upsilon\Gamma$ [66] and KNOT [67] carefully.

Self-synchronizing stream ciphers have, in comparison to synchronous stream ciphers, the appealing property that they also secure the ciphertext against active attacks, i.e. you do not need an additional message authentication code (MAC). However, for the moment I would not recommend them, due the problems with the design process. If you really need a self-synchronizing stream cipher you can still run a block cipher in CFB-mode, but even this has problems. It is probably best not to mix different security goals. Always use two crypto-primitives: a cipher for data security and a MAC to ensure the data integrity. (High level block-cipher modes like CWC [163] or OCB [199] use the same block cipher for two purposes, once as a cipher and once as a MAC, but even these modes clearly separate the two security goals.)

Chapter 11
The Blum-Blum-Shub Generator and Related Ciphers

This chapter covers a class of highly unusual stream ciphers. Instead of being designed to fit real hardware, these ciphers borrow ideas from public key cryptography to enable security proofs. The development of these ciphers started shortly after the development of public key cryptography [33, 240]. The advantage is that we can be very sure that an attacker has no chance of breaking the cipher. The disadvantage is that these ciphers are much slower and need much more memory than normal stream ciphers. They even have problems which come into the range of modern block ciphers. Thus these ciphers are rarely applied as stream ciphers, but they can be applied as part of a key generation protocol.

11.1 Cryptographically Secure Pseudo-random Generators

Every pseudo-random generator can be broken by a brute force search over its internal state. So we cannot expect a pseudo-random generator to be unconditionally secure in the same sense as the one-time pad is unconditionally secure. The best we can expect is that the complexity of breaking the generator is very high. To be able to speak of complexity classes we need a generator family with an arbitrary large internal state. We use the size of the internal state as a parameter to measure the complexity. To avoid additional complications we assume that there is no key scheduling, the key is just the initial state. As usual in complexity theory, we say a problem is easy if it can be solved in polynomial time and it is hard otherwise. This leads us to the following definition.

Definition 11.1 (see [33]) A pseudo-random generator family is said to be *cryptographically secure* if:

1. The time needed to generate the next bit is bounded by a polynomial in the size of the internal state.
2. There exists no polynomial time algorithm which predicts without knowledge of the internal state the next bit with probability $1/2 + \epsilon_n$, where $\epsilon_n \to 0$ as the size n of the internal state tends to infinity.

A. Klein, *Stream Ciphers*, DOI 10.1007/978-1-4471-5079-4_11,
© Springer-Verlag London 2013

Note: A non-deterministic Turing machine can break every generator in polynomial time simply by guessing the internal state and verifying the guess. So the best we can hope for is that breaking the generator is in $\mathcal{NP}\backslash\mathcal{P}$. Since it is still an open problem whether $\mathcal{P} = \mathcal{NP}$ or $\mathcal{P} \neq \mathcal{NP}$ (see Sect. 12.4 for a short introduction to this problem), our proofs will have the form: "If some problem is difficult, then the generator is cryptographically secure".

One should also be aware that not being in $\mathcal{P}$ does not automatically mean that an algorithm is useless in practice. For example, computing Gröbner bases is $\mathcal{EXPSPACE}$-complete, which is several steps above $\mathcal{NP}$ in the hierarchy of complexity classes, nevertheless Gröbner bases are an important and practical tool.

Before we can present some examples of cryptographically secure generators we introduce some notation and terminology.

Definition 11.2 Let $x_1, \ldots, x_n$ be a sequence of pseudo-random bits.

An ε-*distinguisher* is an algorithm which, given the sequence $x_1, \ldots, x_n$ and a true random sequence $y_1, \ldots, y_n$, recognizes the pseudo-random sequence with probability $1/2 + \varepsilon$.

An ε-*next bit predictor* is an algorithm which computes from $x_1, \ldots, x_n$ the next bit x_{n+1} with probability $1/2 + \varepsilon$.

An ε-*previous bit predictor* is an algorithm which computes from $x_1, \ldots, x_n$ the preceding bit x_0 with probability $1/2 + \varepsilon$.

The concepts of distinguisher, next bit predictor and previous bit predictor may look different, but they are essentially the same.

Lemma 11.1 *The following statements are equivalent.*

1. *There exist a distinguisher for a pseudo-random sequence.*
2. *There exist a next bit predictor for a pseudo-random sequence.*
3. *There exist a previous bit predictor for a pseudo-random sequence.*

Proof Assume that you have a predictor for a pseudo-random generator and you want a distinguisher. To decide whether a sequence x was generated by the pseudo-random generator you apply the predictor to a part of x and compare the prediction with the observed bits. If the prediction is correct you accepts the hypothesis that x was generated by the pseudo-random generator. If the prediction in wrong you reject the hypothesis.

On the other hand you can construct a predictor from a distinguisher by applying it to the pair $x_1, \ldots, x_n, 0$ and $x_1, \ldots, x_n, 1$. If x_{n+1} is 0 the distinguisher will tell you that $x_1, \ldots, x_n, 1$ must come from a true random source and vice versa. $\square$

The idea of Blum and Micali [33] is to turn a one-way function into a stream cipher. Before we can state their main result we need some more terminology.

Definition 11.3 Let (S_n) be a series of sets and for each $i \in S_n$ let D_i be some set. We call B a set of *predicates* if

$$B = \{ B_i : D_i \to \{0, 1\} \}.$$

B is an *accessible* set of predicates if there exists a polynomial time algorithm that selects random elements of $I_n = \{(i, x) \mid i \in S_n, x \in D_i\}$. Each element must be selected with probability $\frac{1}{|I_n|}$.

B is *input hard* if there exists no probabilistic polynomial time algorithm that decides $B_i(x) = 1$ with probability $1/2 + \varepsilon$ for some $\varepsilon > 0$.

Theorem 11.2 *Let B be a set of accessible, input hard predicates. For each i let f_i be a permutation of D_i that is computable in polynomial time.*

If $h : D_i \to \{0, 1\}$ with $H : x \mapsto B_i(f_i(x))$ is computable in polynomial time then Algorithm 11.1 is a cryptographically secure pseudo-random generator.

Algorithm 11.1 The Blum-Micali generator

1: **loop**
2:    **output** $B_i(f_i(x))$
3:    $x \leftarrow f_i(x)$
4: **end loop**

Proof By assumption, the computations in Algorithm 11.1 are done in polynomial time.

Assume now that the sequence produced with Algorithm 11.1 can be predicted in polynomial time.

Then there also exists by Lemma 11.1 a previous bit predictor. Apply it to the sequence $s_k = B_i(f_i^k(x))$ $(k = 1, \ldots)$ and you get an efficient probabilistic decision procedure for $B_i(x) = 1$ as $B_i(x) = s_0$.   □

As an example, Blum and Micali give the following generator based on the discrete logarithm problem (see Algorithm 11.2).

Algorithm 11.2 Discrete logarithm generator

Require: p is a prime of the form $2q + 1$, g is an element of $\mathbb{Z}/p\mathbb{Z}$ of order q, $x \in \{g^0, g, \ldots, g^{p-1}\}$

1: **loop**
2:    **output** 1 if $g^x > p/2$ and 0 otherwise
3:    $x \leftarrow g^x$
4: **end loop**

11.2 The Blum-Blum-Shub Generator

We come now to the cipher that gives its name to this chapter. In [32] L. Blum, M. Blum and M. Shub improved the ideas presented in the last section. Their cipher, which is now known as the Blum-Blum-Shub generator, was the first cipher that is cryptographically secure in the sense of Definition 11.1 and also fast enough that it can be applied at least in some special cases in which speed does not have high priority.

The cipher is quite simple (see Algorithm 11.3).

Algorithm 11.3 The Blum-Blum-Shub generator

Require: $n = pq$ where p, q are primes $\equiv 3 \mod 4$
  1: **loop**
  2:     $s \leftarrow s^2 \mod n$
  3:         **output** $s \mod 2$
  4: **end loop**

We reduce the problem of a previous bit predictor for the Blum-Blum-Shub generator to the *Composite Quadratic Residue Problem* which is assumed to be hard (see Sect. 13.5).

Theorem 11.3 *A previous bit predictor for the Blum-Blum-Shub generator gives a Monte Carlo algorithm that solves the composite quadratic residue problem for n. The Monte Carlo algorithm and the predictor have the same time complexity and the same success probability.*

Proof The Blum-Blum-Shub generator is just a particular instance of Algorithm 11.1. Hence we can simply apply Theorem 11.1. But since it is so easy and important, we just repeat the arguments for the special case of the Blum-Blum-Shub generator.

Given s with $\left(\frac{s}{n}\right) = 1$, since -1 is a quadratic non-residue modulo p and q, either s or $-s$ is a quadratic residue modulo n. Since n is odd, s and $n - s$ have different parity.

Compute the Blum-Blum-Shub sequence $x_i = (s^{2^i} \mod n) \mod 2$. The previous bit must be either $x_0 = s \mod 2$ if s is a quadratic residue modulo n or $x_0 = (n - s) \mod 2$ if s is a quadratic non-residue modulo n and hence $n - s$ is a quadratic residue modulo n. So a previous bit predictor solves the quadratic residue problem modulo n. $\square$

Theorem 11.3 transforms a previous bit predictor for the Blum-Blum-Shub generator into a Monte Carlo algorithm for the composite quadratic residue problem, but the success probability could be very low. In the next step we show how to enhance the success probability to any value we like.

Algorithm 11.4 Enhancing the success probability

Require: $n = pq$ with $p \equiv q \equiv 3 \mod 4$ and $\left(\frac{s}{n}\right) = 1$

1: **repeat**
2:     Select x with $\gcd(x, n)$ at random.
3:     Select $e \in \{\pm 1\}$ at random.
4:     $s' \leftarrow ex^2s$
5:     Call the basic Monte Carlo algorithm to decide if s' is a quadratic residue modulo n.
6:     **if** s' is guessed to be a quadratic residue and $e = 1$ or s' is guessed to be a quadratic non-residue and $e = 0$ **then**
7:         This is a hint for s to be a quadratic residue.
8:     **else**
9:         This is a hint for s to be a quadratic non-residue.
10:    **end if**
11: **until** you collected enough hints {Sequential tests could be used to control the error probability of the first and second kind simultaneously}

What Algorithm 11.4 does is select a random s' with $\left(\frac{s'}{n}\right) = 1$ (in lines 2–4), then it decides whether or not s' is a quadratic residue modulo n. The result can be used to decide whether or not s is a quadratic residue (lines 6–10). Each decision is correct with probability $1/2 + \varepsilon$. Standard statistics (see Sect. 15.3.3) can be used to combine the single tests into a result which is correct with arbitrary high probability.

By Lemma 11.1, we know at this point that if the composite quadratic residue problem is hard that there cannot exist an efficient predictor or distinguisher for the Blum-Blum-Shub generator. The absence of an efficient distinguisher also implies that the Blum-Blum-Shub generator must have good pseudo-random properties.

It is also possible to reduce the security of the Blum-Blum-Shub generator directly to the factoring problem. This was first done in [270]. In the following we use the approach from [247]. The key is the following lemma.

Lemma 11.4 *Let $x_i = 2^i a \mod n$ and $s_i = x_i \mod 2$. If $2^{k+1} \geq n$ one can reconstruct a from the sequence $s_0, \ldots, s_k$ with $O(k)$ arithmetic operations.*

Proof Let $x \in \{0, \ldots, n-1\}$ and $y = 2x \mod n$. Since n is odd $y \mod 2 = 0$ implies $y = 2x$ and if $y \mod 2 = 1$ then $y = 2x - n$. If $\hat{y}$ is an approximation of y with $|\hat{y} - y| \leq B$, then

$$\hat{x} = \begin{cases} \frac{\hat{y}}{2} & \text{if } y \mod 2 = 0, \\ \frac{\hat{y}+n}{2} & \text{if } y \mod 2 = 1 \end{cases} \tag{11.1}$$

is an approximation of x with $|\hat{x} - x| \leq B/2$.

Successive application of Eq. (11.1) to the sequence x_i starting with the approximation $\hat{x}_k = \frac{n}{2}$ yields an approximation $\hat{x}_0$ of x_0 with $|\hat{x}_0 - x_0| \leq 1$. Since x_0 is an integer this means that x_0 is determined. $\quad\square$

Theorem 11.5 *A previous bit predictor for the Blum-Blum-Shub generator leads to an efficient factoring algorithm for the modulus n.*

Proof To simplify the presentation of the reduction algorithm we assume that the predictor is always correct and that 2 is a square in $\mathbb{Z}/n\mathbb{Z}$. If the predictor is not perfect the success probability of the factoring algorithm will be less than 1, but with techniques similar to Algorithm 11.4 one can get it very close to 1. For details, and for the case where 2 is not a square, see [247].

Choose x with $\left(\frac{x}{n}\right) = -1$. Without loss of generality we assume that $\left(\frac{x}{p}\right) = 1$ and $\left(\frac{x}{q}\right) = -1$. Compute the Blum-Blum-Shub sequence starting with $y \equiv x^2 \mod n$. y has four roots in $\mathbb{Z}/n\mathbb{Z}$, but only one of these roots $\hat{x}$ is also a square in $\mathbb{Z}/n\mathbb{Z}$. This is the preceding element of the Blum-Blum-Shub sequence.

The predictor will give us the bit $\hat{x} \mod 2$. Applying the predictor to the Blum-Blum-Shub sequence starting with $4y$ will give us the bit $(2\hat{x} \mod n) \mod 2$, because 2 is a square modulo n. Continuing in this fashion we get a sequence of bits which is, by Lemma 11.4, enough to determine $\hat{x}$.

Since $\hat{x} \equiv \pm x \mod p$ and $\left(\frac{\hat{x}}{p}\right) = -\left(\frac{x}{p}\right) = 1$ we get $\hat{x} \equiv -x \mod p$ and similarly we have $\hat{x} \equiv x \mod q$. Thus $\hat{x} + x$ is a multiple of p but not of q. We obtain p by computing $\gcd(\hat{x} + x, n)$.   □

Now that we have completed the security proof for the Blum-Blum-Shub generator, this is an appropriate moment to stop for while and ask ourselves what such a proof means. Does it mean that an attacker should simple give up? Of course not.

If you want to attack a Blum-Blum-Shub generator you can simply polish your social skills. Perhaps in a local pub a clerk will tell you the secret information after the third glass of beer. Most of the time humans are the weakest part of the security chain. See Chap. 7 of [210] for an introduction to social engineering and further references.

Another approach is to put aside the mathematics behind the Blum-Blum-Shub generator and instead try side channel attacks. Since the Blum-Blum-Shub generator is a complex algorithm it will most likely be implemented on a general purpose computer. This opens up a lot of new possibilities. You can try to play around with the branch prediction unit or the cache. Or perhaps you may find a security hole in the underling big-integer library. There exists a lot of work on side channel attacks against the famous RSA cipher [2, 160, 161] and the Blum-Blum-Shub generator is similar enough that most of these methods can be transferred. This is a very real trend. In recent years implementations of RSA-like ciphers have failed regularly to side channel or hacking attacks.

Now let us go back to the mathematical part. We have proved that a distinguisher or predictor for the Blum-Blum-Shub generator will lead to a solution of the composite quadratic residue problem or factorization problem. Let us assume for the moment that the conjectures are true and the best way to solve the composite quadratic residue problem is to compute the factorization, and factorization is indeed a hard problem. This mean that the worst-case complexity of factorization is not polynomial or, if we are lucky, even the average-case complexity is not polynomial. But the attacker is not interested in the worst-case or the average-case, he wants to factor just the special number used in the special system that is under attack. There is still hope that, due to some protocol failure, someone has chosen a special n for which a fast factorization algorithm exists. Again the RSA literature provides a lot of material you can try (see [128, 183]). History shows that this is a real possibility.

In any case, factorization may not be such a hard problem. After the development of RSA, a lot of progress had been made in factorization (quadratic sieve, number field sieve), see [138, 196] for an introduction to this topic. So even if everything else fails, an attacker can still hope that the modulus n was chosen too small and he is able to factorize n.

Finally, the Blum-Blum-Shub generator is only a part of the security chain. It will usually be used to generate a pseudo-random seed for a faster cipher. Perhaps if you can break this cipher, then breaking the Blum-Blum-Shub generator will no longer be necessary. The seed for the Blum-Blum-Shub generator must come from some (external) entropy source. Perhaps you can compromise this source.

All these considerations do not mean that security proofs are useless, but they are a warning that the proofs cover only a small aspect of the true implementation. Even if a cipher is secure, the implementation may have flaws. History has provided enough examples. There is no reason to become too fixated on security proofs [159].

11.3 Implementation Aspects

An efficient implementation of the Blum-Blum-Shub generator is essentially an efficient implementation of large integer arithmetic. This is a classical problem of computer algebra, so we give only a brief introduction to this topic (for further information, see [100, 152]).

A large number is stored as an array of digits with respect to some base B. On a binary computer we will usually work with the base $B = 2^{32}$ or whatever is the internal word size.

Addition and subtraction of n-digit numbers is not a problem. We needs just n elementary additions (subtractions) and we keep track of the carry (which is automatically done by standard hardware).

Multiplication is more difficult. The normal multiplication algorithm that we learn at school needs n^2 elementary multiplications.

A better solution was found by Karatsuba [143]:

$$\left(a_1 B^n + a_0\right)\left(c_1 B^n + c_0\right) = (a_1 c_1) B^{2n} + \left[(a_0 + a_1)(b_0 + b_1) - a_1 b_1 - a_0 b_0\right] B^n + a_0 c_0. \tag{11.2}$$

As one can see from Eq. (11.2), one can reduce the problem of multiplying two $2n$-digit number to that of multiplying three n-digit numbers and some n-digit additions.

The trick is to apply Eq. (11.2) recursively, which leads to the following recursion for the complexity $M(n)$ of multiplying two n-digit numbers

$$M(2n) = 3M(n) + 4n. \tag{11.3}$$

Equations like (11.3) appear quite often in the analysis of recursive algorithms. It is not hard to solve them in general.

Theorem 11.6 *Let $f(x) = O(x^e)$ and let M satisfy the recurrence $M(1) = 1$ and*

$$M(q^k) = rM(q^{k-1}) + f(q^{k-1})$$

then

$$M(x) = \begin{cases} O(x^{\log_q r}) & \text{if } e < \log_q r, \\ O(x^e \log x) & \text{if } e \geq \log_q r. \end{cases}$$

Proof Let $r' = q^e$.

By assumption we have $f(q^k) = O(r'^k)$ and hence

$$f(q^k) \leq cr'^k$$

for some $c > 0$.

In the case $e < \log_q r$ we have $r' < r$. Let $C = \max\{\frac{r'c}{r^3}, (r - r')^{-1}\}$. We claim that

$$M(q^k) \leq C(r^{k+1} - r'^{k+1})$$

for $k \geq 0$.

For $k = 0$ we have $C(r - r') \geq (r - r')^{-1}(r - r') = 1$.

Now let $k > 0$. Then

$$M(q^k) = rM(q^{k-1}) + f(q^{k-1})$$
$$\leq rC(r^k - r'^k) + cr'^{k-1}$$
$$= Cr^{k+1} - C\left(\frac{r}{r'} - \frac{c}{Cr^2}\right)r'^{k+1}$$
$$= C(r^{k+1} - r'^{k+1}).$$

Now we deal with the case $e \geq \log_q r$.

Let $C = \max\{\frac{c}{r'^2}, (r - r')^{-1}\}$. We claim that

$$M(q^k) \leq C(k+1)r'^{k+1}$$

for $k \geq 0$.

For $k = 0$ we have $C(r - r') \geq (r - r')^{-1}(0 + 1)(r - r') = 1$.

Now let $k > 0$ then

$$M(q^k) = rM(q^{k-1}) + f(q^{k-1})$$

$$\leq rCkr'^k + cr'^{k-1}$$

$$= \left(k\frac{rC}{r'} + \frac{c}{r'^2}\right)r'^{k+1}$$

$$\leq \left(kC + \frac{c}{r'^2}\right)r'^{k+1}$$

$$= C(k+1)r'^{k+1}.$$

This proves the theorem. $\qquad\square$

By Theorem 11.6 we get that Karatsuba's algorithm multiplies two n-digit numbers in $O(n^{\log_2(3)}) = O(n^{1.58})$ steps, which is much better than the "school algorithm". The constant hidden in the O-term is very small and even for a few digits Karatsuba's method is faster than the naive multiplication.

With a variation of Karatsuba's idea one can show that multiplication of n-digit numbers can be done in $O(n^{1+\epsilon})$ for each $\epsilon > 0$. The Toom-Cook algorithm, see [56, 265], does exactly this.

The asymptotically fastest multiplication algorithm interprets n-digit numbers as polynomials of degree n in B. The goal is to compute a convolution. This is done by computing the Fourier transform, performing a pointwise multiplication and transforming the result back. Problems arise since we must switch into a ring with nth roots of unity. A careful analysis of the necessary steps shows that multiplication of n-digit numbers can be done in $O(n \log n \log \log n)$ steps. This algorithm was designed by Schönhage and Strassen [236].

For the size used in typical cryptographic applications (a few hundred to a few thousand bits) one uses Karatsuba's algorithm or a variation of the Toom-Cook algorithm, but not the Schönhage-Strassen algorithm. Good free libraries exist that implement these algorithms (for example the GNU multiple precision library [105]). The developers of these libraries always take care to select the best algorithm for the given size of input.

The last operation we need is integer division with remainder. This operation is a bit tricky. The normal "school algorithm" for division involves some guessing steps, which are difficult to automate (see, for example, [152] for a full description of the algorithm). However the "school algorithm" is quite slow (quadratic in the number of digits). The asymptotically fastest division algorithm uses Newton's method to iteratively compute a good approximation of m^{-1} and then computes n/m as nm^{-1} and rounds the result to integers. This proves that division of n-bit integers can be done in $O(M(n))$ time, where $M(n)$ denotes the time needed for multiplication of n-bit integers, i.e. if we use the Schönhage-Strassen multiplication we get the asymptotic bound $O(n \log n \log \log n)$.

However the constant hidden in the O-term is quite big. In the case where we always work modulo the same number p, Montgomery [195] found a way to replace the expensive division by p with a cheap division by a power of 2.

The core of Montgomery's method is Algorithm 11.5.

Algorithm 11.5 Montgomery reduction

Require: $y < pR$
Ensure: $x = yR^{-1} \bmod p,\, 0 \le x < p - 1$
  1: $u \leftarrow -yp^{-1} \bmod R$ {$p^{-1} \bmod R$ can be computed in a pre-processing step}
  2: $x \leftarrow (y + up)/R$
  3: **if** $x \ge p$ **then**
  4:     $x \leftarrow x - p$
  5: **end if**
  6: **return** x

Theorem 11.7 *Given as input a non-negative integer $y < pR$, Algorithm 11.5 computes x with $0 \le x < p$ and $x \equiv yR^{-1} \bmod p$.*

Proof As $u \equiv -yp^{-1} \bmod p$ we have $(y + up) \equiv y - y \equiv 0 \bmod R$. Hence the division in line 2 produces an integer.

We have $x \equiv (y + up)/R \equiv yR^{-1} \bmod p$. The only thing that remains is to prove that $0 \le x < 2p$ and hence the steps in lines 3–5 guarantee that the result of Algorithm 11.5 lies in the interval $[0, p - 1]$.

To see that $(y + up)/R < 2p$ note that $y < pR$ and $u < R$, hence $y + up < pR + R < 2pR$.   □

Algorithm 11.5 needs only two integer multiplications (one in line 1 and one in line 2). In addition it needs a division by R and a reduction modulo R. If R is a power of 2, a division by R is just a shift, which is very cheap. Lines 3–5 contain only one comparison and perhaps a subtraction. So the cost of Algorithm 11.5 is approximately the same as for two integer multiplications, but it performs a reduction modulo p.

Instead of working with the representatives x of the remaining classes directly, we work with their Montgomery transforms $\langle x \rangle = xN \bmod p$.

Adding two Montgomery transforms is easy: $\langle x + y \rangle = \langle x \rangle + \langle y \rangle \bmod p$. Since $\langle x \rangle + \langle y \rangle < 2p$, the reduction modulo p can be done by a comparison and one subtraction (see lines 3–5 of Algorithm 11.5).

To compute $\langle xy \rangle$ one simply applies Montgomery reduction to the product $\langle x \rangle \langle y \rangle$.

The Blum-Blum-Shub Algorithm is well-suited for using Montgomery multiplication. One can do the internal computations with the Montgomery transforms. For the output, the Montgomery transforms must be converted back into the usual representation, but this can be done in parallel, especially if dedicated hardware is used [205].

Even better, one can use Algorithm 11.6 and save the conversion of the output.

Algorithm 11.6 Variation of the Blum-Blum-Shub generator for use with Montgomery reduction

---

Require: $n = pq$ where p, q are primes $\equiv 3 \mod 4$
1: **loop**
2:     $s \leftarrow s^2 R^{-1} \mod n$ {Compute this with Montgomery reduction}
3:     **output** $s \mod 2$
4: **end loop**

If R is a square, the proof of Theorem 11.3 works for Algorithm 11.6 as well as for the original Blum-Blum-Shub generator. (If R is not a square, some extra work is needed to prove the security.) The advantage of Algorithm 11.6 is not only the speed, but also the avoidance of an implementation of a complex integer division algorithm and hence some possible hacking attacks.

Another computational issue is how to generate the primes necessary to set up the Blum-Blum-Shub generator. Here we can find much material common to the RSA cipher. One approach is to generate random numbers and test them for primality. There now exists a deterministic polynomial time algorithm for primality testing [3]. In practice we normally use a faster probabilistic test (like the Solovay-Strassen test [254] or the Miller-Rabin test [191, 217]).

Also noteworthy are the sieving algorithms which find all primes in a given interval. Even with the ancient sieve of Eratosthenes one can do amazing things [214]. The newer Atkin-Bernstein sieve [10] is also worth knowing.

Particularly interesting is the idea of making factoring even harder by using cryptographically strong primes (which provide higher resistance against some special factoring methods that work fast if $n - 1$ or $n + 1$ is smooth) [252], but this is probably not worth the effort [222].

For most purposes it is not necessary to implement our own prime generation, since good free implementations are available, for example [201].

11.4 Extracting Several Bits per Step

With Algorithm 11.6 we have already seen an example which demonstrates that the Blum-Blum-Shub generator must not be based on the least significant bit. This leads us directly to the questions of which bits and how many bits of the Blum-Blum-Shub sequence can be used without weakening the generator. This problem was studied in [5, 270].

We return to the general setting of Theorem 11.2. This time we have j accessible sets of predicates $B^{(k)} = \{B_i^{(k)} : D_i \rightarrow \{0, 1\}\}$ $(1 \leq k \leq j)$. Modify Algorithm 11.1

by outputting in line 2 the j bits $B_i^{(1)}(f_i(x)), \ldots, B_i^{(j)}(f_i(x))$. The question is what are necessary and sufficient conditions for the j-bit Blum-Micali generator to be secure.

A linear transformation of a pseudo-random sequence must still be pseudo-random. Thus an obvious necessary condition is that for each S with $\emptyset \neq S \subseteq \{1, \ldots, j\}$ the predicate $B^S = \{B_i^S : D_i \to \{0, 1\}\}$ where B_i^S is defined by

$$B_i^S(X) = \bigoplus_{k \in S} B_i^{(k)}(x) \tag{11.4}$$

satisfies all the assumptions of Theorem 11.2.

This necessary condition is also sufficient.

Theorem 11.8 (XOR-Condition, see [270]) *Let $B^{(k)} = \{B_i^{(k)} : D_i \to \{0, 1\}\}$ ($1 \leq k \leq j$) be a set of accessible predicates and f_i be a permutation of D_i such that $f_i(x)$ and $B_i^{(k)}(f_i(x))$ are computable in polynomial time.*

If for each non-empty subset S of $\{1, \ldots, j\}$ the predicate B^S defined by Eq. (11.4) is input hard then Algorithm 11.7 is a cryptographically secure pseudo-random generator.

Algorithm 11.7 A variation of the Blum-Micali generator that outputs j bits per step

---

1: **loop**
2:     **output** $B_i^{(1)}(f_i(x)), \ldots, B_i^{(j)}(f_i(x))$
3:        $x \leftarrow f_i(x)$
4: **end loop**

Proof First note that Algorithm 11.7 runs in polynomial time.

Assume that Algorithm 11.7 is not cryptographically secure. Then there exists a previous bit predictor. For some known starting value x one can compute the output of Algorithm 11.7 and use the previous bit predictor to get a probabilistic algorithm T for $B_i^{(j)}(x)$ given $B_i^{(1)}(f_i(x)), \ldots, B_i^{(j-1)}(f_i(x))$.

For all possible $(j-2)$-tuples $b_1, \ldots, b_{j-2}$ compute $T(b_1, \ldots, b_{j-2}, 0)$ and $T(b_1, \ldots, b_{j-2}, 1)$. If both values are equal, you have a good guess for $B_i^{(j)}(x)$ given $B_i^{(1)}(f_i(x)), \ldots, B_i^{(j-2)}(f_i(x))$. If the output differs, there is still a chance to guess $B_i^{(j)}(x)$.

Similarly if $T(b_1, \ldots, b_{j-2}, 0)$ and $T(b_1, \ldots, b_{j-2}, 1)$ are different, we get a good guess for $B_i^{(j)}(x) \oplus B_i^{(j-1)}(x)$ given $B_i^{(1)}(f_i(x)), \ldots, B_i^{(j-2)}(f_i(x))$ and we have still a chance to guess $B_i^{(j)}(x) \oplus B_i^{(j-1)}(x)$ if the outputs are equal.

At this point we have a good probabilistic algorithm T' for $B_i^{(j)}(x)$ or $B_i^{(j)}(x) \oplus B_i^{(j-1)}(x)$, depending on which case is more likely. In any case, T' depends only on $j - 2$ bits instead of $j - 1$ bits.

Continuing in this fashion we finally reach a probabilistic algorithm that depends on no bits and predicts some XOR-sum $B_i^S(X) = \bigoplus_{k \in S} B_i^{(k)}(x)$.

But we assumed that such an algorithm does not exist. Hence Algorithm 11.7 must be cryptographically secure.   □

With arguments similar to the one used in Sect. 11.2 one can prove the XOR-condition for the Blum-Blum-Shub generator and show that extracting $\log \log n$ bits leaves the Blum-Blum-Shub generator asymptotically secure (see [270]). However, this is only half of the truth: the probability of a possible distinguisher still tends to zero, but more slowly when you extract more bits. A more detailed analysis appears in [247]. The authors prove inequalities which let you select, for a given distinguishing advantage ε (the security level) and the number m of bits which should be generated, the size of the modulus n and the number j of bits extracted per step which lead to an optimal speed. The examples given in [247] indicate that you should not extract many bits, often one reaches a higher speed for a given security level by using a smaller n and extracting only one bit.

On the other hand Coppersmith's method (see Sect. 13.6) shows that if we extract more than $n - \sqrt{n}$ bits, the Blum-Blum-Shub generator becomes insecure even if the attacker can observe only two successive time steps.

11.5 The RSA Generator and the Power Generator

We have already mentioned several times that the Blum-Blum-Shub generator has many similarities to the RSA public key system. One can make it even more similar, which leads us to Algorithm 11.8.

Algorithm 11.8 The RSA generator

Require: $n = pq$ where p, q are primes, e is an integer with $\gcd(e, \varphi(n)) = 1$
 1: **loop**
 2:     $s \leftarrow s^e \bmod n$
 3:     **output** $s \bmod 2$
 4: **end loop**

The *power generator* is a generalization of the Blum-Blum-Shub generator and the RSA generator. It is based on the recursion

$$s_{i+1} = s_i^e \bmod n$$

without any restrictions to e.

We now prove a result on the period of the power generator.

Theorem 11.9 (see [97]) *For any positive integer m and any numbers $K_1, K_2 \geq 1$, let W denote the number of seeds (s, e) of the power generator that lead to a period of at most $\lambda(\lambda(n))/(K_1 K_2)$. Then W is bounded by*

$$W \leq \varphi(n)\varphi(\lambda(n)) \left[\frac{\tau(\lambda(n))}{K_1} + \frac{\tau(\lambda(\lambda(n)))}{K_2} \right].$$

Proof In Sect. 13.2 Eq. (13.3) we derive a bound for the number of elements in $\mathbb{Z}/n\mathbb{Z}$ of small order. Theorem 11.9 is essentially a two-fold application of this bound.

The number of elements $s \in \mathbb{Z}/n\mathbb{Z}$ with $\mathrm{ord}_n\, s > \lambda(n)/K_1$ is at least $\varphi(n)(1 - \tau(\lambda(n)))/K_1$. For the moment fix s with $\mathrm{ord}_n\, s = \lambda(n)$. The discrete logarithm with base s of an element in $\langle s \rangle$ is an element of $\mathbb{Z}/\lambda(n)\mathbb{Z}$. The discrete logarithms of the series s_i generated by the power generator are powers of e. The period of the power generator is $\mathrm{ord}_{\lambda(n)}\, e$. The number of elements $e \in \mathbb{Z}/\lambda(n)\mathbb{Z}$ with order $\mathrm{ord}_{\lambda(n)}\, e > \lambda(\lambda(n))/K_2$ is at least $\varphi(\lambda(n))(1 - \tau(\lambda(\lambda(n))))/K_2$. If the order of s is $\lambda(n)/K$ then the period will decrease to $\mathrm{ord}_{\lambda(n)/K}\, e \geq (\mathrm{ord}_{\lambda(n)}\, e)/K$.

Thus there are at least

$$\left[\varphi(n)\left(1 - \frac{\tau(\lambda(n))}{K_1}\right) \right] \left[\varphi(\lambda(n))\left(1 - \frac{\tau(\lambda(\lambda(n)))}{K_2}\right) \right]$$

seeds (s, e) of the power generator that lead to a period of at least $\lambda(\lambda(n))/K_1 K_2$. □

With similar arguments one can also prove a lower bound for the period of the Blum-Blum-Shub generator. Since the Blum-Blum-Shub generator has fewer degrees of freedom the lower bound will be a bit weaker, see [245] Theorem 23.7.

Since the Blum-Blum Shub generator uses only squaring it is the fastest variant of the power generator family.

11.6 Generators Based on Other Hard Problems

Up to now we have essentially two problems as a basis for security proofs: factoring integers and discrete logarithms. These two problems were the first problems that were used in cryptography and stand behind most applications. However, there are other hard problems that can serve as basis for security proofs. They are not as successful as factoring. Lattice based attacks are a real threat for these systems, for example the knapsack-based public key cryptosystem is broken [200, 243], the basic HFE-system (based on multivariate quadratic equations) is broken [241], and the original McEliece system (based on error-correcting codes) is broken [257]. Whenever you see a security proof that uses one of these problems, you should check in detail if it avoids the mistakes made in the broken systems. Caution is advisable. The original knapsack-based system was proposed by M.E. Hellman [190] who is famous for being one of the inventors of public key cryptography. HFE [206] was proposed as an improvement of an older scheme after the author of HFE was able to break it. The construction of correct security proofs is clearly incredibly difficult if even such skilled cryptologists have make mistakes in that respect.

An early attempt at using the knapsack problem for stream ciphers is the knapsack or subset sum generator [229].

The idea is to compute an LFSR $x_1, \ldots, x_n$ and output

$$y_i = \left\lfloor \frac{x_i w_0 + \cdots + x_{i-k} w_w}{B} \right\rfloor$$

for some weight $w_1, \ldots, w_k$ and some divisor B. Usually one chooses B as a power of 2, i.e. one discards the lower bits of $x_i w_0 + \cdots + x_{i-k} w_w$. The idea behind the generator is that addition is highly non-linear when interpreted as a Boolean function.

The knapsack generator has no rigorous security proof and several instances of the knapsack generator are broken (see [101]).

From coding theory we know that, in general, decoding a linear code is a hard problem. J.-B. Fisher and J. Stern [92] suggest Algorithm 11.9 as a stream cipher based on the decoding problem.

Let $\rho < 1$. Given $M \in \mathbb{F}_2^{\lfloor \rho n \rfloor \times n}$ and $y = Mx$ with $\delta n \leq w(x) \leq \delta' n$, determine x. If the possible weight of x is too small, the problem is easy (there are simply not enough possibilities for x). If the weight of x is about $\frac{n}{2}$ the problem is again easy (there are too many possible solutions). Fisher and Stern suggest choosing δ and δ' near the Gilbert-Warshamov bound.

Algorithm 11.9 The Fisher-Stern generator

Require: $M \in \mathbb{F}_2^{\lfloor \rho n \rfloor \times n}$, $x \in \mathbb{F}_2^n$, $w(x) = \delta n$
  1: **loop**
  2:   $y \leftarrow Mx$
  3:   Split y into y_1 and y_2 with y_1 has $\lceil \log_2(\binom{n}{\delta n}) \rceil$ bits.
  4:   Set x to the y_1-th word in $\mathbb{F}_2^n$ of weight δn.
  5:   **output** y_2
  6: **end loop**

At the moment no attacks against the Fisher-Stern generator are known. The security proof is similar to Theorem 11.2 (for details see [92]). Despite its foundation on a (presumably) hard problem, I am personally not quite convinced. Since $\binom{n}{\delta n}$ is not a power of 2, one has to accept a small bias in the mapping performed in line 4. Can an attacker really make no use of this bias? The order of the words of weight δn used in line 4 is not used in the security proof. It is no problem to construct an order which leads to a cycle of length 1 and hence a weak cipher.

Another hard problem that is used as a basis for security proofs is the solution of multivariate quadratic equations over finite fields. A stream cipher that is based on this problem is QUAD [19]. The structure of QUAD is simple (Algorithm 11.10).

Algorithm 11.10 The QUAD cipher

Require: $S = (Q_1, \ldots, Q_{kn})$ is a secret system of kn random multivariate
   quadratic equations over $\mathbb{F}_q$ in n variables.
1: **loop**
2:    Compute $S(x) = (Q_1(x), \ldots, Q_{kn}(x))$ where x is the current internal state
3:    **output** $Q_{n+1}(x), \ldots, Q_{kn}(x)$
4:    Set the new internal state to $Q_1(x), \ldots, Q_n(x)$.
5: **end loop**

QUAD is very fast, but the parameter mentioned in the QUAD design paper
is definitely too small, as shown in [286]. If even the designers of QUAD have
difficulties in choosing the right parameters, it seems very risky to use it in practice.

So if you want to go for maximal security and use a stream cipher with a security
proof, I recommend the Blum-Blum-Shub generator. Its simple structure makes it
unlikely to miss some hidden requirements. Its speed is fast enough for the generation of pseudo-random keys, and for all other purposes normal ciphers are better
than the available ciphers with security proofs. There is definitely no need to use
a more complicated cipher; you will have enough problems securing your implementation of the Blum-Blum-Shub generator against side channel attacks and other
programming errors. Do not forget to plan a big security margin. One never knows
what future progress will be in factoring algorithms.

For the same reason I recommend RSA as a public key cipher. It is simple, which
helps to avoid mistakes in the implementation. Even if it is slower than the alternatives, this does not matter, since in most applications you use it only as a signature
scheme or in the key exchange protocol for a fast symmetric cipher. Simplicity is
an advantage at this point. Don't forget that it has been unbroken for more than 30
years and is widely used. Thus if, contrary to common belief, factoring is not hard,
you will not miss the new developments and hopefully will be able to replace RSA
by a new and better cipher before someone breaks your system.

11.7 Unconditionally Secure Pseudo-random Sequences

We have seen that it is possible to construct pseudo-random generators that are
computationally secure under the assumption that the attacker only has access to
a small number of bits. It is a natural question whether one can also construct unconditionally secure pseudo-random sequences. The answer is yes, as was shown
by U.M. Maurer and J.L. Massey [180].

Definition 11.4 A function $f : \mathbb{F}_2^k \to \mathbb{F}_2^n$ is a (k, n, e) *perfect local randomizer* if
whenever $X_1, \ldots, X_k$ are independent random variables with $p(X_i = 1) = 1/2$ then
every subset of size e of the n random variables $f(X_1, \ldots, X_k)$ is a set of independent random variables.

The idea of Definition 11.4 is that we expand a small true random sequence of length k to a pseudo-random sequence of length n. An attacker that only has access to e bits of the pseudo-random sequence should not be able to distinguish it from a true random sequence.

Theorem 11.10 *A linear function* $f : \mathbb{F}_2^k \to \mathbb{F}_2^n$ *is a* (k, n, e) *perfect local randomizer if and only if it is an* $(e + 1)$-*error detecting code.*

Proof Let $f(x) = Gx$.

G is the generator matrix of an $e + 1$ error detecting code if and only if each set of e columns of G is linearly independent.

If G has e columns that are linearly dependent then the corresponding e-bits of the pseudo-random sequence satisfies a linear equation, which clearly distinguishes them from a true random sequence.

If, on the other hand, each set of e columns of G is linearly independent then for every projection P onto e coordinates, $f_P : x \mapsto PGx$ is a surjective linear mapping from $\mathbb{F}_2^n$ to $\mathbb{F}_2^e$. Hence $|f_p^{-1}(y)| = 2^{n-e}$ for all $y \in \mathbb{F}_2^e$. This proves that every e-tuple has the same probability 2^{-e}, i.e. f is a (k, n, e) perfect local randomizer. $\qquad\square$

Linear perfect local randomizers are, by Theorem 11.10, just linear codes. We can apply all bounds from coding theory. For example, the Gilber-Varshamov bound shows that a (k, n, e) perfect local randomizer exists if $e \le \frac{k}{\log_2 n}$.

One can view Theorem 11.10 as a warning. Linear functions are weak cryptographic functions, but even they can be locally unpredictable. If you see a security proof for a cipher you should always check if the assumptions are plausible. Can the attacker really access only a small (polynomial) part of the pseudo-random bits? What happens if the attacker is more powerful? Is the cipher robust against a more powerful attacker? The Blum-Blum-Shub generator still seems to be secure if the attacker has access to exponentially many bits, but can use only polynomial time. This is not part of the security proof, but it is a good sign and makes us trust the Blum-Blum-Shub generator even more. Even better, it is not realistic to assume that the attacker will be able to access exponentially many bits, because this would mean that the sender can compute exponentially many steps. However, if this were possible, the attacker could buy the same hardware and break the generator by factoring.

In contrast, the linear codes of Theorem 11.10 are broken immediately if the attacker can access $e + 1$ bits and not only e bits as assumed in the security proof. This is a bad feature which rules linear codes out as cryptographic codes. It is also very plausible that an attacker can observe more bits than we expect.

Part III
Mathematical Background

Chapter 12
Computational Aspects

12.1 Bit Tricks

We often need to work with bitwise operations. The performance of many correlation attacks depends on choosing efficient bit manipulating algorithms.

Many bit tricks are not commonly known. To help the reader, this section collects all the bit operation material we need. For a more extensive treatment, see the excellent book [156].

We denote by $a \mathbin{\&} b$ the bitwise AND operation, $a \mid b$ is the bitwise OR and $a \oplus b$ is the bitwise XOR. All three operations satisfy the commutative and associative law. In addition we have the following distributive laws:

$$(a \mid b) \mathbin{\&} c = (a \mathbin{\&} c) \mid (b \mathbin{\&} c), \tag{12.1}$$

$$(a \mathbin{\&} b) \mid c = (a \mid c) \mathbin{\&} (b \mid c), \tag{12.2}$$

$$(a \oplus b) \mathbin{\&} c = (a \mathbin{\&} c) \oplus (b \mathbin{\&} c). \tag{12.3}$$

By $\bar{a}$ we denote the bitwise complement of a. We have the De Morgan laws:

$$\overline{a \mathbin{\&} b} = \bar{a} \mid \bar{b}, \qquad \overline{a \mid b} = \bar{a} \mathbin{\&} \bar{b}, \qquad \overline{a \oplus b} = \bar{a} \oplus \bar{b}. \tag{12.4}$$

We denote the left-shift by k-bits by $a \ll k$ and the right-shift by $a \gg k$.

12.1.1 Infinite 2-adic Expansions

It is often useful to work with infinite 2-adic numbers $a = (\ldots a_3 a_2 a_1 a_0)_2$. In this representation we have $-1 = (\ldots 1111)_2$ and we have fractions like $-1/3 = (\ldots 010101)_2$.

The equation $x + \bar{x} = (\ldots 111)_2 = -1$ connects addition with the bitwise complement.

A. Klein, *Stream Ciphers*, DOI 10.1007/978-1-4471-5079-4_12,
© Springer-Verlag London 2013

We will often use the masks

$$\mu_0 = (\ldots 0101010101010101)_2 = -1/3,$$
$$\mu_1 = (\ldots 0011001100110011)_2 = -1/5,$$
$$\mu_2 = (\ldots 0000111100001111)_2 = -1/17$$

and so on. In general we have $\mu_k = -1/(2^{2^k} + 1)$ and $\mu_k \ll 2^k = \overline{\mu_k}$.

On a computer we only need finite precision. In the case of a 2^d-bit register we can compute the truncated constant μ_k by $(2^{2^d} - 1)/(2^{2^k} + 1)$.

12.1.2 Sideway Addition

An operation that is extremely useful for cryptography is the sideway addition. Some correlation attacks (see Sect. 4.1.6.1) are effectively just a repeated application of the sideway addition, hence optimizing this operation is very important for the efficiency of these attacks.

For a word $w = (w_{n-1} \ldots w_1 w_0)$, we define

$$\text{SADD}(w) = \sum_{i=1}^{n-1} w_i.$$

The sideway addition is also known as the population count or Hamming weight of w.

Because it is so important, many processors have a special machine instruction for this operation. For example, on a Mark I or Knuth's MMIX, it is named SADD, in the SSE4.2 instruction set it is POPCNT, and on a NEC SX-4 it is VPCNT. Unfortunately the widespread Intel ix86 platform has no sideway addition.

On computers that do not have a sideway addition instruction there are essentially two algorithms to perform sideway addition. The first is based on table look-up (Algorithm 12.1).

Algorithm 12.1 Sideway addition based on table look-up

Require: w is a k-byte word $w = (w_{k-1}, \ldots, w_0)_{256}$.
Ensure: $S = \text{SADD}(w)$
1: $T := [\text{SADD}(0), \text{SADD}(1), \ldots, \text{SADD}(255)]$ $\{T$ is computed at compile time$\}$
2: $S := 0$
3: **for** $j := 0$ **to** $k - 1$ **do**
4:      $S := S + T[w_j]$
5: **end for**

Alternatively we can use Algorithm 12.2, which is due to D.B. Gillies and J.C.P. Miller ([281]). We present the algorithm in its version for 64-bit numbers (originally it was developed for the 35-bit numbers on EDSAC).

Algorithm 12.2 Sideway addition (64 bit words)

1: $S := w - ((w \gg 1) \,\&\, \mu_0)$
2: $S := (S \,\&\, \mu_1) + ((S \gg 2) \,\&\, \mu_1)$
3: $S := (S + (S \gg 4)) \,\&\, \mu_2$
4: $S := (aS \mod 2^{64}) \gg 56$ {where $a = (11111111)_{256}$}

When Algorithm 12.2 reaches line 4. We have $S = (s_7 \ldots s_0)_8$ with $s_0 + \cdots + s_7 = \mathrm{SADD}(w)$. Since $\mathrm{SADD}(w)$ fits without difficulty into 8 bits we can do the last summation by a multiplication. If multiplication is slow on our CPU, we can replace line 4 by $S := S + (S \gg 8)$, $S := S + (S \gg 16)$, $S := (S + (S \gg 32)) \,\&\, 255$.

In general Algorithm 12.1 is faster on 32-bit CPUs and Algorithm 12.2 is the best on 64-bit CPUs.

If we know that $\mathrm{SADD}(w)$ is small (less than 4) there is a special algorithm due to P. Wegner (see Exercise 17.23).

12.1.3 Sideway Addition for Arrays

Computing the sideway addition for a large array of words is a special problem. The naive method simply calls one of the algorithms of the previous subsection for every word and adds the results.

A better method was found by Harley and later improved by Seal [237, 277]. It is based on a circuit called a *carry save adder* (CSA). A carry save adder is just a sequence of independent full adders. A full adder takes the input bits a, b, c and outputs the low-bit (sum) l and high-bit (carry) h. In Boolean notation:

$$l \leftarrow (a \oplus b) \oplus c, \qquad h \leftarrow (a \wedge b) \oplus \big[(a \oplus b) \wedge c\big].$$

Thus a carry save adder can be implemented with five operations and a standard processor.

The simplest way to use the CSA is to group the elements of the array into groups of three and apply one CSA and two word size sideway additions. The number of ones is twice the output of the sideway addition of the high word plus the sideway addition of the low word. This is Harley's original idea. Seal made some improvements and simplifications (see Algorithm 12.3).

An even more elaborate algorithm for this problem is developed in [82].

Table 12.1 (taken from [82]) shows how specialized algorithms can speed up the sideway addition of arrays.

Algorithm 12.3 Sideway addition Harley-Seal method

1: $S \leftarrow 0, l \leftarrow 0$ {initialize}
2: **for** i from 0 to $\lfloor (n-1)/2 \rfloor$ **do**
3:     $(h, l) \leftarrow \text{CSA}(w_{2i}, w_{2i+1}, l)$
4:     $S \leftarrow S + \text{SADD}(h)$
5: **end for**
6: $S \leftarrow 2S + \text{SADD}(l)$ {the high bits count double}
7: **if** n even **then**
8:     $S \leftarrow S + \text{SADD}(w_n)$ {If n is even we must treat the last word separately}
9: **end if**
10: **return** S

Table 12.1 Comparing sideway addition algorithms for arrays

Algorithm	Relative speed
Simple loop (table look-up)	100 %
Harley-Seal	73 %
Third iteration of Harley-Seal	59 %
Algorithm developed in [82]	39 %

12.2 Binary Decision Diagrams, Implementation Aspects

In Chap. 5 we described the idea of binary decision diagrams and related basic algorithms. In this Appendix we have a closer look at the implementation details.

12.2.1 Memory Management

There are two strategies to maintain a BDD in memory. One uses a reference counter per node and the other stores all nodes of a BDD in a continuous region. Both have their pros and cons.

12.2.1.1 Nodes with Reference Counter

In this implementation we store for each node, its label, two pointers to the successor nodes and a reference counter, which counts how many pointers point to this node.

If the reference counter reaches zero, we know that the node is no longer reachable and can be removed from the memory. In the C++ programming language reference counters must be implemented explicitly, but there are languages like Java which adds a reference counter to each object implicitly.

An advantage of this memory model is that nodes can be allocated one by one and that the BDD algorithms needn't care about memory management, since this

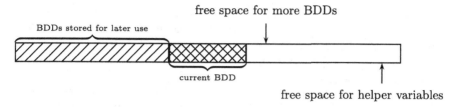

Fig. 12.1 Memory layout

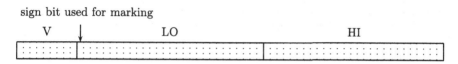

Fig. 12.2 A BDD node in memory

is done by the node structure. The nodes with reference counter model works well with top-down algorithms and shows its full strength when it comes to automatic reordering strategies.

A disadvantage is that the reference counters (and other required data structures such as unique tables) generally need a lot of memory. A typical implementation will use about 200 bits per node. Another drawback is that with this memory model it is difficult to make use of the cache.

12.2.1.2 An RPN Calculator for BDDs

For the cryptographic applications, I prefer an alternative memory model. The idea is to implement a reverse polish notation (RPN) style calculator for BDDs. The primary data structure is a stack of BDD-nodes.

Figure 12.1 shows the basic memory layout of the package.

One advantage of the stack-based approach is the memory management. There is none. At the beginning the program allocates space for the stack and afterwards it just updates the pointer NTOP that indicates the current top.

All BDDs on the stack are stored as a sequence of triples (v, l, h) of value, low pointer and high pointer. According to the rules:

- The first two nodes are $(v_{max} + 1, 0, 0)$ and $(v_{max} + 1, 1, 1)$ representing false and true, respectively.
- The node of the BDD must be ordered by the index V of the variables. The nodes with the highest value V comes first.
- The last node must be the root of the BDD.

This allows a very compact representation (Fig. 12.2). A 64-bit word can hold a node with two 27-bit pointers and a 9-bit value and still leaves us an extra status bit which we interpret as a sign bit of the low pointer.

With this memory layout we can store a BDD with 2^{27} nodes (about 1 GB of memory). Taking into account that we need additional space to manipulate the BDD, this is enough for a standard desktop machine with only 2–4 GB RAM. (This implementation needs roughly only half of the memory in comparison with implementations based on the nodes with reference counter model.)

In addition, the sequential storage of the nodes works perfectly with breadth-first algorithms and makes the most out of the memory hierarchy of the computer system.

Of course the scheme also has disadvantages. The entire memory management moves into the BDD manipulation algorithms, resulting in code with very involved pointer operations (which is hard to maintain). Dynamic variable reordering especially requires a lot work, but can also be implemented very efficiently (see Sect. 12.2.3).

Another drawback is that one can manipulate only the top BDD on the stack and that different BDDs cannot share nodes (see Sect. 12.2.4 for a partial solution). In the remainder of the section we describe how to implement an RPN calculator for BDDs.

12.2.2 Implementation of the Basic Operations

In the stack-based setting the implementation of the most basic BDD operations is straightforward.

Counting the number of solutions requires us to loop over all BDD nodes. This algorithm profits directly from the fact that the nodes are continuous in memory. Additional storage is taken from the top of the allocated memory (again continuous). Everything behaves well with respect to cache effects. In this way one can also implement similar algorithms (determining the generating function, Boolean optimization, evaluating the reliability polynomial, etc.).

Evaluating the BDD requires us to follow a path from the root to a sink. Here we cannot make use of the continuous memory layout, but the stack-based approach is no worse than a classical depth first approach. Similarly there are algorithms for finding the lexicographic least solution, enumerating all solutions, generating random solutions, etc.

Simple transforms (complement and dual) can also make little use of the continuous memory. Computing the complement is an especially interesting case. In a classical pointer based approach we can just exchange the two sinks (provided there are no overlaps) and get the complement in $O(1)$ time. In our stack-based approach we must keep the order of the sinks $(v_{max} + 1, 0, 0)$ and $(v_{max} + 1, 1, 1)$ unchanged, thus we have to loop over all BDD nodes and exchange 0 pointers by 1 pointers and vice versa ($O(B)$ time).

The reduction algorithm is more complicated. Here we pay the price for the continuous memory layout. See Algorithm 5.2 for a full description of the pointer operations.

The reduction algorithm is the core of the package and all other complicated algorithms are only variations. In Sect. 5.1.1 we have already seen how to apply a

Boolean operation to two BDDs. The idea is to build the generic melt and then apply reduction.

Another situation in which we can use reduction is the following.

Consider a BDD for the function $f(x_1, \ldots, x_n, y, z_1, \ldots, z_m)$ with variable order $x_1, \ldots, x_n, y, z_1, \ldots, z_n$. Assume that in every solution of f the variable y satisfies $y = g(x_1, \ldots, x_n)$. We call y a *check variable*.

This is quite often the case. For example, one method to compute the BDD for the functional composition

$$f\big(h_1(x_1, \ldots, x_n), \ldots, h_k(x_1, \ldots, x_n)\big)$$

is to compute

$$f(y_1, \ldots, y_k) \wedge \big(y_1 = h_1(x_1, \ldots, x_n)\big) \wedge \cdots \wedge \big(y_k = h_k(x_1, \ldots, x_n)\big).$$

BDD nodes labeled by a *check variable* have a very special form: either the low or the high pointer must point to the sink $\perp$.

Computing the BDD $\exists y : f$ for a check variable y is simple. For every node labeled by y we give the pointer pointing to $\perp$ the value of the other pointer. This yields a non-reduced BDD (all nodes labeled y are now superfluous since both pointers point to the same node) for $\exists y : f$.

Reducing this BDD is faster than applying the unspecialized quantifier $\exists$ to the BDD (variation of the melting process).

The *restricted existential operator* can be applied quite frequently, so it deserves a specialized algorithm.

12.2.3 Implementation of Reordering Algorithms

Dynamic variable reordering [227] is an important technique that automatically finds a good variable order for the given problem.

The elementary step in the variable reorder algorithm is swapping two adjacent variables. Figure 12.3 illustrates the swapping operation. There are four types of nodes. Nodes in level i are either tangled or solitary depending on whether or not they have a j node as descendant. (In Fig. 12.3 a and c are tangled and b is solitary.) Nodes of level j are either hidden or visible, depending on whether or not they can be reached only from i nodes. (In Fig. 12.3 d and e are hidden but f is visible.)

The swap moves all solitary nodes one level down and all visible nodes one level up. The tangled nodes change their labels from i to j. The hidden nodes are deleted and new nodes have to be created to represent the calculation behind tangled nodes.

Several aspects of variable swapping do not cooperate well with the representation of the BDD chosen in our stack-based approach.

- The order of the nodes is changed. (In the figure the node order was a, b, c, d, e, f and it became a, c, f, new, new, b, new.)

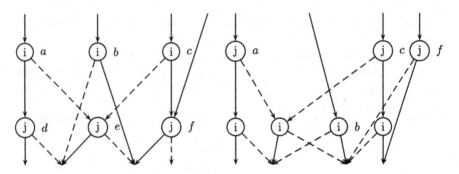

Fig. 12.3 Variable swapping

This is in general a big disadvantage for breadth-first algorithms since it destroys locality of the nodes, with negative effects with respect to caching.

In our special approach we would have to copy a node from one location to another, which leads to the problem that all pointers pointing to that node would have to be updated.

- Without reference pointers it is difficult to determine the visible nodes. In fact, it is impossible without scanning all the levels above i.

In [218] several strategies were considered to integrate dynamic reordering into breadth-first packages. The best solution of [218] was to temporarily change the data structure, use conventional depth-first algorithms for reordering and then change the data structure back.

In the next sections another solution which is better suited to the breadth-first setting, and which can make more use from the cache, will be developed.

12.2.3.1 Jumping up

The easiest reordering operation is the "jump up" operation (one variable moves to the top).

Figure 12.4 shows how to build a (non-reduced) BDD for the "jump up" operation. One simply doubles every node above the variable that jumps up and adds a new node at the top.

Doubling the nodes is quite cheap and behaves well with respect to the cache. The complexity of the reduction algorithm is not too bad.

12.2.3.2 Sifting down

Testing all possible variable orders is far too expensive. Instead of testing all possible orders one usually uses sifting [227]. The idea is to temporarily fix the order for all but one variable and search the best possible position for the remaining variable.

Fig. 12.4 The variable 3 jumps up

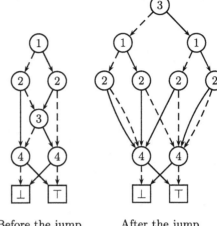

Before the jump   After the jump

Sifting down is the key algorithm in our dynamic reordering tool set. We explain the idea for the special case where the variable x_1 should move to the bottom.

Consider the simple BDD shown in Fig. 12.5 (a). It is easy to construct a BDD for the THEN-case (x_1 equals true) with x_1 moved down. Similarly we can construct a BDD for the ELSE-case with x_1 moved down. These two BDDs are shown in Fig. 12.5 (b). (Note that it is only a coincidence that the THEN-BDD and the ELSE-BDD are disjoint. We do not need this property.) To get the full BDD with x_1 moved down we must compute the OR of the THEN-BDD and the ELSE-BDD.

To do this we use the melting algorithm described in Sect. 5.1.1. It builds the BDD one level at a time, which fits perfectly with our continuous memory layout and makes good use of the cache.

Note that the last level in melt must always consist of the three nodes $1T \diamond \perp$, $\perp \diamond 1F$ and $1T \diamond 1F$ independent from the BDD. We immediately see one possible reduction: The node $1T \diamond 1F$ is superfluous and can be replaced by $\top$. No other reduction steps are possible since the melt itself is reduced and replacing a link to $1T \diamond 1F$ by a link to $\top$ cannot result in a new superfluous node, since the original diagram for the melt contains no direct links to $\top$.

The basic form of the algorithm as described above moves the variable x_1 down into the last level. Now we extend the algorithm to sifting.

- Instead of choosing the THEN-BDD and the ELSE-BDD with respect to the first variable, we can use any other variable to construct the two sub-BDDs, i.e. we can jump down with any variable.
- In the same way, if we construct in Fig. 12.5 the THEN-BDD with x_1 at the last level, we can get the THEN-BDD with x_1 at any other level. The only difference is that we must place in the THEN-BDD and the ELSE-BDD the check nodes for the variable x_1 not in the last level, but at some higher levels. Figure 12.6 shows, for example, the THEN-BDD and the ELSE-BDD for the BDD of Fig. 12.5 (a) with x_1 moved to level 2.

Fig. 12.5 Moving a variable
from the top to the bottom

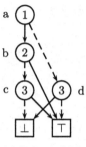

(a) A simple BBD

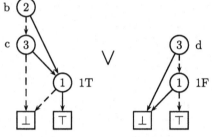

(b) the two sub-BBDs (x_1 moved down)

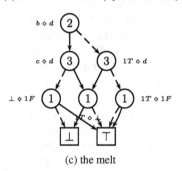

(c) the melt

Fig. 12.6 Sifting down

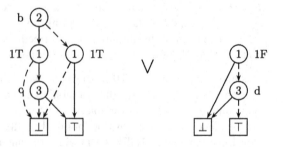

- At this point we can let a variable jump from any level to any lower level. This is still not quite what we need for dynamic variable reordering, but look a little closer at the process of jumping down.

When we let $a = (1 \; ? \; b : c)$ be the first node of the BDD (as in Fig. 12.5)—for simplicity assume that the nodes b and c both lie at level 2—then the first node of the BDD in the THEN-case will be b (no matter how many levels the variable x_1 jumps down) and the first node of the ELSE-case will be c. So, no matter how many levels we jump down, the first level of the melt will be $b \diamond c$.

In the case $V(b) < V(c)$ the LO pointer of $b \diamond c$ will point to $LO(b) \diamond c$. Thus it does not matter if we call the node $b \diamond c$ or $b \diamond 1F$; the structure of the melt is unchanged. The case $V(b) > V(c)$ is similar.

This leads to a variation of the sifting algorithm. We start the melting to process for x_1 jumping from the top to the bottom as shown in Fig. 12.5. Melt is computed level wise as described in Sect. 5.1.1. After each level we compute the size of the BDD under the condition that x_1 jumps to that level. The size is the number of nodes $\alpha \diamond \alpha'$ we have already constructed in the melt plus the number of nodes below the current level in the original BDD plus the number of nodes labeled 1 we would have to create when stopping at this level.

By this procedure we find for each level l the size $S_B(l)$ of the BDD under the condition that x_1 jumps to level l. Then we can select the l for which the BDD size is the smallest and let x_1 jump directly to that place. This is quite cheap: we still have the original BDD that contains all nodes that lie below the new level of x_1 and the melt which contains all nodes which lie above the new level of x_1, so all we have to do is to compute the new nodes labeled with 1. This is again just a variation of the melting algorithm.

Since the new sifting algorithm is a variation of the melting algorithm that works level wise, it fits well into the breadth-first style of our package. Most memory access will be made in a small memory range which is good for cache effects.

• It remains to explain how to compute the size $S_B(l)$ of the BDD where x_1 jumps to level l. The problem is how to compute the number of nodes labeled with 1.

In the package we take an extreme approach to this problem. The algorithm doesn't simply calculate the number of 1 nodes. As an effect, the sifting algorithm will sometimes place a variable one or two positions out of the optimal place. However, in practice the speed-up that comes from the simpler algorithms outweighs the small errors.

A way to get the correct number is the following. For each node $\alpha \diamond \alpha'$ in the melt, let l be the level of the node and l' be the smallest level of a node that has $\alpha \diamond \alpha'$ as successor. If x_1 jumps at some level between l' and l then the node $\alpha \diamond \alpha'$ would appear as a node with label 1 as result. So one can determine the exact values $S_B(l)$ without computing the 1-nodes at every level.

12.2.4 Emulating a BDD Base

The biggest drawback of the implementation of a BDD package as an RPN-calculator is that the different BDDs cannot share states.

Here is an idea to deal with this problem. To store two overlapping BDDs f and g, add a special variable S and store the BDD for $S?f : g$. By adding further special variables one can store more than two overlapping BDDs. To support this style of storing several BDDs, the package should have a lot of exotic quantifiers such as $\lambda x_i : f = \overline{f[x_i/0]} \wedge f[x_i/1]$ which replace, in this case, the normal Boolean operations on two different BDDs.

This emulation of a BDD base by a single BDD works quite well in several applications. For example, it is not difficult to compute the BDDs representing a binary product (in [156] this problem was used as an example explaining the necessity of BDD bases).

Nevertheless this style of emulating a BDD base has its limitations since we need to handle all BDDs in the base simultaneously.

So if one often works with several overlapping BDDs, the node with reference counter implementation is the best, but then one must accept the bigger memory overhead and one cannot make as much use of the cache as the stack-based implementation.

12.3 The O-Notation

The O-notation is a convenient but somewhat ambiguous way to express the asymptotic behavior of a function. We use it mainly to express the asymptotic behavior of an algorithm, hence its appearance in this chapter about computer science.

Definition 12.1 For functions f and g from $\mathbb{N}$ or $\mathbb{R}^+$ to $\mathbb{R}^+$ we write:

1. $f(n) = O(g(n))$ if

$$\limsup_{n \to \infty} \frac{f(n)}{g(n)} < \infty.$$

2. $f(n) = o(g(n))$ if

$$\lim_{n \to \infty} \frac{f(n)}{g(n)} = 0.$$

3. $f(n) = \Omega(g(n))$ if

$$\limsup_{n \to \infty} \frac{f(n)}{g(n)} > 0.$$

4. $f(n) = \omega(g(n))$ if

$$\lim_{n \to \infty} \frac{f(n)}{g(n)} = \infty.$$

5. $f(n) = \Theta(g(n))$ if $f(n) = O(n)$ and $f(n) = \Omega(n)$, i.e. if

$$0 < \limsup_{n \to \infty} \frac{f(n)}{g(n)} < \infty.$$

What makes the O-notation ambiguous is the fact that there is no way to express which variable should tend to infinity. We may want to say $f(m) = \frac{m^2+1}{m+1} = O(m)$, in which case we mean that $\limsup_{m \to \infty} f(m)/m < \infty$. Or we say $f(n)$ is polynomial if there exists a constant c for which $f(n) = O(n^c)$, and here we clearly want $n \to \infty$ and not $c \to \infty$. Which variable is intended becomes clear in context. In this book we follow the convention that the variable that tends to infinity is called n, N, m or similar, while c, d or similar denote constants.

To make the notation even more ambiguous it is a convention that if we use x, y or h as a variable in the O-notation we mean that x, y or h should tend to zero instead of infinity. So we can write "A differentiable function f satisfies $f(x) = f(0) + xf'(0) + o(x)$" and mean $\lim_{x \to 0} \frac{f(x)-f(0)-xf'(0)}{x} = 0$.

Another aspect of the O-notation is how it is used in equality chains. If we write $O(f(n)) = O(g(n))$ we mean that $h(n) = O(f(n))$ implies $h(n) = O(g(n))$. In particular, in the context of the O-notation the equality is not symmetric. $O(f(n)) = O(g(n))$ and $O(g(n)) = O(f(n))$ mean totally different things. For example we can write $f(n) = O(n \log n) = o(n^2)$ and mean that $f(n) = O(n \log n)$ and that implies $f(n) = o(n^2)$. In contrast $o(n^2) = O(n \log n)$ is wrong. For example $n^{1.5} = o(n^2)$, but $n^{1.5} = O(n \log n)$ is false.

12.4 The Complexity Classes $\mathcal{P}$ and $\mathcal{NP}$

Turing machines are a very simple abstract computer model which are commonly used in complexity theory. To really understand the complexity classes $\mathcal{P}$ and $\mathcal{NP}$ which feature in the basic theory behind most of the security proofs in cryptography we need at least some knowledge of Turing machines. We present here a very short introduction. For further reading, see [130] (first edition; the section edition was adapted to the lower lever of present day students and skips a lot of the interesting material) or [219].

Definition 12.2 A *non-deterministic Turing machine* is an abstract computer which consists of the following components:

- A control unit which can take its states from a finite set Q. Some states $Q_t \subseteq Q$ are marked as *terminal states*. The machine stops its computation if it reaches a terminal state. One state q_0 is marked as the initial state.
- A finite number k of memory tapes. Each position on a memory tape can contain a symbol from a finite memory alphabet Σ. One element $\sqcup \in \Sigma$ is a black symbol.
- A write-protected input tape which contain a word over the finite input alphabet Σ_I. The special symbol $\$ \in \Sigma_I$ is used to mark the end of the input.
- An output tape which contains the output which will be a word over the finite output tape Σ_O.
- A program or *transition relation*

$$\delta \subseteq \left(Q \times \Sigma_I \times \Sigma^k\right) \times \left(Q \times (\Sigma_I \times M_I) \times (\Sigma \times M)^k \times (\Sigma_O \times M_O)\right)$$

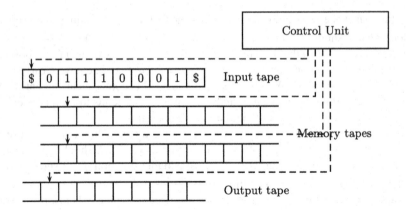

Fig. 12.7 A Turing machine

where M_I, M, M_O are the possible movements of input-head, memory-heads and output-head. Possible movements are left, right and don't move. The transition function must ensure that the input head never leaves the finite input tape.

Figure 12.7 illustrates a Turing machine.

The Turing machine is said to be *deterministic* if δ is a function

$$\left(Q \times \Sigma_I \times \Sigma^k\right) \rightarrow \left(Q \times (\Sigma_I \times M_I) \times (\Sigma \times M)^k \times (\Sigma_O \times M_O)\right).$$

The Turing machine starts the computation in the following state. The finite control is in the state q_0. If the input is w, then $\$w\$$ is written on the input tape and the input head is on the left $\$$. All memory tapes and the output tape are filled with the blank symbol.

The computation is a sequence of steps in which the machine performs the following actions:

- It changes the current state of the control.
- It changes the content of the memory tapes and the output tape at the current head position.
- It moves all heads.

Allowed changes are given by the transition relation δ.

The computation stops if the machine reaches a terminal state. Whenever this happens the content of the output tape is said to be the result of the computation.

Note that for deterministic Turing machines there exists a unique sequence of allowed computation steps, but a non-deterministic Turing machine can have many possible computation paths. It is possible that a non-deterministic Turing machine can produce more than one result for an input. It is also possible that a Turing machine might produce no output at all, either because it runs in a infinite loop and never reaches a terminal state or, in the case of a non-deterministic Turing machine, that there is simply no valid computation path.

It is rather difficult to get an intuition of what computation in a non-deterministic Turing machine means. A typical computation of a non-deterministic Turing machine is divided into two phases. In the first phase the machine uses non-deterministic steps to write down a possible solution of the problem. In the second phase the machine uses deterministic steps to verify the guess from the first phase. If the guess was wrong, then the computation path will end without reaching a terminal state. If the guess was correct it will be written to the output tape before the machine switches to a terminal state.

The model 12.2 can be varied in several aspects. Here are some common variants you can find in the literature:

- The Turing machine has no special input tape. The input is just written on the first memory tape. (The special input tape is just a technical detail which allows us to define sublinear tape complexity classes.)
- In many parts of theoretical computer science it is convenient to restrict the focus to language recognition instead of computing the value of a function. In this case the only possible outputs are "Yes" and "No" and we can remove the output tape and signal the output by two special terminal states q_{Yes} and q_{No}.
- One can restrict the model to only one memory tape. There are several tape reduction theorems that prove that the exact number of tapes does not affect the classical complexity classes $\mathcal{P}$ and $\mathcal{NP}$.
- Several small variations may restrict the possible tape alphabets (for example one can require $\Sigma = \Sigma_I = \Sigma_O$) or the movement of the heads (the input head may be only read from left to right). Sometimes the transition relation of a deterministic Turing machine is allowed to be a partial function.

In the following we will restrict our attention to language recognition, i.e. Turing machines which produce either the output "Yes" or "No".

Definition 12.3 For a finite alphabet Σ we denote by Σ^* the set of all finite words over Σ.

A formal language L over a finite alphabet Σ is a subset of Σ^*.

For a Turing machine M the formal language recognized by M is the set of all words $x \in \Sigma_I^*$ for which at least one computation path produces the output "Yes". (Note that in the case of a non-deterministic machine, it is perfectly acceptable that some computation paths produce no output at all or even the output "No". The word can be accepted anyway.)

Let $f : \mathbb{N} \to \mathbb{N}$.

The formal language L lies in the complexity class $\text{NTime}(f)$ ($\text{DTime}(f)$) if there exists a non-deterministic (deterministic) Turing machine that accepts L and for each word in $w \in L$ there exists an accepting computation path of at most $f(|w|)$ steps. L lies in $\text{NSpace}(f)$ ($\text{DSpace}(f)$) if there exists a non-deterministic (deterministic) Turing machine that accepts L and for each word in $w \in L$ there exists an accepting computation path which uses at most $f(|w|)$ memory cells.

Classes like $\text{Dtime}(f)$ are model dependent; small changes of the model (such as restricting the number of tapes) already have an affect. To get model independent classes we take unions. This leads to the classical complexity hierarchy that starts with:

- $\mathcal{L} = \bigcup_{k \in \mathbb{N}} \mathrm{DSpace}(k \log n)$
- $\mathcal{NL} = \bigcup_{k \in \mathbb{N}} \mathrm{NSpace}(k \log n)$
- $\mathcal{P} = \bigcup_{k \in \mathbb{N}} \mathrm{DTime}(n^k)$
- $\mathcal{NP} = \bigcup_{k \in \mathbb{N}} \mathrm{NTime}(n^k)$
- $\mathcal{PSPACE} = \bigcup_{k \in \mathbb{N}} \mathrm{DSpace}(n^k) = \bigcup_{k \in \mathbb{N}} \mathrm{NSpace}(n^k)$
- $\mathcal{EXP} = \bigcup_{k \in \mathbb{N}} \mathrm{DTime}(k^n)$
- $\mathcal{NEXP} = \bigcup_{k \in \mathbb{N}} \mathrm{NTime}(k^n)$
- $\mathcal{EXPSPACE} = \bigcup_{k \in \mathbb{N}} \mathrm{DSpace}(k^n) = \bigcup_{k \in \mathbb{N}} \mathrm{NSpace}(k^n)$

and then continues indefinitely via double-exponentiation, and so on.

Hidden in the definition of the hierarchy there are several results from complexity theory:

- The union theorem, which states that the classes are complexity classes in the sense of Definition 12.3, i.e. there exists a function $f : \mathbb{N} \to \mathbb{N}$ with $\mathcal{P} = \mathrm{DTime}(f)$, and so on.
- Hierarchy theorems which prove that each class is a subclass of the next one and the inclusions $\mathcal{NL} \subsetneq \mathcal{PSPACE}, \mathcal{P} \subsetneq \mathcal{EXP}, \mathcal{NP} \subsetneq \mathcal{NEXP}$ and so on are proper inclusions.

  In particular, the hierarchy is infinite: there is no maximal complexity class.
- The theorem $\mathrm{DSpace}(f) \subseteq \mathrm{NSpace}(f) \subseteq \mathrm{DSpace}(f^2)$ which proves that $\mathcal{PSPACE}$ is the same for deterministic and non-deterministic machines.

Most problems we face in mathematics belong to the lower classes of the hierarchy. Almost all numerical methods belong to $\mathcal{P}$. $\mathcal{NP}$ is the typical complexity of combinatorial problems. In cryptography $\mathcal{NP}$ is the highest complexity an attack can have. As the encryption algorithm should be in $\mathcal{P}$, a non-deterministic Turing machine can break the cipher in polynomial time by guessing the key and verifying the guess. $\mathcal{PSPACE}$ and $\mathcal{EXP}$ are typical complexities for backtracking algorithms as they appear in the analysis of games. Gröbner bases are one of the rare examples in which we have to go up to $\mathcal{EXPSPACE}$. Higher classes can appear in formal logic. For example, deciding if a sentence in *Presburger arithmetic* (everything you can write down with $\forall, \exists, \wedge, \vee, \neg, +$ and $\leq$ where the variables are integers) is true needs double exponential non-deterministic time.

Up to now no one has been able to prove that two neighbors in the hierarchy are distinct. We know that $\mathcal{NL} \subsetneq \mathcal{PSPACE}$ and hence at least one of the inclusions $\mathcal{NL} \subseteq \mathcal{P} \subseteq \mathcal{NP} \subseteq \mathcal{PSPACE}$ must be a proper one. It is conjectured that all classes in the hierarchy are distinct. $\mathcal{P} = \mathcal{NP}$ would have a dramatic impact on several areas, including cryptography. To prove that $\mathcal{P} \neq \mathcal{NP}$ is considered the most important open problem in complexity theory.

The concepts of reduction and complete problems give some insight into the nature of the complexity classes. These notions were originally defined for $\mathcal{NP}$, but we can also introduce them in their full generality.

Definition 12.4 Let $L_1 \subseteq \Sigma_1^*$ and $L_2 \subseteq \Sigma_2^*$. L_1 is said to be *reducible* to L_2 with respect to some complexity class $\mathcal{C}$ if there exists a function $f : \Sigma_1^* \to \Sigma_2^*$ in $\mathcal{C}$ with

$$w \in L_1 \quad \Longleftrightarrow \quad f(w) \in L_2$$

for all $w \in \Sigma_1$. We write this as $L_1 \leq_C L_2$.

We say L_1 and L_2 are *equivalent* with respect to C-reduction if L_1 can be reduced to L_2 and vice versa:

$$L_1 \equiv_C L_2 \quad \Longleftrightarrow \quad L_1 \leq_C L_2 \wedge L_2 \leq_C L_1.$$

From now, fix some reduction relation $\leq$. A language L is called *hard* for the class C' if for all $L' \in C'$ we have $L' \leq L$ and L is called *complete* for the class C' if $L \in C'$ and is hard.

When we speak of $\mathcal{NP}$-completeness or $\mathcal{PSPACE}$-completeness we implicitly assume that $\leq$ is reduction with respect to deterministic polynomial time.

When we speak of $\mathcal{NL}$-completeness or $\mathcal{P}$-completeness we also implicitly assume that $\leq$ is reduction with respect to deterministic logarithmic space.

Reduction is a central concept, not only in complexity theory: all security proofs in cryptography are reduction theorems. Reduction can also be used to design algorithms (see, for example, Sect. 12.5.2).

Here is a small list of $\mathcal{NP}$-complete problems:

- The *satisfiability problem*: Given a Boolean expression, does there exist an allocation of the variables such that it evaluates to true? This is usually the first problem which is proved to be $\mathcal{NP}$-complete. All other $\mathcal{NP}$-complete problems are reduced either directly or indirectly to this problem.
- The *knapsack problem*: Given a set S of integers and an integer n, does there exists $S' \subset S$ with $\sum_{s \in S'} s = n$? This problem is infamous in cryptography because several broken cryptography systems tried to base their security on this problem [200].
- The *linear decoding problem*: Given a generator matrix G of a linear code C and a word w, find the codeword $c \in C$ with minimal distance to w.
- *Integer linear programming*: Given a matrix $A \in \mathbb{Z}^{n \times m}$ and a vector $b \in \mathbb{Z}^n$, does there exist a vector $x \in \mathbb{Z}^m$ with $Ax \leq b$? This problem is interesting in two respects. Firstly, it is one of the rare cases for which it is much easier to prove that the problem is $\mathcal{NP}$-hard than to prove that it is in $\mathcal{NP}$. Secondly, the analogous problem for rational linear programming is in $\mathcal{P}$.
- Given a lattice L, determine the shortest vector in L.

As mentioned before, it is still an open problem to prove $\mathcal{P} \neq \mathcal{NP}$. Here is an explanation of why it is so difficult. One can extend the Turing machine model by oracles. An oracle gives a Turing machine an answer to a specific question in one step. Almost all proof techniques that work for normal Turing machines also work for Turing machines with oracles and give the same result. However, there exists an oracle O_1 such that $\mathcal{P}^{O_1} = \mathcal{NP}^{O_1}$ and another oracle O_2 with $\mathcal{P}^{O_2} \neq \mathcal{NP}^{O_2}$.

So a proof of $\mathcal{P} \neq \mathcal{NP}$ must make use of the fact that no oracle is present, which requires new proof techniques.

It is not known whether or not $\mathcal{NP}$ is closed under complements. The class of problems for which the complement is in $\mathcal{NP}$ is denoted by co-$\mathcal{NP}$. It is conjectured that $\mathcal{NP} \neq co$-$\mathcal{NP}$, which would also imply $\mathcal{P} \neq \mathcal{NP}$.

The breaking of a cryptosystem is a problem in $co\text{-}\mathcal{NP} \cap \mathcal{NP}$—just guess the key and you can verify everything in polynomial time.

A big problem for security proofs is that no problem is known to be complete for the class $co\text{-}\mathcal{NP} \cap \mathcal{NP}$. There are essentially two strategies to deal with the problem:

- Choose a hopefully difficult problem in $co\text{-}\mathcal{NP} \cap \mathcal{NP}$ (factoring, discrete logarithm) and reduce the security of the cipher to it.
- Choose an $\mathcal{NP}$-complete problem. Restrict it somehow so that the resulting problem lies in $co\text{-}\mathcal{NP} \cap \mathcal{NP}$ and reduce the security of the cipher to it.

The first strategy seems more reasonable and most systems that use this strategy remain unbroken. The second strategy is attractive since there is a wide variety of $\mathcal{NP}$-complete problems to choose from, but restricting an $\mathcal{NP}$-complete problem does not guarantee that the restricted problem is still hard. Many systems that have followed the second strategy are now broken (especially if they start with the knapsack problem [200]).

12.5 Fast Linear Algebra

In algebraic attacks (see Sect. 6.1.2) we have to deal with very large systems of linear equations. For systems of that size the usual methods from linear algebra are too slow. The attacks benefit a lot from advanced methods from linear algebra. (Some of these methods are infamous because they are numerically not as stable as the classical methods, which is of course irrelevant to cryptography, where we deal only with finite fields.)

12.5.1 Matrix Multiplication

12.5.1.1 Simple Algorithms

Let R be a ring and let $A, B \in R^{n \times n}$ be two $n \times n$ matrices. Our goal is to compute the product $C = AB$.

By definition of matrix multiplication

$$c_{i,j} = \sum_{k=1}^{n} a_{i,k} b_{k,j}.$$

Converting this definition into a computer program easy. We summarize it as follows.

Theorem 12.1 *The naive matrix multiplication algorithm needs n^3 ring multiplications and $n^3 - n^2$ ring additions.*

Until the computer era arrived, no one had ever considered if there could be other ways to perform matrix multiplication. Before we start with the asymptotically fast algorithms, we give a small variation of the naive algorithm (see also [282]).

Algorithm 12.4 Winograd's algorithm for multiplying small matrices

Require: R commutative ring, n even, $A, B \in R^{n \times n}$.
Ensure: $C = AB$
1: **for** i **from** 1 **to** n **do**
2:     $A_i = \sum_{k=1}^{n/2} a_{i,2k-1} a_{i,2k}$
3: **end for**
4: **for** j **from** 1 **to** n **do**
5:     $B_j = \sum_{k=1}^{n/2} b_{2k-1,j} b_{2k,j}$
6: **end for**
7: **for** i **from** 1 **to** n **do**
8:     **for** j **from** 1 **to** n **do**
9:        $c_{i,j} = \sum_{k=1}^{n/2} (a_{i,2k-1} + b_{2k,j})(a_{i,2k} + b_{2k-1,j}) - A_i - B_j$
10:     **end for**
11: **end for**

Theorem 12.2 *If R is a commutative ring, Algorithm 12.4 is correct and needs $n^3/2 + n^2$ ring multiplications and $\frac{3}{2}n^3 + 2n^2 - n$ ring additions.*

Proof To prove the correctness we must just expand the expression in line 9 and get

$$
c_{i,j} = \sum_{k=1}^{n/2} (a_{i,2k-1} + b_{2k,j})(a_{i,2k} + b_{2k-1,j}) - A_i - B_j
$$

$$
= \sum_{k=1}^{n/2} a_{i,2k-1} a_{i,2k} + \sum_{k=1}^{n/2} b_{2k,j} a_{i,2k}
$$

$$
+ \sum_{k=1}^{n/2} a_{i,2k-1} b_{2k-1,j} + \sum_{k=1}^{n/2} b_{2k,j} b_{2k-1,j} - A_i - B_j
$$

$$
= \sum_{k=1}^{n/2} b_{2k,j} a_{i,2k} + \sum_{k=1}^{n/2} a_{i,2k-1} b_{2k-1,j}
$$

$$
= \sum_{k=1}^{n} a_{i,k} b_{k,j}.
$$

Note how we used that the ring is commutative in the last equation.

Algorithm 12.4 needs $n(n/2)$ multiplications to compute the A_i and $n(n/2)$ to compute the B_j. The main loop in line 9 needs $n^2(n/2)$ multiplications. Counting the additions is similar. $\qquad\square$

Algorithm 12.4 trades multiplications for additions. Since addition is normally faster than multiplication we can expect Algorithm 12.4 to be better than the naive algorithm on most machines.

12.5.1.2 Strassen's Algorithm

The naive matrix multiplication algorithm needs 8 multiplications and 4 additions to multiply 2×2 matrices.

In 1969 Strassen found a method to multiply two 2×2 matrices (over non-commutative rings) with only 7 multiplications and 18 additions [260].

Later Winograd improved this to 7 multiplications and only 15 additions (see Algorithm 12.5).

Algorithm 12.5 Strassen's algorithm to multiply 2×2 matrices

$$A = \begin{pmatrix} A_{11} & A_{12} \\ A_{21} & A_{22} \end{pmatrix} \qquad\qquad B = \begin{pmatrix} B_{11} & B_{12} \\ B_{21} & B_{22} \end{pmatrix}$$

$$S_1 \leftarrow A_{21} + A_{22} \qquad\qquad T_1 \leftarrow B_{12} - B_{11}$$

$$S_2 \leftarrow S_1 - A_{11} \qquad\qquad T_2 \leftarrow B_{22} - T_1$$

$$S_3 \leftarrow A_{11} - A_{21} \qquad\qquad T_3 \leftarrow B_{22} - B_{12}$$

$$S_4 \leftarrow A_{12} - S_2 \qquad\qquad T_4 \leftarrow T_2 - B_{21}$$

$$P_1 \leftarrow A_{11}B_{11} \qquad\qquad P_5 \leftarrow S_1 T_1$$

$$P_2 \leftarrow A_{12}B_{21} \qquad\qquad P_6 \leftarrow S_2 T_2$$

$$P_3 \leftarrow S_4 B_{22} \qquad\qquad P_7 \leftarrow S_3 T_3$$

$$P_4 \leftarrow A_{22}T_4$$

$$U_1 \leftarrow P_1 + P_2 \qquad\qquad U_5 \leftarrow U_4 + P_3$$

$$U_2 \leftarrow P_1 + P_6 \qquad\qquad U_6 \leftarrow U_3 - P_4$$

$$U_3 \leftarrow U_2 + P_7 \qquad\qquad U_7 \leftarrow U_3 + P_5$$

$$U_4 \leftarrow U_2 + P_5$$

$$C = AB = \begin{pmatrix} U_1 & U_5 \\ U_6 & U_7 \end{pmatrix}$$

The trade-off of one multiplication for 11 additions in a 2×2 matrix multiplication may seem insignificant. However, Algorithm 12.5 holds for non-commutative rings, and so we can apply recursion. Interpret a $2^k \times 2^k$ matrix as a 2×2 block matrix with block size $2^{k-1} \times 2^{k-1}$. So one can perform the multiplication of $2^k \times 2^k$ matrices by 7 multiplications and 15 additions of $2^{k-1} \times 2^{k-1}$ matrices. To multiply the $2^{k-1} \times 2^{k-1}$ matrices you use Strassen's formula again. This is very similar to Karatsuba's algorithm for multiplying n-digit numbers (see Sect. 11.3). Let $M(n)$ be the number of operations needed to compute the product of two $n \times n$ matrices by Strassen's algorithm. Then

$$M(2^k) = 7M(2^{k-1}) + 15(2^{k-1})^2. \tag{12.5}$$

Applying Theorem 11.6 to Eq. (12.5) we get that for $n = 2^k$, the multiplication of two $n \times n$ matrices takes $O(7^k) = O(n^{\log_2(7)})$ operations. Since $\log_2 7 \approx 2.807 < 3$ this is much faster than the naive algorithm.

To extend Strassen's algorithm to matrix dimensions different from 2^k one has to pad the matrices with zeros. The asymptotic complexity stays the same but the constant gets slightly worse if n is not a power of 2.

How practical is Strassen's algorithm? It trades one multiplication of $n/2 \times n/2$ matrices for 11 additions. Assuming that we multiply the $n/2 \times n/2$ matrices using the naive algorithm, for $n \geq 22$ we get $(n/2)^3 \geq 11(n/2)^2$, so one may expect that Strassen recursion is favorable for matrix dimensions above 22. In real world implementations there is some additional overhead, but Strassen's algorithm is already favorable for small matrix dimensions. The crossover point lies between 32 and 128, depending on the hardware [13, 127].

It should be mentioned that Strassen's algorithm is numerically less stable than the naive algorithm. This is a general disadvantage of the advanced algorithms. For applications in cryptography, these considerations are irrelevant, since we work only with finite fields.

12.5.1.3 Pan's Algorithm

After Strassen's discovery it was natural to ask what the smallest ω is such that an $O(n^\omega)$ algorithm for $n \times n$ matrix multiplication exists. The naive algorithm shows $\omega \leq 3$ and Strassen's algorithm shows $\omega \leq \log_2 7$. In 1978 Pan was able to improve Strassen's algorithm [203]. The current world record is $\omega \leq 2.376$ (an algorithm of Coppersmith and Winograd [59]). The interesting field of fast matrix multiplications is treated in detail in the books [40, 204].

The only lower bound on ω is the trivial lower bound $\omega \geq 2$. It is conjectured that for every $\epsilon > 0$ there is an $O(n^{2+\epsilon})$ algorithm for matrix multiplication. Perhaps there is an $O(n^2 \log n)$ algorithm for matrix multiplication (somewhat analogous to the famous $O(n \log(n))$ algorithm for polynomial multiplication).

Most textbooks on computer algebra describe Strassen's algorithm in detail and then mention the advanced algorithms as we did above. Then they note that the constants in the algorithm of Coppersmith and Winograd are so large that the algorithm is only advantageous for astronomical large n. This leaves the reader with impression that, from the practical point of view, Strassen's algorithm is the fastest matrix multiplication algorithm, i.e. for all practical purposes $\omega \approx 2.808$. Indeed it is the only algorithm implemented in most systems.

However, Pan gives in his book [204] a class of algorithms that achieve $\omega \approx 2.775$ and which are superior to Strassen's algorithm for matrices of dimension above a few thousand. Since we can easily reach such dimensions in algebraic attacks, these algorithms are important to cryptology.

In the remaining section we will describe Pan's algorithm and, in the process, also introduce techniques which are important for the understanding of more theoretical results like the current world record of Coppersmith and Winograd.

The first ingredient is the notion of bilinear algorithms as trilinear forms.

The naive matrix multiplication is described by the equations

$$c_{i,j} = \sum_{k=1}^{n} a_{i,k} b_{k,j}.$$

To transform these n^2 equations into a single algebraic expression we multiply the right-hand side by the formal variables $c_{i,j}$ and add the equations together. Thus matrix multiplication is described by the single expression

$$\sum_{i=1}^{n}\sum_{j=1}^{n}\sum_{k=1}^{n} a_{i,k} b_{k,j} c_{i,j}. \tag{12.6}$$

Using this notation, Strassen's algorithm translates to the following identity.

$$\sum_{i=1}^{2}\sum_{j=1}^{2}\sum_{k=1}^{2} a_{i,k} b_{k,j} c_{i,j}$$

$$= a_{1,1} b_{1,1}(c_{11} + c_{1,2} + c_{2,1} + c_{2,2})$$

$$+ a_{1,2} b_{1,2} c_{1,1}$$

$$+ (a_{1,2} - a_{2,1} - a_{2,2} + a_{1,1}) b_{2,2} c_{1,2}$$

$$- a_{2,2}(b_{2,2} - b_{1,2} + b_{1,1} - b_{2,1}) c_{2,1}$$

$$+ (a_{2,1} + a_{2,2})(b_{1,2} - b_{1,1})(c_{1,2} + c_{2,2})$$

$$+ (a_{2,1} + a_{2,2} - a_{1,1})(b_{2,2} - b_{1,2} + b_{1,1})(c_{1,2} + c_{2,1} + c_{2,2})$$

$$+ (a_{1,1} - a_{2,1})(b_{2,2} - b_{1,2})(c_{2,1} + c_{2,2}). \tag{12.7}$$

To obtain the algorithm from Eq. (12.7), one must collect the coefficients of the formal variables $c_{i,j}$ on both sides.

That Strassen's algorithm needs only 7 multiplications to obtain the product of 2×2 matrices means for the trilinear Eq. (12.7) simply that the right hand side has only 7 terms. Thus the search for fast algorithms is equivalent to the problem of decomposing a given trilinear form into few terms.

Definition 12.5 The *rank* of a trilinear form $T(A, B, C)$ is the minimal number of terms in all possible decompositions.

To make the trilinear form in Eq. (12.6) more symmetric, we swap the indices of $c_{i,j}$ and write:

$$T(A, B, C) = \sum_{i=1}^{n} \sum_{j=1}^{n} \sum_{k=1}^{n} a_{i,k} b_{k,j} c_{j,i}. \tag{12.8}$$

The goal is to find good upper bounds for the rank of $T(A, B, C)$.

For the following let R be a (non commutative) ring with 1. The second ingredient for Pan's algorithm is Lemma 12.3.

Lemma 12.3 *Let $n \in \mathbb{N}$ and assume $n + 1$ is invertible in R. For $a_1, \ldots, a_n, b_1, \ldots, b_n \in R$, let*

$$a_0 = -\sum_{i=1}^{n} a_i,$$

$$b_0' = -\frac{\sum_{i=1}^{n} b_i}{n + 1},$$

$$b_i' = b_i + b_0.$$

Then $\sum_{i=0}^{n} a_i = \sum_{i=0}^{n} b_i' = 0$ and

$$\sum_{i=1}^{n} a_i b_i = \sum_{i=0}^{n} a_i b_i'.$$

Proof Let $A = \sum_{i=1}^{n} a_i$ and $B = \sum_{i=1}^{n} b_i$ then

$$\sum_{i=0}^{n} a_i b_i' = A \frac{B}{n+1} + \sum_{i=1}^{n} a_i \left(b_i - \frac{B}{n+1} \right)$$

$$= \sum_{i=1}^{n} a_i b_i + \sum_{i=1}^{n} a_i \left(-\frac{B}{n+1} \right) + \frac{AB}{n+1}$$

$$= \sum_{i=1}^{n} a_i b_i.$$

$\square$

Lemma 12.3 allows us to transform two $n \times n$ matrices A and B into two $(n + 1) \times (n+1)$ matrices A' and B' with the property that AB is just a submatrix of $A'B'$ and that the row sums and column sums of A' and B' are 0. Having the matrices in this special form is advantageous, since many terms will cancel automatically.

Pan's algorithm is best described as the problem of computing three matrix products $C = AB$, $W = UV$ and $Z = XY$ simultaneously.

In the notation with trilinear forms we have to compute

$$\sum_{i,j,k} a_{i,j} b_{j,k} c_{k,i} + u_{j,k} v_{k,i} w_{i,j} + x_{k,i} y_{i,j} z_{j,k} \tag{12.9}$$

($3n^3$ terms). Note the way we have chosen the indices.

We assume that the row and column sums are 0, i.e. $\sum_i a_{i,j} = \sum_j a_{i,j} = 0$ and so on. Lemma 12.3 shows how to bring the input matrices A, B, U, V, X and Y into this form, and we will see later how to ensure this form for the output matrices C, W, Z.

Now consider the following trilinear form:

$$\sum_{i,j,k} (a_{i,j} + u_{j,k} + x_{k,j})(b_{j,k} + v_{k,i} + y_{i,j})(c_{k,i} + w_{i,j} + z_{j,k})$$

(n^3 terms).

Pan used the more graphical *aggregating table*

$a_{i,j}$	$b_{j,k}$	$c_{k,i}$
$u_{j,k}$	$v_{k,i}$	$w_{i,j}$
$x_{k,i}$	$y_{i,j}$	$z_{j,k}$

to describe it.

Multiplying out we get all terms of Eq. (12.9) and most other terms cancel, for example

$$\sum_{i,j,k} a_{i,j} \otimes b_{j,k} \otimes w_{i,j} = \sum_{i,j} a_{i,j} \otimes \left(\sum_k b_{j,k} \right) \otimes w_{i,j} = \sum_{i,j} a_{i,j} \otimes 0 \otimes w_{i,j} = 0.$$

The only remaining terms are the *diagonal correction* terms $a_{i,j} v_{k,i} z_{j,k}$, $u_{j,k} y_{i,j} c_{k,i}$ and $x_{k,i} b_{j,k} w_{i,j}$ and the *anti-diagonal correction* terms $x_{k,i} v_{k,i} c_{k,i}$, $u_{j,k} b_{j,k} z_{j,k}$ and $a_{i,j} y_{i,j} w_{i,j}$.

First consider the anti-diagonal correction terms. In each term only two different indices appear. Thus there are just $3n^2$ different anti-diagonal correction terms. So computing this correction is cheap in comparison to computing the main terms.

We need a way to deal with the diagonal correction terms. Here comes the third ingredient for Pan's algorithm: double the size of the matrices.

We now have $2n \times 2n$ matrices with the property that

$$\sum_{i=1}^{n} a_{i,j} = \sum_{i=n+1}^{2n} a_{ij} = \sum_{j=1}^{n} a_{i,j} = \sum_{j=n+1}^{2n} a_{ij} = 0.$$

We say that such matrices have *Pan-normal-form*.

Instead of one aggregating table, we now have 8 aggregating tables. In the tables we use $\tilde{i}$ as an abbreviation for $i + n$ and similarly $\tilde{j} = j + n$, $\tilde{k} = k + n$

$$
\begin{array}{c|c|c}
a_{i,j} & b_{j,k} & c_{k,i} \\
u_{j,k} & v_{k,i} & w_{i,j} \\
x_{k,i} & y_{i,j} & z_{j,k}
\end{array}
\qquad
\begin{array}{c|c|c}
-a_{i,j} & b_{j,\bar k} & -c_{\bar k,i} \\
u_{\bar j,k} & v_{k,i} & w_{i,\bar j} \\
x_{k,\bar i} & y_{\bar i,j} & z_{j,k}
\end{array}
$$

$$
\begin{array}{c|c|c}
a_{i,\bar j} & b_{\bar j,k} & c_{k,i} \\
-u_{j,k} & v_{k,\bar i} & -w_{\bar i,j} \\
x_{\bar k,i} & y_{i,j} & z_{j,\bar k}
\end{array}
\qquad
\begin{array}{c|c|c}
a_{\bar i,j} & b_{j,k} & c_{k,\bar i} \\
u_{j,\bar k} & v_{\bar k,i} & w_{i,j} \\
-x_{k,i} & y_{i,\bar j} & -z_{\bar j,k}
\end{array}
$$

$$
\begin{array}{c|c|c}
a_{\bar i,\bar j} & b_{\bar j,k} & c_{\bar k,\bar i} \\
u_{j,\bar k} & v_{\bar k,\bar i} & w_{\bar i,j} \\
x_{\bar k,\bar i} & y_{\bar i,j} & z_{j,\bar k}
\end{array}
\qquad
\begin{array}{c|c|c}
-a_{\bar i,\bar j} & b_{\bar j,k} & -c_{k,\bar i} \\
u_{j,\bar k} & v_{\bar k,\bar i} & w_{\bar i,j} \\
x_{\bar k,i} & y_{i,\bar j} & z_{j,\bar k}
\end{array}
$$

$$
\begin{array}{c|c|c}
a_{\bar i,j} & b_{j,\bar k} & c_{\bar k,\bar i} \\
-u_{\bar j,\bar k} & v_{\bar k,i} & -w_{i,j} \\
x_{k,\bar i} & y_{\bar i,\bar j} & z_{\bar j,k}
\end{array}
\qquad
\begin{array}{c|c|c}
a_{i,\bar j} & b_{j,\bar k} & c_{\bar k,i} \\
u_{\bar j,k} & v_{k,\bar i} & w_{\bar i,\bar j} \\
-x_{\bar k,\bar i} & y_{i,j} & -z_{j,\bar k}
\end{array}
$$

Observe that every diagonal correction term appears exactly twice, once with positive sign and once with negative sign. Hence the diagonal correction terms cancel.

Now have a closer look at the remaining $24n^2$ anti-diagonal correction terms. Consider the following 8 anti-diagonal correction terms:

$$
n(a_{ij}y_{ij}w_{ij} - a_{ij}y_{\bar i j}w_{\bar i j} - a_{i\bar j}y_{ij}w_{\bar i j} + a_{\bar j}y_{ij}w_{ij}
$$
$$
+ a_{i\bar j}y_{\bar i\bar j}w_{\bar i\bar j} - a_{i\bar j}y_{ij}w_{\bar i\bar j} - a_{i\bar j}y_{ij}w_{ij} + a_{ij}y_{\bar i}w_{\bar i\bar j}).
$$

This is essentially the same trilinear form as the product of 2×2 matrices. So one can apply Strassen's trick to replace the eight terms by only 7 terms. This reduces the number of anti-diagonal correction terms from $24n^2$ to $21n^2$.

At this point we have decomposed the trilinear form of three matrix products of Pan normalized $2n \times 2n$ matrices into only $8n^3 + 21n^2$ terms instead of $24n^3$ terms (naive implementation). This is a significant speed-up.

Lemma 12.3 explains how to transform the input matrices to Pan normal form. We must still deal with the output matrices. If one computes all products of the aggregation tables and the anti-diagonal correction terms and collect the results in the variables indicated by the trilinear form, we do not get the pan normalized output $C = \begin{pmatrix} C_{11} & C_{12} \\ C_{21} & C_{22} \end{pmatrix}$ but a matrix $\hat C = \begin{pmatrix} \hat C_{11} & \hat C_{12} \\ \hat C_{21} & \hat C_{22} \end{pmatrix}$.

The relation between $C_{11} = (c_{ij})_{1 \le i,j \le n}$ and $\hat C_{11} = (\hat c_{ij})_{1 \le i,j \le n}$ is

$$
\hat c_{i,j} = c_{i,j} + d_i + e_j. \tag{12.10}
$$

Similar relations hold between C_{12} and $\hat C_{12}$, and so on.

To compute the Pan-normalized result one has to find d_i and e_j in Eq. (12.10). Note that with e_i and d_j, $e'_i = e_i - x$ and $d'_j = d_j + x$ are also solutions of (12.10). So one can assume without loss of generality that $d_1 = 0$. Then

$$\sum_{j=1}^{n} e_j = \sum_{j=1}^{n} \hat{c}_{0,j},$$

$$d_i = n^{-1} \left(\sum_{j=1}^{n} \hat{c}_{i,j} - \sum_{j=1}^{n} e_j \right),$$

$$e_j = n^{-1} \left(\sum_{i=1}^{n} \hat{c}_{i,j} - \sum_{i=1}^{n} d_i \right).$$

So one can compute correction values d_i and e_i and hence the Pan-normalized result C with $O(n^2)$ operations.

Pan's algorithm can also be adapted to compute just one matrix product instead of three matrix products simultaneously. One just has to specialize the nine matrices to $A = U = X$, $B = V = Y$ and $C = W = Z$.

The algorithm becomes less symmetric. We omit the details, which can be found in Chap. 31 of [203] or in our fully documented reference implementation. Theorem 12.4 summarizes the results (using the same terminology as Lemma 12.3).

Theorem 12.4 *If $n \in N$ is even and $\frac{n+2}{2}$ is invertible in R, then the product of two matrices in $R^{n \times n}$ can be computed by using only*

$$\frac{1}{3}n^3 + \frac{15}{4}n^2 + \frac{97}{6}n + 20$$

ring multiplications.

Algorithm 12.6 shows a bird's eye view of Pan's algorithm.

Algorithm 12.6 Pan's matrix multiplication

Input: $A, B \in R^{2n \times 2n}$, $n + 1$ must be invertible in R

1. Convert A, B to A', $B' \in R^{(2n+2) \times (2n+2)}$ into Pan-normal-form using Lemma 12.3. This takes $O(n^2)$ operations (only scalar multiplications, no ring multiplication).
2. Compute the 8 aggregating tables and the anti-diagonal corrections terms and collect the results. There are $n^3/3 + \frac{15}{4}n^2 + \frac{97}{6}n + 20$ ring multiplications.
3. Renormalize the result to get $C' = A'B'$. ($O(n^2)$ operations, again no ring multiplication).

Table 12.2 Speed of Pan's algorithm

Size of the matrix $(2n)$	44	46	48	50	52
Number of ring multiplication	36386	41144	46300	51870	57870
Exponent in asymptotic algorithm	2.77522	2.77509	2.77508	2.7752	2.77532

Of course, as in the case of Strassen's algorithm, Pan's algorithm shows its full strength only if applied recursively to block matrices. The optimal speed is achieved if one chooses $n \approx 25$ in Algorithm 12.6. In Table 12.2 one can see the achievable speed for different sizes of n. As one can see, Pan's algorithm needs asymptotically $n^{2.775}$ operations to multiply $n \times n$ matrices, which is better than Strassen's algorithm ($n^{2.807}$ operations). The crossover point is quite low, between 1000 and 10000, so Pan's algorithm is really interesting in practice.

12.5.1.4 Binary Matrices

When multiplying binary matrices we must take care to make use of the inherent parallelism of bitwise operations.

In this section we always work with 64-bit words.

One can store a 64×64 binary matrix as an array of 64 words. Each word represents a whole row. Algorithm 12.7 shows how to multiply two such 64×64 binary matrices.

Algorithm 12.7 Multiplication of 64×64 binary matrices

Require: A, B two arrays of 64 words, each word represent a matrix row (index numeration starts with 0).

Ensure: $C = AB$

1: **for** i **from** 0 **to** 63 **do**
2:      $x \leftarrow A[i]$
3:      **if** x odd **then**
4:          $C[i] \leftarrow B[0]$
5:      **else**
6:          $C[i] \leftarrow 0$
7:      **end if**
8:      **for** j **from** 1 **to** 63 **do**
9:          $X \leftarrow X/2$ {Division by 2 is just a right shift}
10:         **if** x odd **then**
11:            $C[i] \leftarrow C[i] \oplus B[j]$
12:         **end if**
13:      **end for**
14: **end for**

Algorithm 12.7 uses just the definition of matrix multiplication, but note how we use the bitwise XOR in line 11 to do 64 additions in parallel. The multiplication is hidden in the IF-statements; since $\mathbb{F}_2$ has only 0 and 1 as elements an IF-statement is a good replacement for a multiplication.

Algorithm 12.7 is very efficient with respect to its use of the inherent parallelism of bitwise operations, but it makes no use of the mathematics that stands behind Strassen's algorithm and the other advanced matrix multiplications. Addition of two 64×64 binary matrices needs only 64 word operations, but multiplication with Algorithm 12.7 needs more than 10000 operations. This is a clear sign that we should switch from the naive algorithm to Strassen's algorithm earlier.

The right idea is to interpret a 64-bit word as an 8×8 binary matrix, i.e. we interpret the word $(a_{63} \ldots a_1 a_0)_2$ as the binary matrix

$$A = \begin{pmatrix} a_{63} & a_{62} & \cdots & a_{56} \\ a_{55} & a_{54} & \cdots & a_{48} \\ \vdots & \vdots & \ddots & \vdots \\ a_7 & a_6 & \cdots & a_0 \end{pmatrix}.$$

The bitwise XOR is matrix addition. In [155] Knuth describes the abstract machine MMIX which he uses in his books "The Art of Computer Programming". The MMIX machine has the instruction MXOR, which is the multiplication of two 64-bit words as 8×8 binary matrices. Knuth notes that the MXOR instruction has many applications (rearranging bytes, multiplication in $\mathbb{F}_{256}$, etc.), but at the moment real machines do not include it. Instead real machines "have a dozen or so ad hoc instructions that handle only the most common special cases".

As long as we do not have an MMIX machine we can use Algorithm 12.8 to simulate MXOR by common instructions (see also [154], which describes a program simulating an MMIX machine).

Algorithm 12.8 Multiplication of 8×8 binary matrices (MXOR)

Require: A, B two 64-bit words, representing 8×8 binary matrices
Ensure: C is a 64-bit word representing the 8×8 binary matrix $C = AB$
 1: last_column $\leftarrow 0x0101010101010101$
 2: last_row $\leftarrow 0x00000000000000FF$
 3: $C \leftarrow 0$
 4: **for** i **from** 0 **to** 7 **do**
 5:     $C \leftarrow C \oplus (((a \gg i) \,\&\, \text{last\_column}) \cdot \text{last\_row}) \,\&$
                    $(((b \gg (8i)) \,\&\, \text{last\_row}) \cdot \text{last\_column})$
 6: **end for**

Algorithm 12.8 looks very involved, since it makes heavy use of bit manipulations, but it is just the naive matrix multiplication. Let's have a closer look at the individual operations. When multiplying

$$A = \begin{pmatrix} a_{00} & \cdots & a_{07} \\ \vdots & \ddots & \vdots \\ a_{70} & \cdots & a_{77} \end{pmatrix} \quad \text{and} \quad B = \begin{pmatrix} b_{00} & \cdots & b_{07} \\ \vdots & \ddots & \vdots \\ b_{70} & \cdots & b_{77} \end{pmatrix}$$

we need

$$C' = \begin{pmatrix} a_{07}b_{70} & \cdots & a_{07}b_{77} \\ \vdots & \ddots & \vdots \\ a_{77}b_{70} & \cdots & a_{77}b_{77} \end{pmatrix}$$

as part of the result.

To get C' we need

$$A' = \begin{pmatrix} a_{07} & \cdots & a_{07} \\ \vdots & \ddots & \vdots \\ a_{77} & \cdots & a_{77} \end{pmatrix}.$$

We obtain A' by extracting the last column of A (done by the bit operation $A \& 0x0101010101010101$) and copy the elements to the other columns (multiplication by $0xFF$, the multiplication will not produce a carry).

Similarly we obtain

$$B' = \begin{pmatrix} b_{70} & \cdots & b_{77} \\ \vdots & \ddots & \vdots \\ b_{70} & \cdots & b_{77} \end{pmatrix}$$

by extracting the last row of B ($B \& 0xFF$) and copying this row to the other rows (multiplication by $0x0101010101010101$). This is exactly what is done in line 5 of Algorithm 12.8.

With Algorithm 12.8 and the bitwise XOR we have all we need to deal efficiently with 8×8 binary matrices. Larger binary matrices are built from these basic blocks.

12.5.2 Other Matrix Operations

In the previous section we focused on matrix multiplication algorithms. In this section we want to explain why.

Let

$$\omega(MM) = \inf\{\omega | \text{there is an } O(n^{\omega}) \text{ algorithm for } n \times n \text{ matrix multiplication}\}.$$

As we learned in the previous section $\omega(MM) \leq 2.376$ (see [59]). It is still an open problem whether the exponent $\omega(MM)$ is independent of the ground field. For the moment all we know is that if $\omega(MM)$ is at all dependent on the ground field then it depends only on the characteristic of the ground field. We will write $\omega_F(MM)$ if the ground field F is important.

Similarly we define $\omega(MI)$ as the infimum over all ω for which an $O(n^\omega)$ algorithm for inverting an $n \times n$ matrix exists. The exponent for the problem of computing the determinant of an $n \times n$ matrix is denoted by $\omega(DET)$ and $\omega(SLE)$ is the exponent for solving a system of n linear equations in n variables.

Theorem 12.5 *For an arbitrary field F we have*

$$\omega_F(SLE) \le \omega_F(MM) = \omega_F(MI) = \omega_F(DET).$$

Proof Consider the following identity for block matrices.

Let

$$X = \begin{pmatrix} X_{11} & X_{12} \\ X_{21} & X_{22} \end{pmatrix}$$

and assume that X_{11} is invertible. This is no restriction, just choose a suitable basis. Let

$$Z = X_{22} - X_{21} X_{11}^{-1} X_{12}$$

then

$$X = \begin{pmatrix} I & 0 \\ X_{21} X_{11}^{-1} & I \end{pmatrix} \begin{pmatrix} X_{11} & 0 \\ 0 & Z \end{pmatrix} \begin{pmatrix} I & X_{11}^{-1} X_{12} \\ 0 & I \end{pmatrix}, \qquad (12.11)$$

$$X^{-1} = \begin{pmatrix} I & -X_{11}^{-1} X_{12} \\ 0 & I \end{pmatrix} \begin{pmatrix} X_{11}^{-1} & 0 \\ 0 & Z^{-1} \end{pmatrix} \begin{pmatrix} I & 0 \\ -X_{21} X_{11}^{-1} & I \end{pmatrix}. \qquad (12.12)$$

Let $T(MI(n))$ be the number of ring operations needed to compute the inverse of an $n \times n$ matrix. Equation (12.12) shows that inverting a $2n \times 2n$ matrix can be done by two inversions of $n \times n$ matrices (computing X_{11}^{-1} and Z^{-1}) and 6 multiplications of $n \times n$ matrices (two multiplications to compute Z and 4 multiplications to evaluate Eq. (12.12)). Furthermore, we need two matrix additions. Altogether

$$T(MI(2n)) \le 2T(MI(2n)) + 6O(n^{\omega(MM)}) + 2n^2,$$

which proves $T(MI(2n)) = O(n^{\omega(MM)})$ or $\omega(MI) \le \omega(MM)$.

From Eq. (12.11) we see $\det(X) = \det X_{11} \det Z$. Hence the number $T(DET(n))$ of operations needed to compute the determinant of an $n \times n$ matrix satisfies

$$T(DET(2n)) \le 2T(DET(n)) + 2O(n^{\omega(MM)}) + n^2 + 1$$

and hence $\omega(DET) \le \omega(MM)$.

To prove $\omega(MM) \le \omega(MI)$ choose $X_{11} = I$. Then the lower right submatrix of X^{-1} is $Z^{-1} = (X_{22} - X_{21} X_{21})^{-1}$. We obtain the product $X_{21} X_{21}$ by first inverting the $2n \times 2n$ matrix X and then inverting the $n \times n$ matrix Z^{-1}. Therefore

$$T(MM(n)) \le T(MI(2n)) + T(MI(n)) + n^2$$

which proves $\omega(MM) \leq \omega(MI)$.

For the proof that $\omega(MI) \leq \omega(DET)$, see [18].

Solving a linear system of equations can be done by inverting the matrix and multiplying the inverse by a vector which shows

$$T\big(SLE(n)\big) \leq T\big(MI(n)\big) + n^2$$

or $\omega(SLE) \leq \omega(MI)$. It is conjectured that $\omega(SLE) = \omega(MI)$, but at the moment this is an open problem. $\qquad\Box$

Theorem 12.5 is a theorem on the asymptotic behavior of matrix algorithms, but note that the constructions used in the proof all have very small constants, so they give feasible algorithms.

12.5.3 Wiedmann's Algorithm and Black Box Linear Algebra

The problem of solving linear equations is of special importance to cryptography. An interesting algorithm for this problem is due to Wiedmann [280]. It is especially useful for sparse linear systems (which are quite common in cryptography). It is also a good application for the Berlekamp-Massey algorithm (see Sect. 2.4.2). Both points justify a section devoted just to Wiedmann's algorithm.

The main idea is the following. Let $A \in F^{n \times n}$ be a non-singular matrix and let $b \in F^n$. Consider the sequence $a_i = A^i b$. This is a linear recurrence sequence. In contrast to the linear recurrence sequence we have studied in Chap. 2, it is not defined over a field but over the vector space F^n.

The sequence a_i still has a minimal polynomial m which is the unique monic polynomial $m = \sum_{i=0}^{d} m_i x^i$ of least degree for which

$$\sum_{i=0}^{d} m_i A^i b = 0 \qquad (12.13)$$

holds. m is a divisor of the minimal polynomial of A which is in turn a divisor of the characteristic polynomial of A. In particular, the degree d is bounded by n. We will see later how to compute m.

Winograd's idea is to simply multiply Eq. (12.13) by A^{-1} to obtain

$$A^{-1}b = -m_0^{-1} \sum_{i=1}^{d} m_i A^{i-1}b. \qquad (12.14)$$

Equation (12.14) can be efficiently evaluated in a Horner-like fashion. The interesting point is that Winograd's algorithm uses a matrix only to evaluate it at vectors. Thus instead of storing the matrix as an $n \times n$ table, we just need a black box which returns for an input $v \in F^n$ the result $Av \in F^n$. We call this way of doing linear algebra *black box linear algebra* to distinguish it from the traditional explicated linear algebra in which we manipulate the entries of A explicitly. A special case is *sparse linear algebra* in which a matrix is stored as a list of triples $(i, j, m_{i,j})$ with $m_{i,j} \neq 0$. Winograd first proposed his algorithm in the context of sparse linear algebra in which the evaluation Av is very fast.

We still have to explain how to compute the minimal polynomial m of the sequence $a_i = A^i b$. The problem is that for sequences over vector spaces there is no analog of the Berlekamp-Massey algorithm. The idea is to choose a random u and compute the minimal polynomial b_u of the linear recurrence sequence $(u^t A^i b)_{i \in N}$ with the normal Berlekamp-Massey algorithm. m_u is a divisor of the minimal polynomial m and if the field F has enough elements there is a high probability that $m_u = m$. Thus we have an effective probabilistic algorithm to compute m. Again the algorithm needs only a black box to evaluate A at some vectors v, so we stay in the framework of black box linear algebra.

We do not go into the details. See Sect. 12.4 of [100] or [280] for an analysis of the probability that $m_u = m$.

Chapter 13
Number Theory

The Blum-Blum-Shub generator (Chap. 11) uses a lot of number theory. This chapter collects together all the results from number theory that we need.

13.1 Basic Results

We write $a \equiv b \mod n$ if and only if $n|(a - b)$. For each congruence class $b + n\mathbb{Z}$ there exists a unique representative $a \in \{0, \ldots, n - 1\}$. We denote this representative by $a = b \mod n$ (note the different use of $=$ and $\equiv$). Most programming languages provide a built in operator for $a = b \mod n$ (in C-style syntax a = b%n).

By $\tau(n)$ we denote the number of divisors of n. If $n = p_1^{e_1} \cdots p_k^{e_k}$ then $\tau(n) = (e_1 + 1) \cdots (e_k + 1)$.

Recall the *Chinese Remainder Theorem*:

Theorem 13.1 (Chinese Remainder Theorem) *Let $n_1, \ldots, n_k$ be pairwise relatively prime. Then the system*

$$x \equiv a_1 \mod n_1$$

$$\vdots$$

$$x \equiv a_k \mod n_k$$

has a unique solution $x \equiv a \mod n$ with $n = n_1 \cdots n_k$.

With the extended Euclidean algorithm the solution a can be effectively computed.

A. Klein, *Stream Ciphers*, DOI 10.1007/978-1-4471-5079-4_13,
© Springer-Verlag London 2013

13.2 The Group $(\mathbb{Z}/n\mathbb{Z})^\times$

By the Chinese remainder theorem we get for $n = p_1^{e_1} \cdots p_k^{e_k}$ that $(\mathbb{Z}/n\mathbb{Z})^\times$ is isomorphic to the direct product $(\mathbb{Z}/p_1^{e_1}\mathbb{Z})^\times \times \cdots \times (\mathbb{Z}/p_k^{e_k}\mathbb{Z})^\times$.

By $\varphi(n)$ we denote the number of elements in $\{1, \ldots, n-1\}$ relatively prime to n. φ is called the *Eulerian function*. If $n = p_1^{e_1} \cdots p_k^{e_k}$ then

$$\varphi(n) = p_1^{e_1-1}(p_1 - 1) \cdots p_k^{e_k-1}(p_k - 1). \tag{13.1}$$

The order of the group $(\mathbb{Z}/n\mathbb{Z})^\times$ is $\varphi(n)$ and hence we obtain:

Lemma 13.2 *If a is relatively prime to n then*

$$a^{\varphi(n)} \equiv 1 \mod n.$$

Theorem 13.3 *If $n = p_1^{e_1} \cdots p_k^{e_k}$ then*

$$\varphi(n) = p_1^{e_1-1}(p_1 - 1) \cdots p_k^{e_k-1}(p_k - 1).$$

By the *Carmichael function* $\lambda(n)$ we denote the smallest integer m such that $a^m = 1$ for all a in the group $(\mathbb{Z}/n\mathbb{Z})^\times$. More explicitly we have

$$\lambda\left(p^e\right) = \begin{cases} p^{e-1}(p-1) & \text{if } p \geq 3 \text{ or } e \leq 2, \\ 2^{e-2} & \text{if } p = 2 \text{ and } e \geq 3 \end{cases} \tag{13.2}$$

for a prime power p^e and for $n = p_1^{e_1} \cdots p_k^{e_k}$:

$$\lambda(n) = \text{lcm}\left(\lambda\left(p_1^{e_1}\right), \ldots, \lambda\left(p_k^{e_k}\right)\right).$$

To prove the result on the period of the power generator we need the following estimate of the number of elements in $\mathbb{Z}/n\mathbb{Z}$ of small multiplicative order.

Lemma 13.4 (Friedlander, Pomerance, Shparlinski [97]) *Let n be a positive integer and let j be a divisor of $\lambda(n)$. The number $N_j(n)$ of elements in $\mathbb{Z}/n\mathbb{Z}$ with order dividing $\lambda(n)/j$ is bounded by*

$$N_j(n) \leq \frac{\varphi(n)}{j}.$$

Proof Let $n = p_1^{e_1} \cdots p_k^{e_k}$ and let $\lambda(n) = q_1^{d_1} \cdots q_l^{d_l}$. If $\text{ord}_n x | \frac{\lambda(n)}{j}$ then for all $i = 1, \ldots, k$:

$$\text{ord}_{p_i^{e_i}} x \left| \frac{\lambda(p_i^{e_i})}{j_i} \right.$$

where j_i contains all prime factors q_g of j for which $q_g^{d_g} | \lambda(p_i^{e_i})$.

The number of elements of $\mathbb{Z}/p_i^{e_i}\mathbb{Z}$ with $\mathrm{ord}_{p_i^{e_i}} x \mid \frac{\lambda(p_i^{e_i})}{j_i}$ is at most $\varphi(p_i^{e_i})/j_i$. The bound is sharp except for the case $p_i = 2$, $e_i \geq 3$, in which case it can be lowered by a factor of $1/2$.

Using the Chinese remainder theorem to combine the results modulo prime powers we obtain

$$N_j \leq \prod_{i=1}^{k} \frac{\varphi(p_i^{e_i})}{j_i} = \frac{\varphi(n)}{j_1 \cdots j_k} \leq \frac{\varphi(n)}{j}. \qquad \square$$

Let $\tau(n)$ be the number of divisors of n.

Summing over all j with $j \geq K$ (at most $\tau(\lambda(n))$ summands less than $\frac{\varphi(n)}{K}$) we get that the number $S_k(n)$ of elements in $\mathbb{Z}/n\mathbb{Z}$ with order at most $\lambda(n)/K$ is bounded by

$$S_k(n) \leq \frac{\varphi(n)\tau(\lambda(n))}{K}. \qquad (13.3)$$

13.3 The Prime Number Theorem and Its Consequences

To estimate how fast we can generate random primes for cryptographic applications, we need to know the asymptotic density of the primes.

Definition 13.1 For $x > 0$ let

$$\pi(x) = \sum_{p \leq x} 1 \quad and \quad \vartheta(x) = \sum_{p \leq x} \ln p$$

where the sums run over all primes less than or equal to x.

A famous result of number theory is:

Theorem 13.5

$$\pi(x) \sim \frac{x}{\ln x} \qquad (13.4)$$

and

$$\vartheta(x) = \Theta(x). \qquad (13.5)$$

Proof See, for example, [122] Chap. 22. Equation (13.4) is known as the *prime number theorem* and Eq. (13.5) is an important step in the classical proof of the prime number theorem. $\qquad \square$

Other results in this direction are:

Theorem 13.6

$$\sum_{p \leq n} \frac{1}{p} = \ln \ln n + B_1 + o(1) \tag{13.6}$$

and

$$\prod_{p \leq n} \left(1 - \frac{1}{p}\right) \sim \frac{e^{\sigma}}{\ln n} \tag{13.7}$$

for some constant σ.

Proof See, for example, [122] Theorem 427 and Theorem 429. Equation (13.7) is known as Mertens' theorem. □

Equation (13.4) tells us that a random integer of size approximately e^n is prime with probability $1/n$. Choosing random integers and testing them for primality is therefore fast enough for practical purposes. In most cryptographic applications $n \approx 550$ (about 800 bits) is secure enough.

Next we will use Eq. (13.5) to estimate the size of the Eulerian function.

Theorem 13.7

$$\frac{n}{\varphi(n)} = O\left(\ln \ln(n)\right).$$

Proof By Eq. (13.1) we have

$$\varphi(n) = n \prod_{p \mid n} \left(1 - \frac{1}{p}\right)$$

where the product runs over all prime divisors of n.

Let $\nu(n)$ denote the number of prime divisors of n. The factor $(1 - \frac{1}{p})$ is smaller if p is smaller, hence

$$\varphi(n) \geq n \prod_{i=1}^{\nu(n)} \left(1 - \frac{1}{p_i}\right)$$

where the product runs over the $\nu(n)$ smallest primes. By Eq. (13.7) we have

$$\varphi(n) \geq nc \frac{1}{\ln p_{\nu(n)}} \tag{13.8}$$

for some constant c.

On the other hand

$$n \geq \prod_{p \mid n} p \geq \prod_{i=1}^{\nu(n)} p_i = \prod_{p \leq p_{\nu(n)}} p$$

and hence

$$\ln n \geq \sum_{p \leq p_{\nu(n)}} \ln p = \vartheta(p_{\nu(n)}) \geq c' p_{\nu(n)} \tag{13.9}$$

by Eq. (13.5).

By combining Eqs. (13.8) and (13.9) we get

$$\varphi(n) \geq \frac{c}{\ln c'} \frac{n}{\ln \ln n}$$

which is precisely the statement of the theorem. $\square$

Theorem 13.7 has some applications in cryptology. It tells us, for example, that we can efficiently generate random primitive polynomials (see Sect. 14.3).

13.4 Zsigmondy's Theorem

Zsigmondy's theorem is useful in the theory of finite linear groups.

Theorem 13.8 (see [290]) *Let $a, b, n \in \mathbb{Z}$ with $\gcd(a, b) = 1$, $|a| > 1$ and $n \geq 1$. Then $a^n - b^n$ has a prime divisor p that does not divide $a^m - b^m$ for any $0 < m < n$, except in the following cases*:

(a) $n = 1$ and $a - b = 1$.
(b) $n = 2$ and $a + b = \pm 2^{\mu}$.
(c) $n = 3$ and $\{a, b\} = \{2, -1\}$ or $\{a, b\} = \{-2, 1\}$.
(d) $n = 6$ and $\{a, b\} = \{2, 1\}$ or $\{a, b\} = \{-2, -1\}$.

We will not need Theorem 13.8 in its full generality, so we will prove it only for $a > 1$ and $b = 1$. The following simple proof is due to Lüneburg [173].

Like all proofs of Zsigmondy's theorem, Lüneburg's proof is based on cyclotomic polynomials. For $n \geq 1$ the *cyclotomic polynomial* Φ_n is defined as

$$\Phi_n(x) = \prod_{i=1}^{\varphi(n)} (x - \zeta_i),$$

where $\zeta_1, \ldots, \zeta_{\varphi(n)}$ are the primitive roots of unity of order n. We note without proof that $\Phi_n \in \mathbb{Z}[x]$ (see, for example, [140]).

Lemma 13.9 *Let $a > 1$ and $n > 1$ be integers. Let p be a prime dividing $\Phi_n(a)$ and let f be the order of a modulo p. Then $n = fp^i$ for some non-negative integer i. If $i > 0$ then p is the largest prime dividing n. If in addition $p^2 | \Phi_n(a)$ then $n = p = 2$.*

Proof Since $\Phi_n(a)$ divides $a^n - 1$, so does p. Hence $f|n$. Let $n = fp^i w$ with $\gcd(w, p) = 1$. Assume that $w \neq 1$, let $r = fp^i$. Then $p|a^r - 1$ and

$$\frac{a^n - 1}{a^r - 1} = \frac{((a^r - 1) + 1)^w - 1}{a^r - 1} = w + \sum_{j=2}^{w} \binom{w}{j}(a^r - 1)^{j-1} \equiv w \neq 0 \mod p.$$

But for each proper divisor r of n the number $\Phi_n(a)$ (and hence p) is a divisor of $(a^n - 1)/(a^r - 1)$. Thus $w = 1$, i.e. $n = fp^i$.

Now let $i > 0$. Since f is a divisor of $p - 1$ (Fermat's little theorem) p must be the largest prime divisor of n.

Now assume $i > 0$ and $p^2|\Phi_n(a)$. For $p > 2$ we have

$$\frac{a^n - 1}{a^{n/p} - 1} = \frac{((a^{n/p} - 1) + 1)^p - 1}{a^{n/p} - 1}$$

$$= p + \binom{p}{2}(a^{n/p} - 1) + \sum_{j=3}^{p} \binom{p}{j}(a^{n/p} - 1)^{j-1}$$

$$\equiv p \mod p^2$$

and hence $p^2 \nmid \Phi_n(a)$.

If $p = 2$, then $n = 2^i$ since p is the largest prime divisor of n. But then $\Phi_n(a) = (a^n - 1)/(a^{n/2} - 1) = a^{2^{i-1}} + 1$. For $i > 2$ we have $a^{2^{i-1}} + 1 \equiv 2 \mod 4$, i.e. $p^2 = 4 \nmid \Phi_n(a)$. Thus $n = 2$, as stated in the lemma. $\square$

Now we can prove the special case of Zsigmondy's theorem.

Proof of Theorem 13.8 for $a > 1$ and $b = 1$ We first check that the cases mentioned in Theorem 13.8 are indeed exceptions.

Suppose now that there is no prime p such that the order of a modulo p is n.

If $n = 2$, Lemma 13.9 implies $\Phi_n(a) = a + 1 = 2^s$ for some integer s, which is condition (b) of Theorem 13.8.

From now on assume $n > 2$. By Lemma 13.9 $\Phi_n(a) = p$, where p is the largest prime factor of n.

If $p = 2$ then $n = 2^i$ and $\Phi_n(a) = a^{2^{i-1}} + 1 > 2$, a contradiction.

Now let $p > 2$. Then $n = p^i r$ where r divides $p - 1$. Let $\zeta_1, \ldots, \zeta_{\phi(r)}$ be the primitive roots of unity of order r, then

$$p = \Phi_n(a)$$

$$= \frac{\Phi_r(a^{p^i})}{\Phi_r(a^{p^{i-1}})}$$

$$= \frac{\prod_{j=1}^{\phi(r)}(c^p - \zeta_j)}{\prod_{j=1}^{\phi(r)}(c - \zeta_j)} \quad \text{with } c = a^{p^{i-1}}$$

$$> \left(\frac{b^p - 1}{b + 1} \right)^{\phi(r)}$$

$$\geq c^{p-2}.$$

Hence $p = 3$, $c = a = 2$ and $i = 1$. Thus either $n = 3$ or $n = 6$. For $n = 3$ and $a = 2$ we have $7 | a^n - 1$ but $7 \nmid a^k - 1$ for $k < n$. Thus $n = 6$ and $a = 2$, which is case (d) in Theorem 13.8, is the only exception.   □

13.5 Quadratic Residues

An integer a relative prime to n is said to be a *quadratic residue* modulo n if $x^2 \equiv a$ mod n has a solution. We denote the set of all quadratic residues by $QR(n)$. The quadratic residue problem is to decide whether a given number lies in $QR(n)$.

Definition 13.2 (Legendre symbol) Let p be an odd prime. The *Legendre symbol* $\left(\frac{a}{p} \right)$ is defined by:

$$\left(\frac{a}{p} \right) = \begin{cases} 0 & \text{if } a \equiv 0 \mod p \\ 1 & \text{if } a \text{ is a quadratic residue modulo } p \\ -1 & \text{otherwise.} \end{cases}$$

The Legendre symbol can be efficient evaluated.

Lemma 13.10 (Euler's criterion) *Let p be an odd prime then*

$$\left(\frac{a}{p} \right) \equiv a^{(p-1)/2} \mod p.$$

Proof By Fermat's little theorem $a^{p-1} \equiv 1 \mod p$ if $a \not\equiv 0 \mod p$.

Hence if $a \equiv x^2 \mod p$ then $a^{(p-1)/2} \equiv x^{p-1} \equiv 1 \mod p$. Since the equation $a^{(p-1)/2} \equiv 1 \mod p$ has at most $(p-1)/2$ solutions and we have proved that the $(p-1)/2$ quadratic residues are solutions of $a^{(p-1)/2} \equiv 1 \mod p$, we know that the quadratic non-residues must satisfy $a^{(p-1)/2} \equiv -1 \mod p$.   □

One simple consequence of Euler's criterion is:

Corollary 13.11

$$\left(\frac{ab}{p} \right) = \left(\frac{a}{p} \right) \left(\frac{b}{p} \right).$$

Proof By Euler's criterion

$$\left(\frac{ab}{p}\right) \equiv (ab)^{(p-1)/2} \mod p$$

$$\equiv a^{(p-1)/2}b^{(p-1)/2} \mod p$$

$$\equiv \left(\frac{a}{p}\right)\left(\frac{b}{p}\right).$$

$\square$

Lemma 13.10 already describes a way to solve the quadratic residue problem modulo primes, but one can do better:

Theorem 13.12 (quadratic reciprocity law) *If p and q are odd primes then $\left(\frac{p}{q}\right) = -\left(\frac{q}{p}\right)$ if $p \equiv q \equiv 3 \mod 4$ and $\left(\frac{p}{q}\right) = \left(\frac{q}{p}\right)$ otherwise.*
In addition, $\left(\frac{2}{p}\right) = 1$ if and only if $p \equiv \pm 1 \mod 8$ and $\left(\frac{-1}{p}\right) = 1$ if and only if $p \equiv 1 \mod 4$.

Proof The proof of this classical theorem can be found in any number theory book, see for example [122] Theorems 99, 95 and 82. $\square$

Theorem 13.12 allows a Euclidean algorithm style computation of the Legendre symbol.
We generalize the Legendre symbol by:

Definition 13.3 (Jacobi symbol) Let $b \geq 3$ be odd and let $b = p_1^{e_1} \cdots p_k^{e_k}$ be the factorization of b. The *Jacobi symbol* $\left(\frac{a}{b}\right)$ is defined by:

$$\left(\frac{a}{b}\right) = \left(\frac{a}{p_1}\right)^{e_1} \cdots \left(\frac{a}{p_k}\right)^{e_k}.$$

Algorithm 13.1 computes the Jacobi symbol in an efficient way (see also Exercise 17.25).
Next we study the quadratic residue problem modulo a composite number. We deal only with the case of distinct prime factors which is relevant for applications in cryptography (for prime powers, see [122] Sect. 8.3).

Theorem 13.13 *Let $n = p_1 \cdots p_k$ be the product of k distinct odd primes.*
If $\gcd(a, n) = 1$, then the equation $x^2 \equiv a \mod n$ has either 0 or 2^k solutions.

Proof Assume that $x^2 \equiv a \mod n$ has a solution. As $\gcd(a, n) = 1$, for each prime factor p_i the congruence $x^2 \equiv a \mod p_i$ has exactly two solutions $x_{1/2}^{(i)}$. By the Chinese remainder theorem the system of linear congruences $x \equiv x_{j_i}^{(i)} \mod p_i$ has exactly one solution for every $j \in 1, 2^k$. Each such x is a solution of $x^2 \equiv a \mod n$ and on the other hand each solution of $x^2 \equiv a \mod n$ must be of this form. $\square$

Algorithm 13.1 Evaluating the Jacobi symbol

Require: a, b are integers, $a < b$, b odd
Ensure: return $\left(\frac{a}{b}\right)$
 1: **if** $a = 1$ **then**
 2:    **return** 1
 3: **end if**
 4: **if** a even **then**
 5:    Write $a = 2^k a'$ with a' odd.
 6:    **if** k odd and $b \equiv \pm 3 \mod 8$ **then**
 7:       **return** $-\left(\frac{a'}{b}\right)$ {recursive call}
 8:    **else**
 9:       **return** $\left(\frac{a'}{b}\right)$ {recursive call}
10:    **end if**
11: **else**
12:    $b' = b \mod a$
13:    **if** $b' = 0$ **then**
14:       **return** 0
15:    **end if**
16:    **if** $a \equiv b \equiv 3 \mod 4$ **then**
17:       **return** $-\left(\frac{b'}{a}\right)$ {recursive call}
18:    **else**
19:       **return** $\left(\frac{b'}{a}\right)$ {recursive call}
20:    **end if**
21: **end if**

Let $n = pq$ be the product of two distinct primes. The Jacobi symbol gives at least partial information as to whether a number a is a quadratic residue. If $\left(\frac{a}{n}\right) = -1$ then either a is a quadratic non-residue modulo p or modulo q, and in both cases a cannot be a quadratic residue modulo n.

If $\left(\frac{a}{n}\right) = 1$ there are two cases: either $\left(\frac{a}{p}\right) = \left(\frac{a}{q}\right) = 1$ and hence a is a quadratic residue modulo n or $\left(\frac{a}{p}\right) = \left(\frac{a}{q}\right) = -1$ and a is a quadratic non-residue modulo n. Both cases are, with $\frac{(p-1)(q-1)}{4}$ residues, equally likely.

If the factorization $n = pq$ is known it is easy to decide between the cases. At the moment no other method than factoring is known to decide whether or not a with $\left(\frac{a}{n}\right) = 1$ is a quadratic residue. Every method which does this would also determine $\left(\frac{a}{p}\right)$ and $\left(\frac{a}{q}\right)$ and hence leak at least some information about the prime factors.

13.6 Lattice Reduction

Definition 13.4 Let $b_1, \ldots, b_k$ be a set of linearly independent vectors in $\mathbb{R}^n$. Then $L = \mathbb{Z}b_1 + \cdots + \mathbb{Z}b_k$ is a k-dimensional *lattice* in $\mathbb{R}^n$. The vectors $b_1, \ldots, b_k$ form a basis of the lattice L.

In many applications we are interested in a basis consisting of short vectors. The Lenstra, Lenstra, Lovász basis reduction (LLL-reduction, for short) [168] gives us a (partial) solution of the short vector problem. It essentially transfers the Gram-Schmidt orthogonalization to lattices.

Let us recall the Gram-Schmidt orthogonalization algorithm.

Definition 13.5 Let $(\cdot, \cdot)$ be a bilinear form of $\mathbb{R}^n$ and $f_1, \ldots, f_n$ be a basis of $\mathbb{R}^n$. The *Gram-Schmidt* orthogonal basis $f_1^*, \ldots, f_n^*$ consists of the vectors

$$f_i^* = f_i - \sum_{j=1}^{i-1} \mu_{i,j} f_j^*$$

where $\mu_{i,j} = \frac{(f_i, f_j)}{(f_j^*, f_j^*)}$.

The core idea of LLL-reduction is to calculate the Gram-Schmidt orthogonalization and round the coefficients $\mu_{i,j}$ to stay inside the lattice (see Algorithm 13.2).

Algorithm 13.2 LLL basis reduction

Require: $f_1, \ldots, f_n$ is the basis of a lattice $L \subset \mathbb{R}^n$
1:   $i \leftarrow 2$
2:   **while** $i < n$ **do**
3:      Compute the Gram Schmidt orthogonalization {Actually this is just an update step which can use data from the previous computation}
4:      $f_i = f_i - \sum_{j=1}^{k} \lfloor \mu_{i,j} \rceil f_j$
5:      **if** $\|f_{i-1}^*\|^2 > 2\|f_i^*\|^2$ **then**
6:         Swap f_{i-1} and f_i
7:         $i \leftarrow i - 1$ unless $i = 2$
8:      **else**
9:         $i \leftarrow i + 1$
10:    **end if**
11: **end while**

Theorem 13.14 *Algorithm* 13.2 *uses* $O(n^4 \log n)$ *arithmetic operations. It computes a basis* $f_1, \ldots, f_n$ *of the lattice* L *with the following properties:*

(a) $\|f_i^*\|^2 \leq 2\|f_{i-1}^*\|^2$ *for* $2 \leq i \leq n$.
(b) $\|f_1\| \leq 2^{(n-1)/2}\|f\|$ *for each* $f \in L \backslash \{0\}$.

Proof The proof of the LLL-basis reduction can be found in most computer algebra textbooks. See, for example, Chap. 16 of [100].   □

LLL-reduction has several applications in cryptography. In the context of the Blum-Blum-Shub generator, Coppersmith's method for finding small roots of modular equations is of special interest. The problem is: Given a polynomial f and an integer of unknown factorization, find all x with $f(x) \equiv 0 \mod$ and $|x| \le x_0$ for a bound x_0 as large as possible. Coppersmith published in 1997 a nice LLL-based method (Algorithm 13.3) for solving this problem (see [60, 131]). His method is relevant to the bit security of RSA-like ciphers.

Algorithm 13.3 Coppersmith's method (univariate case)

---

(1) Given a polynomial f of degree d and a modulus n of unknown factorization which is a multiple of b.

(2) Choose δ with $b \ge N^\beta$ and $\varepsilon \le \beta$. Let $m = \lceil \frac{\beta^2}{d\varepsilon} \rceil$ and $t = \lfloor \delta m(\frac{1}{\beta} - 1) \rfloor$.

(3) Compute the polynomials

$$g_{i,j}(x) = x^j N^i f^{m-i}(x) \quad \text{for } i = 0, \ldots, m-1, \ j = 0, \ldots, d-1$$

$$h_i(x) = x^i f^m(x) \quad \text{for } i = 0, \ldots, t-1$$

(4) Let $X = \lceil N^{\beta^2/d - \varepsilon} \rceil$. Construct a lattice with basis consisting of the coefficients of $g_{i,j}(Xx)$ and $h_i(xX)$.

(5) Find a short vector in this lattice by using the LLL method. You have found a polynomial $\hat{f}$ which is a linear combination of the polynomial $g_{i,j}(Xx)$ and $h_i(xX)$. Hence for $f(x_0) \equiv 0 \mod n$ implies $\hat{f}(x_0) \equiv 0 \mod b$.

(6) As the norm of $\hat{f}$ is small, $x_0 \le X$ and $\hat{f}(x_0) \equiv 0 \mod n^m$ implies $\hat{f}(x_0) = 0$. Compute all integer solutions of $\hat{f}$ using standard methods. For every root x_0 check if $f(x_0) \equiv 0 \mod b$.

(7) You have now found all solutions of $f(x) \equiv 0 \mod b$ with $|x| \le X$.

We do not go deeper into the mathematics of Coppersmith's method. Instead we mention without proof:

Theorem 13.15 *Algorithm 13.3 works correctly and needs at most $O(\varepsilon^{-7}\delta^5 \log^2 N)$ arithmetic operations.*

Proof See [60] or [131]. □

Note that the essential point of Coppersmith's method is that it works for integers with unknown factorization. If the factorization of n is known, there are better methods for solving $f(x) \equiv 0 \mod n$ which use the Chinese remainder theorem and Hensel lifting.

Chapter 14
Finite Fields

Finite fields are the basis for the mathematics of shift registers. We will especially need the constructive parts of the theory of finite fields. In the next subsections we summarize the most important facts. For further references see, for example, [140].

14.1 Basic Properties

- The characteristic of a (finite) field is a prime.
- For every prime power q there exists up to isomorphism only one finite field. We denote the finite field of size q by $\mathbb{F}_q$.
- The multiplicative group $\mathbb{F}_q^\times$ of $\mathbb{F}_q$ is cyclic.
- $\mathbb{F}_{p^e}$ is a subfield of $\mathbb{F}_{p^d}$ if and only if $e \mid d$.
- The Galois group $GL(\mathbb{F}_{q^e}/\mathbb{F}_q)$ is cyclic and is generated by the *Frobenius automorphism* $x \mapsto x^q$.
- An element of $\mathbb{F}_q$ is *primitive* if it generates the multiplicative group. $\mathbb{F}_q$ contains $\varphi(q-1)$ primitive elements.

14.2 Irreducible Polynomials

Definition 14.1 A polynomial is called *monic* if its leading coefficient is 1.

We want to count the number of irreducible monic polynomials of degree n over $\mathbb{F}_q$.

Let us recall the *Möbius inversion*. Let A and B be functions from $\mathbb{N}^+$ into an additive group G, then

$$B(n) = \sum_{d \mid n} A(d)$$

A. Klein, *Stream Ciphers*, DOI 10.1007/978-1-4471-5079-4_14,
© Springer-Verlag London 2013

if and only if

$$A(n) = \sum_{d|b} \mu(d) B\left(\frac{n}{d}\right)$$

where μ is the *Möbius function* defined by

$$\mu(n) = \begin{cases} 1 & \text{if } n = 1, \\ (-1)^k & \text{if } n \text{ is the product of } k \text{ distinct primes}, \\ 0 & \text{otherwise.} \end{cases}$$

Theorem 14.1 *The number of irreducible monic polynomials of degree n over* $\mathbb{F}_q$ *is*

$$\frac{1}{n} \sum_{d|n} \mu(d) q^{n/d} = \frac{q^n}{n} + O\left(q^{n/2}/n\right).$$

Proof Let A be the number of elements ζ in $\mathbb{F}_{q^n}$ with $\mathbb{F}_q(\zeta) = \mathbb{F}_q$ and let $B(n) = |\mathbb{F}_{q^n}|$.

Then $B(n) = q^n = \sum_{d|n} A(d)$, since every element of $\mathbb{F}_{q^n}$ generates a subfield $\mathbb{F}_{q^d}$ of $\mathbb{F}_{q^n}$, with $d|n$.

By the Möbius inversion we have $A(n) = \sum_{d|n} \mu(d) q^{n/d}$. Since every irreducible monic polynomial of degree n has n roots, $A(n)/n$ is the number of irreducible monic polynomials of degree n.

The bounds $B(n) = q^n \geq A(n) = \sum_{d|n} \mu(d) q^{n/d} \geq q^n - \sum_{j=1}^{n/2} q^j$ give $A(n) = q^n + O(q^{n/2})$. □

Since the number of monic polynomials of degree n over $\mathbb{F}_q$ is q^n, the probability that a random monic polynomial of degree n is irreducible is about $1/n$, which is quite high. So repeatedly testing monic polynomials until we find an irreducible one is an attractive probabilistic algorithm. To make it effective we need a criterion which decides if a given polynomial is irreducible. Theorem 14.2 is such a criterion.

Theorem 14.2 *Let* $f \in \mathbb{F}_q[x]$ *be a monic polynomial of degree n. Then f is irreducible if and only if* $\gcd(x^{q^d} - x, f) = 1$ *for all proper divisors d of n.*

Proof Suppose f is not irreducible, then f has a zero ζ which lies in $\mathbb{F}_{q^d}$ for some proper divisor of n.

Since the multiplicative group of $\mathbb{F}_{q^d}$ is cyclic, $\zeta^{q^d} - \zeta = 0$ and hence $\gcd(x^{q^d} - x, f) \neq 1$.

On the other hand, $\gcd(x^{q^d} - x, f) \neq 1$ implies that f has a zero ζ which satisfies $\zeta^{q^d} - \zeta = 0$. Hence $\zeta \in F_{q^d}$, which is the splitting field of $x^{q^d} - x$ over $\mathbb{F}_q$. □

For a given polynomial f we can compute $x^{q^d} \mod f$ in only $O(\log q^d)$ steps by using the square-and-multiply algorithm, so Theorem 14.2 is indeed an efficient test for irreducibility.

14.3 Primitive Polynomials

We call a polynomial $f \in \mathbb{F}_q[x]$ of degree n *primitive* if its roots have order $q^n - 1$ (see Definition 2.5). Since the multiplicative group of $\mathbb{F}_{q^n}$ is cyclic, the number of primitive elements in $\mathbb{F}_{q^n}$ is $\phi(q^n - 1)$ and hence $\mathbb{F}_q[x]$ contains $\phi(q^n - 1)/n$ primitive polynomials of degree n.

An obvious way to generate a random primitive element (and, with that element, also a random primitive polynomial) is Algorithm 14.1.

Algorithm 14.1 Choosing a random primitive element of $\mathbb{F}_q$

1: Compute the factorization $q - 1 = p_1^{e_1} \ldots p_s^{e_s}$.
2: **repeat**
3:     Choose c random in $\mathbb{F}_q^\times$.
4: **until** $c^{(q-1)/p_i} \neq 1$ for $i \in \{1, \ldots, s\}$

The problematic part of Algorithm 14.1 is the factorization of $q - 1$, but there are specialized algorithms for factoring numbers of the form $p^k - 1$, so we can run it even for large q. At the time of writing $2^{827} - 1$ is the smallest Mersenne number which is not completely factored and $2^{1061} - 1$ is the smallest Mersenne number which is not a prime, but no factor is known.

Another influence on the performance of Algorithm 14.1 is the probability that a random element is primitive.

Note that it possible to find a sequence of prime numbers p_k such that the chance that a random element of $\mathbb{F}_{p_k}^\times$ is primitive approaches 0 for $k \to \infty$. One simply chooses $p_k \equiv 1 \mod k!$, which is possible due to Dirichlet's theorem. Then

$$\frac{\phi(p_k - 1)}{p_k - 1} \leq \prod_{\substack{p \leq k \\ p\,\text{prime}}} \left(1 - \frac{1}{p}\right) \to 0 \quad \text{for } k \to \infty.$$

However, for all practical sizes of q the probability that a random element is primitive is quite high. We recall the asymptotic behavior of the Eulerian φ-function (see Theorem 13.7) $\frac{m}{\varphi(m)} = O(\ln \ln(m))$.

Thus the chance that a random element of $\mathbb{F}_q^\times$ is primitive is $\Omega(1/\ln \ln q)$, which is for all reasonable sizes of q large enough.

At the moment no algorithm is known that finds a primitive element in polynomial time (see [50] for an overview of what has been done in that direction).

14.4 Trinomials

Linear feedback shift registers with sparse feedback polynomials can be implemented very efficiently (see Algorithm 2.10). So there is special interest in irreducible and primitive polynomials of low weight. In this section we collect some facts about trinomials.

Theorem 14.3 *The trinomial $x^n + x^k + 1$ is irreducible over $\mathbb{F}_2$ if and only if $x^n + x^{n-k} + 1$ is irreducible over $\mathbb{F}_2$.*

Proof If $f(x) = x^n + x^k + 1$ then $g(x) = x^n + x^{n-k} + 1 = x^n f(1/x)$. So every factorization of f gives a factorization of g.   □

Richard G. Swan [261] proved a very strong result on the number of irreducible factors of a trinomial.

Theorem 14.4 *Let f be a monic polynomial with integral coefficients over $\mathbb{Q}_p$. Assume that the reduction $\bar{f}$ of f over $\mathbb{F}_p$ has no multiple roots. Let r be the number of irreducible factors of $\bar{f}$. Then $r \equiv n \bmod 2$ if and only if the discriminant $D(f)$ is a square in $\mathbb{Q}_p$.*

In particular, for trinomials over $\mathbb{F}_2$ we get the following. Let $n > k > 0$ and let exactly one of n and k be odd. Then $x^n + x^k + 1 \in \mathbb{F}_2[x]$ has an even number of irreducible factors if and only if one of the following cases hold:

(a) *n is even, k is odd and $n \neq 2k$ and $nk/2 \equiv 0$ or $1 \bmod 4$.*
(b) *$n \equiv \pm 3 \bmod 8$, k is even, $k \nmid 2n$.*
(c) *$n \equiv \pm 1 \bmod 8$, k is even, $k \mid 2n$.*

Proof See [261] Theorem 1 and Corollary 5.   □

Two corollaries of Swan's Theorem are:

Corollary 14.5 *If n is a multiple of 8, then every trinomial of degree n is reducible over $\mathbb{F}_2$.*

Proof First note that for even n and k, $x^n + x^k + 1 = (x^{n/2} + x^{k/2} + 1)^2$ is not irreducible. If n is a multiple of 8 and k is odd then $2k \neq n$ and $nk/2 \equiv 0 \bmod 4$, i.e. by Theorem 14.4 $x^n + x^k + 1$ has an even number of factors. In particular, it is reducible.   □

Corollary 14.6 *If $n \equiv 3 \bmod 8$ or $n \equiv 5 \bmod 8$ and $x^n + x^{2d} + 1 \in \mathbb{F}_2[x]$ is irreducible, then d is divisor of n.*

Proof By Theorem 14.4 $x^n + x^{2d} + 1$ contains an even number of irreducible factors if $d \nmid n$.   □

There are also very efficient algorithms for testing a trinomial for irreducibility (see [29, 30]). This allows us to generate irreducible trinomials of high degree.

We close this section with a table of primitive and irreducible polynomials of low weight (Table 14.1).

14.5 The Algebraic Normal Form

Consider the polynomial ring $R = \mathbb{F}[x_1, \ldots, x_n]$. With a polynomial $p \in R$ we associate the polynomial mapping $x \mapsto p(x)$. For infinite fields this is a one-to-one correspondence and there exist functions which are not polynomial. For finite fields the situation is different:

Theorem 14.7 *For any function* $f : \mathbb{F}_q^n \to \mathbb{F}_q$ *there exists a unique polynomial* p *with* $\deg_i(p) < q$ *for* $1 \le i \le n$ *and* $f(x) = p(x)$ *for all* $x \in \mathbb{F}_q^n$.

Proof Let $f : \mathbb{F}_q^n \to \mathbb{F}_q$. By Lagrange interpolation we have

$$f(x) = \sum_{\alpha \in \mathbb{F}_q^n} f(\alpha) \prod_{j=1}^n \frac{(x_i^q - x_i)}{x_i - \alpha_i}.$$

Hence f can be written as a polynomial p with $\deg_i(p) < q$ for $1 \le i \le n$.

To prove the uniqueness of p, note that there are q^{q^n} functions from $f : \mathbb{F}_q^n \to \mathbb{F}_q$ and the same number of polynomials with $\deg_i(p) < q$ for $1 \le i \le n$. $\qquad\square$

Definition 14.2 The polynomial p of Theorem 14.7 is called the *algebraic normal form* of f.

The algebraic normal form allows us to speak of the algebraic degree of a function over finite fields.

Table 14.1 Primitive and irreducible polynomials over $\mathbb{F}_2$ of low weight

Degree	Primitive polynomial	Irreducible polynomial
2	$x^2 + x + 1$	$x^2 + x + 1$
3	$x^3 + x + 1$	$x^3 + x + 1$
4	$x^4 + x + 1$	$x^4 + x + 1$
5	$x^5 + x^2 + 1$	$x^5 + x^2 + 1$
6	$x^6 + x + 1$	$x^6 + x + 1$
7	$x^7 + x + 1$	$x^7 + x + 1$
8	$x^8 + x^4 + x^3 + x^2 + 1$	$x^8 + x^4 + x^3 + x^2 + 1$
9	$x^9 + x^4 + 1$	$x^9 + x^4 + 1$
10	$x^{10} + x^3 + 1$	$x^{10} + x^3 + 1$
11	$x^{11} + x^2 + 1$	$x^{10} + x^3 + 1$
12	$x^{12} + x^6 + x^4 + x + 1$	$x^{12} + x^3 + 1$
13	$x^{13} + x^4 + x^3 + x + 1$	$x^{13} + x^4 + x^3 + x + 1$
14	$x^{14} + x^5 + x^3 + x + 1$	$x^{14} + x^5 + 1$
15	$x^{15} + x + 1$	$x^{15} + x + 1$
16	$x^{16} + x^5 + x^3 + x^2 + 1$	$x^{16} + x^5 + x^3 + x^2 + 1$
17	$x^{17} + x^3 + 1$	$x^{17} + x^3 + 1$
18	$x^{18} + x^7 + 1$	$x^{18} + x^7 + 1$
19	$x^{19} + x^5 + x^2 + x + 1$	$x^{19} + x^5 + x^2 + x + 1$
20	$x^{20} + x^3 + 1$	$x^{20} + x^3 + 1$
21	$x^{21} + x^2 + 1$	$x^{21} + x^2 + 1$
22	$x^{22} + x + 1$	$x^{22} + x + 1$
23	$x^{23} + x^5 + 1$	$x^{23} + x^5 + 1$
24	$x^{24} + x^4 + x^3 + x + 1$	$x^{24} + x^4 + x^3 + x + 1$
25	$x^{25} + x^3 + 1$	$x^{25} + x^3 + 1$
26	$x^{26} + x^6 + x^2 + x + 1$	$x^{26} + x^6 + x^2 + x + 1$
27	$x^{27} + x^5 + x^2 + x + 1$	$x^{27} + x^5 + x^2 + x + 1$
28	$x^{28} + x^3 + 1$	$x^{28} + x^3 + 1$
29	$x^{29} + x^2 + 1$	$x^{29} + x^2 + 1$
30	$x^{30} + x^6 + x^4 + x + 1$	$x^{30} + x + 1$
31	$x^{31} + x^3 + 1$	$x^{31} + x^3 + 1$
32	$x^{32} + x^7 + x^6 + x^2 + 1$	$x^{32} + x^7 + x^6 + x^2 + 1$
33	$x^{33} + x^{13} + 1$	$x^{33} + x^{13} + 1$
34	$x^{34} + x^8 + x^4 + x^3 + 1$	$x^{34} + x^7 + 1$
35	$x^{35} + x^2 + 1$	$x^{35} + x^2 + 1$

Chapter 15
Statistics

15.1 Measure Theory

Although most probabilities we meet in cryptography are simple discrete probabilities, it is convenient to describe the statistical background in the more general measure theoretic context. In this section we collect all the measure theoretic notions we need.

For a finite set Ω we can assign to every subset of Ω a probability, but for infinite sets this is no longer possible. So we assign a probability measure only to "nice" sets. The right definition of a nice set is given in the following.

Definition 15.1 $\mathcal{F} \subseteq \mathcal{P}(\Omega)$ is called σ-algebra if:

(a) $\emptyset \in \mathcal{F}$.
(b) If $A \in \mathcal{F}$ then $\bar{A} = \Omega \setminus A \in \mathcal{F}$.
(c) If $A_i \in \mathcal{F}$ for $i \in \mathbb{N}$ then $\bigcup_{i=0}^{\infty} A_i \in \mathcal{F}$.

The pair $(\Omega, \mathcal{F})$ is called a *measurable space*.

Definition 15.2 Let $(\Omega, \mathcal{F})$ be a measurable space, then $\mu : \mathcal{F} \to \mathbb{R}^{\geq 0} \cup \{\infty\}$ is a *measure* if:

(a) $\mu(\emptyset) = 0$.
(a) If $A_i \in \mathcal{F}$ for $i \in \mathbb{N}$ with $A_i \cap A_j = \emptyset$ for $i \neq j$ then $\mu(\bigcup_{i=0}^{\infty} A_i) = \sum_{j=0}^{\infty} \mu(A_i)$.

μ is a *probability measure* if $\mu(\Omega) = 1$.

From analysis we know the special case of the Borel σ-algebra $\mathcal{B}$ on $\mathbb{R}$, which is the smallest σ-algebra that contains all intervals. Once we have a measurable space and a measure, we can proceed as in analysis and define the measure integral, which we will denote by $\int_A f(\omega)d\mu$. The famous Lebesgue integral is just the special case with $(\mathbb{R}, \mathcal{B})$ as measurable space and the measure μ with $\mu([a, b]) = b - a$.

Let $(\Omega, \mathcal{F})$ and $(\Omega', \mathcal{F}')$ be two measurable spaces. Then a function $f : \Omega \to \Omega'$ is called *measurable* (with respect to $\mathcal{F}$ and $\mathcal{F}'$) if $f^{-1}(F') \in \mathcal{F}$ for all $F' \in \mathcal{F}'$.

A. Klein, *Stream Ciphers*, DOI 10.1007/978-1-4471-5079-4_15,
© Springer-Verlag London 2013

If $f : \Omega \to \mathbb{R}^+$ is measurable then $\mu'(A) = \int_A f(\omega)\mathrm{d}\mu$ defines a new measure μ'. We say μ' has *μ-density* f.

15.2 Simple Tests

15.2.1 The Variation Distance

In cryptography we often observe a random variable X which is either distributed by P or by Q and we must decide which of the two possible cases hold.

Definition 15.3 Let P and Q be two probability distributions on a common measurable space $(\Omega, \mathcal{F})$. A *distinguisher* is a measurable function $D : \Omega \to \{0, 1\}$. The probability of success is $|P(D = 1) - Q(D = 1)|$.

Of course we are interested in the best possible distinguisher. This leads us to the definition of the variation distance.

Definition 15.4 Let P and Q be two probability measures on $(\Omega, \mathcal{F})$. The *variation distance* between P and Q is

$$\|P - Q\| = \sup_{F \in \mathcal{F}} |P(F) - Q(F)|.$$

The variation distance is a useful and common measure for the distance between two probability distributions. Theorem 15.1 shows how to compute the variation distance for probability measures with densities.

Theorem 15.1 *Let μ be a measure on $(\Omega, \mathcal{F})$. If P has μ-density ϕ_1 and P has μ-density ϕ_2, then*

$$\|P - Q\| = \frac{1}{2} \int |\phi_1(\omega) - \phi_2(\omega)| \mathrm{d}\mu.$$

In particular, if $\Omega = \{\omega_1, \ldots, \omega_n\}$ is a finite set, then

$$\|P - Q\| = \frac{1}{2} \sum_{j=1}^{n} |P(\omega_i) - Q(\omega_i)|.$$

Proof Let $F = \{\omega \mid \phi_1(\omega) \geq \phi_2(\omega)\}$, then

$$|P(F) - Q(F)| = \int_F \phi_1(\omega) - \phi_2(\omega)\mathrm{d}\mu = \int_F |\phi_1(\omega) - \phi_2(\omega)| \mathrm{d}\mu$$

and similarly

$$\left|P(F) - Q(F)\right| = \left(1 - \int_{\bar{F}} \phi_1(\omega) \mathrm{d}\mu\right) - \left(1 - \int_{\bar{F}} \phi_2(\omega) \mathrm{d}\mu\right)$$
$$= \int_{\bar{F}} \left|\phi_1(\omega) - \phi_2(\omega)\right| \mathrm{d}\mu,$$

hence

$$\left|P(F) - Q(F)\right| = \frac{1}{2} \int \left|\phi_1(\omega) - \phi_2(\omega)\right| \mathrm{d}\mu.$$

Now let $F' \neq F$. Let $F_1 = F \backslash F'$ and $F_2 = F' \backslash F$, then

$$\left|P(F') - Q(F')\right|$$
$$= \left(P(F) - Q(F)\right) - \left(P(F_1) - Q(F_1)\right) + \left(Q(F_2) - F(F_2)\right)$$
$$\leq P(F) - Q(F),$$

i.e. $|P(F) - Q(F)| = \sup_{F' \in \mathcal{F}} |P(F') - Q(F')|$. $\qquad\square$

15.2.2 The Test Problem

In this section we will give a short introduction to statistical tests. The basic setting in statistics is that we have a measurable space $(\Omega, \mathcal{F})$ and a parameter space Θ. For each parameter $\theta \in \Theta$ we have probability measures P_θ on $(\Omega, \mathcal{F})$. A partition of Θ into two sets H_0 and H_1 defines a test problem. H_0 is called the *hypothesis* and H_1 is called the *alternative*.

Every measurable function $\phi : \Omega \to [0, 1]$ is a test. The value $\phi(x)$ is the probability that the test rejects the hypothesis H_0 given the observation x. If ϕ takes only the values 0 and 1 it is a *deterministic test*. Otherwise we speak of a *randomized test*. We do not use randomized tests in this book, but the theory does not become simpler if we restrict ourselves to deterministic tests.

The quality of a test is measured by:

Definition 15.5 Let $(\Omega, \mathcal{F}, (P_\theta)_{\theta \in \Theta}, H_0, H_1)$ be a test problem and ϕ be a test. The function $\beta_\phi : \Theta \to [0, 1]$ with

$$\beta_\phi(\theta) = \int \phi_x \mathrm{d}P_\theta$$

is the *quality function* of ϕ (given the test problem $(\Omega, \mathcal{F}, (P_\theta)_{\theta \in \Theta}, H_0, H_1)$).

For a test ϕ we call $\sup_{\theta \in H_0} \beta_\phi(\theta)$ the *error probability of the first kind*. ϕ is a test at level α if the error probability of the first kind is at most α. The set of all level α tests is denoted by Φ_α.

For $\theta \in H_1$ we call $1 - \beta_\phi(\theta)$ the *error probability of the second kind* of ϕ at point θ.

15.2.3 Optimal Tests

Once we have a notion of a general statistical test, it is natural to ask how we can find good tests.

Definition 15.6 For a set of tests Φ the test $\phi \in \Phi$ is *most powerful* if for every $\phi' \in \Phi$ and every $\theta \in H_1$ the inequality

$$\beta_\phi(\theta) \geq \beta_{\phi'}(\theta)$$

holds.

Note that it is not guaranteed that a most powerful test exists if H_1 contains more than one element. We are interested in a most powerful level α test.

For the cases that H_0 and H_1 are simple, i.e. if both contain only one element, the Neyman-Pearson Lemma characterizes these tests.

Theorem 15.2 (Neyman-Pearson Lemma) *Let P_0 and P_1 be two probability measures with densities f and g, respectively, with respect to some measure μ.*

Then a most powerful α test for $H_0 = \{f\}$ against $H_1 = \{g\}$ exists and every such test ϕ satisfies

$$\phi(\omega) = \begin{cases} 1 & g(\omega) > cf(\omega), \\ 0 & g(\omega) < cf(\omega), \end{cases} \tag{15.1}$$

where c is the smallest value for which $\int_{g(\omega)>cf(\omega)} f(\omega)\mathrm{d}\mu \leq \alpha$, and

$$\beta_\phi(f) = \alpha \tag{15.2}$$

if $c > 0$.

Conversely, every test that satisfies (15.1) and (15.2) is most powerful for the class Φ_α of level α tests.

Proof Let ϕ be a test that satisfies (15.1) and (15.2) and let ϕ' be any other level α test. Then $(\phi(\omega) - \phi'(\omega))(g(\omega) - cf(\omega)) \geq 0$ for all $\omega \in \Omega$. Hence

$$0 \leq \int (\phi(\omega) - \phi'(\omega))(g(\omega) - cf(\omega))\mathrm{d}\mu = \beta_\phi(g) - \beta_{\phi'}(g) + c\beta_{\phi'}(f) - c\beta_\phi(f). \tag{15.3}$$

Since $\beta_\phi(f) = \alpha$ (condition (15.2)) and $\beta_{\phi'} \leq \alpha$ since ϕ' is a level α test, we get

$$0 \leq \beta_\phi(g) - \beta_{\phi'}(g),$$

i.e. ϕ is a most powerful test in the class Φ_α.

If, on the other hand, ϕ' does not satisfy (15.1), then $(\phi(\omega) - \phi'(\omega))(g(\omega) - cf(\omega)) > 0$ on a set of measure > 0 and the inequality in Eq. (15.3) is strict. If $c > 0$ and ϕ' does not satisfy (15.1) then $c\beta_{\phi'}(f) - c\beta_\phi(f) > 0$ and hence $\beta_{\phi'}(g) < \beta_\phi(g)$.

So every most powerful test in Φ_α must satisfy (15.1) and (15.2). $\square$

Note that the optimal level α test is non-deterministic if $\int_{g(\omega)=cf(\omega)} f(\omega)\mathrm{d}\mu \neq 0$. So even if almost all tests that are used in practice are deterministic tests, we need non-deterministic tests to get a nice theory.

Normally we apply Theorem 15.2 in the following form:

Let P_0 and P_1 be two probability measures with densities f and g, respectively, with respect to some measure μ. We observe iid. random variables $X_1, \ldots, X_n$ and want to test the hypothesis $H_0 : X_i \sim f$ against the alternative $H_1 : X_i \sim g$.

Then the most powerful level α-test is to compute the likelihood ratio

$$L_n = \prod_{i=1}^n \frac{g(X_i)}{f(X_i)} \tag{15.4}$$

and reject H_0 in favor of H_1 if L_n exceeds a critical value c_α. This kind of test is also called a *likelihood quotient test*.

15.2.4 Bayesian Statistics

In statistics the usual assumption is that we have no *a priori* information on the unknown parameter. In most cases this is quite reasonable. The assumption: "Nature selects the mortality rate of a disease uniformly at random in the range between 40 % and 60 %" does not make sense. (Even more crazy is the assumption that a universe selects it at creation with probability 50 %, using a god for that purpose, which was made in [269].)

In cryptography we are in the fortunate situation of having *a priori* information on the unknown parameter. The unknown parameter is the key, or a part of it, and we know the key generation protocol (most times it selects the key uniformly at random).

In this case we can utilize Bayes' Theorem for our statistics. In the simple case of discrete probabilities we observe an event A and compute for each possible parameter θ the *a posteriori* probability of θ by

$$P(\theta|A) = \frac{P(\theta)P_\theta(A)}{P(A)}.$$

The *a posteriori* probability describes all the information we have about θ after the observation of A. Thus the best we can do is to select the θ which maximizes $P(\theta \mid A)$.

15.3 Sequential Tests

15.3.1 Introduction to Sequential Analysis

Classical statistics deals with the following problem: We observe n (identical and independently distributed) random variables $X_1, \ldots, X_n$. We have a hypothesis H_0 and an alternative H_1 and must decide from the observation $X_1, \ldots, X_n$ whether we accept H_0 or reject H_0 and favor H_1.

Sequential analysis extends this classical model in the following way: We have a potentially infinite sequence of random variables $X_1, X_2, \ldots$. At each time step t we examine the observed random variables $X_1, \ldots, X_t$ and must decide between the following possibilities:

- we accept the hypothesis H_0 and stop;
- we reject the hypothesis H_0 and stop; or
- we decide that we have not enough data and wait until we observe the next random variable X_{t+1}.

Classical statistics is just a special case of sequential statistics in which we always choose the third alternative if $t < n$ and choose one of the first alternatives for $t = n$.

The idea of sequential statistics fits well with stream ciphers. A stream cipher generates a potentially infinite key stream and after each observed bit the attacker can decide if he has seen enough data or if he has to wait for more data.

Sequential analysis was developed in the 1940s by Wald. His book [273] is still a good introduction. For a more recent introduction to sequential analysis I recommend [249]. The remaining part of this section will introduce the parts of sequential analysis we need.

15.3.2 Martingales

The theoretical framework needed to formulate sequential tests uses martingales. Martingales are a standard topic in probability theory that can be found in many books (in the following I will always refer to the nice book of P. Billingsley [27]).

We start with the definition of conditional expectation.

Definition 15.7 Let X be a random variable on $(\Omega, \mathcal{F}, P)$ and let $\mathcal{G} \subseteq \mathcal{F}$ be a σ-field. Then $E[X \| \mathcal{G}]$ is called the *conditional expected value* of X given $\mathcal{G}$ if:

1. $E[X \| \mathcal{G}]$ is $\mathcal{G}$ measurable.
2. $E[X \| \mathcal{G}]$ satisfies

$$\int_G E[X \| \mathcal{G}] dP = \int_G X dP$$

for all $G \in \mathcal{G}$.

It is easy to see that the conditional expectation is unique up to values on a set of measure 0, but it is difficult to prove that the conditional expectation always exists (see the Radon-Nikodym Theorem, Sect. 32 in [27]).

Example 15.1 Suppose $(\Omega, \mathcal{F}, P)$ is a discrete probability space with $P(\omega_i) = p_i$.
Choose $\mathcal{G} = \{\emptyset, A, \bar{A}, \Omega\}$. Then $E[X\|\mathcal{G}]$ is constant on A. Let $E[X\|\mathcal{G}](\omega) = E[X\|A]$ for $\omega \in A$.
The equation

$$\int_A E[X\|\mathcal{G}]dP = \int_A XdP$$

simplifies to

$$\sum_{\omega \in A} p_i X(\omega_i) = P(A)E[X\|A]$$

and hence

$$E[X\|A] = \frac{\sum_{\omega \in A} p_i X(\omega_i)}{P(A)}.$$

Thus in the case of discrete probability spaces the complicated Definition 15.7 simplifies to the notion of conditional expectation we know from school.

Now we can introduce martingales.

Definition 15.8 Let $(\Omega, \mathcal{F}, P)$ be a probability space. A *filtration* of σ-algebras is a sequence $\mathcal{F}_1, \mathcal{F}_2, \ldots$ of σ-algebras with $\mathcal{F}_i \subseteq \mathcal{F}_{i+1}$.
A *martingale* adapted to the filtration is a sequence $X_1, X_n, \ldots$ of random variables with:

1. X_i is $\mathcal{F}_i$ measurable.
2. $E[|X_i|] < \infty$.
3. With probability 1

$$E[X_{i+1}\|\mathcal{F}_i] = X_i. \tag{15.5}$$

Sometimes the filtration $\mathcal{F}_1, \mathcal{F}_2, \ldots$ is not important. We say that $X_1, X_n, \ldots$ is a martingale if it is a martingale with respect to some filtration $\mathcal{F}_1, \mathcal{F}_2, \ldots$. In this case the filtration $\mathcal{F}_n = \sigma(X_1, \ldots, X_n)$ always works.
If we replace Eq. (15.5) in the definition of a martingale by the inequality

$$E[X_{i+1}\|\mathcal{F}_i] \geq X_i \tag{15.6}$$

we get the definition of a *submartingale* and if we replace it by the opposite inequality

$$E[X_{i+1}\|\mathcal{F}_i] \leq X_i \tag{15.7}$$

we get a *supermartingale*.
An important special case of a submartingale is the following example.

Example 15.2 Let $\Delta_1, \Delta_2, \ldots$ be a sequence of independent random variables with $E[\Delta_n] \geq 0$ and $E[|\Delta_n|] < \infty$ for all $n \in \mathbb{N}$. Let $X_n = \sum_{i=1}^{n} \Delta_i$. Then

$$E[X_{n+1}\|X_1, \ldots, X_n] = E\left[\sum_{i=1}^{n+1} \Delta_i \,\Big\|\, \Delta_1, \ldots, \Delta_n\right]$$

$$= \sum_{i=1}^{n} \Delta_i + E[\Delta_{n+1}]$$

$$\geq X_n$$

and hence $X_1, X_2, \ldots$ is a submartingale.

In our applications we always have submartingales of this type.

The next important concept is the stopping time.

Definition 15.9 Let τ be a random variable which takes values in $\mathbb{N} \cup \{\infty\}$. It is a *stopping time* with respect to the filtration $\mathcal{F}_1 \subseteq \mathcal{F}_2 \subseteq \cdots$ if $\{\tau = n\} \in \mathcal{F}_n$ for all $n \in \mathbb{N}$.

For a stopping time τ we define the σ-field $\mathcal{F}_\tau$ by

$$\mathcal{F}_\tau = \left\{F \in \mathcal{F} \,|\, F \cap \{\tau = n\} \in \mathcal{F}_n \text{ for all } n \in \mathbb{N}\right\}$$

$$= \left\{F \in \mathcal{F} \,|\, F \cap \{\tau \leq n\} \in \mathcal{F}_n \text{ for all } n \in \mathbb{N}\right\}. \tag{15.8}$$

Supermartingales describe typical casino games. Whatever the player does, the expected value is negative. As many players find out the hard way, this does not change if you play several games and follow some stopping strategy. The optimal sampling theorem is a precise formulation of this observation.

Theorem 15.3 (Optimal sampling theorem) *Let* $X_1, \ldots, X_n$ *be a (sub)martingale with respect to* $\mathcal{F}_1, \ldots, \mathcal{F}_n$ *and let* τ_1, τ_2 *be stopping times which satisfy* $1 \leq \tau_1 \leq \tau_2 \leq n$, *then* X_{τ_1}, X_{τ_2} *is a (sub)martingale with respect to* $\mathcal{F}_{\tau_1}, \mathcal{F}_{\tau_2}$.

Proof For every $n \in \mathbb{N}$ we have $\{\tau_2 \leq n\} \subseteq \{\tau_1 \leq n\}$, since $\tau_1 \leq \tau_2$. Hence for every $F \in \mathcal{F}_{\tau_1}$ we have $F \cap \{\tau_2 \leq n\} = (F \cap \{\tau_1 \leq n\}) \cap \{\tau_2 \leq n\} \in \mathcal{F}_n$, i.e. $F \in \mathcal{F}_{\tau_2}$.

Assume that $X_1, \ldots, X_n$ is a submartingale with respect to $\mathcal{F}_1, \ldots, \mathcal{F}_n$.

Since $|X_{\tau_i}| \leq \sum_{j=1}^{n} |X_j|$ and $\int \sum_{j=1}^{n} |X_j| < \infty$, X_{τ_i} is integrable. We must prove $E[X_{\tau_2}|\mathcal{F}_1] \geq X_{\tau_1}$ or equivalently $\int_A X_{\tau_2} - X_{\tau_1} \geq 0$ for all $A \in \mathcal{F}_{\tau_1}$.

$F \in \mathcal{F}_1$ implies $F \cap \{\tau_1 < k \leq \tau_2\} = [F \cap \{\tau_1 \leq k - 1\}] \cap \{\tau_2 \leq k - 1\}^C$ lies in $\mathcal{F}_{k-1}$. Hence

$$\int_F X_{\tau_2} - X_{\tau_1} dp = \int_F \sum_{k=2}^{n} I_{\{\tau_1 < k \leq \tau_2\}} X_k - X_{k-1} \quad \text{telescope sum}$$

$$= \sum_{k=2}^{n} \int_{F \cap \{\tau_1 < k \leq \tau_2\}} X_k - X_{k-1}$$

$$\geq 0 \quad \text{by the submartingale property.}$$

If $X_1, \ldots, X_n$ is a martingale with respect to $\mathcal{F}_1, \ldots, \mathcal{F}_n$ we apply the same argument to the martingale $-X_1, \ldots, -X_n$ and get $E[X_{\tau_2} | \mathcal{F}_1] = X_{\tau_1}$. □

For two stopping times τ_1 and τ_2 we let $\tau_1 \wedge \tau_2 = \min \tau_1, \tau_2$. A short calculation proves that $\tau_1 \wedge \tau_2$ is again a stooping time. An important special case of the optimal sampling theorem is:

Corollary 15.1 *Let* $X_1, X_2, \ldots$ *be a (sub)martingale and let* τ *be a stopping time. Then* $X_{\tau \wedge 1}, X_{\tau \wedge 2}, \ldots$ *is also a (sub)martingale.*

Proof $\tau_1 = \tau \wedge (n - 1)$ and $\tau_2 = \tau \wedge n$ satisfy the assumptions of Theorem 15.3. □

Finally we mention without proof an important limit theorem for martingales.

Theorem 15.4 (Doob's martingale convergence theorem) *Let* $X_1, X_2, \ldots$ *be a submartingale. If* $K = \sup_n E[|X_n|] < \infty$ *then* $X_n \to X$ *with probability 1, where* X *is a random variable satisfying* $E[|X|] < K$.

Proof See Theorem 35.5 in [27]. □

15.3.3 Wald's Sequential Likelihood Ratio Test

Remember that for the (non-sequential) test problem with simple hypotheses against a simple alternative, the likelihood quotient test (Neyman-Pearson test) described in Sect. 15.2.3 is optimal.

With classical tests we can test to minimize the type-I error, but we have no control over the size of the type-II error. Wald's sequential likelihood ratio test (Algorithm 15.1) lets us control both error types.

We will prove that Algorithm 15.1 controls both error types in several steps.

Lemma 15.1 *If* P *is absolutely continuous with respect to* Q *then* L_n, $n \geq 1$, *is a martingale with respect to the filtration* $\mathcal{F}_n = \sigma(X_1, \ldots, X_n)$ *under* H_0.

Algorithm 15.1 Wald's sequential test

- Let $X_1, \ldots$ be iid. Let H_0 be the hypothesis "X_i has density f" and H_1 be the alternative "X_i has density g".
- Let the type-I error probability be α and the type-II error probability be β.
- Set $A = \frac{\beta}{1-\alpha}$ and $B = \frac{1-\beta}{\alpha}$.
- Let τ be the stopping time

$$\tau = \inf_n \{n \geq 1 : L_n \leq A \text{ or } L_n \geq B\}$$

where $L_n = \prod_{i=1}^n \frac{g(X_i)}{f(X_i)}$ is the likelihood quotient.
- Accept H_0 if $L_\tau \leq A$ and reject H_0 in favor of H_1 if $L_\tau \geq B$.

Proof

$$E_0[L_{n+1}\|\mathcal{F}_n] = L_n E_0\left[\frac{g(X_{n+1})}{f(X_{n+1})}\,\bigg\|\,\mathcal{F}_n\right]$$

$$= L_n \int \frac{g(x)}{f(x)} f(x)d\mu$$

$$= L_n \int g(x)d\mu = L_n.$$

Implicitly we have shown $E_0[L_n] = E_0[L_0] = E_0[1] = 1.$ $\qquad\square$

Obviously $L_n \geq 0$ and hence $E_0[|L_n|] = E_0[L_n] = 1$ and by Doob's Theorem (Theorem 15.4) L_n converges almost surely under E_0. We will now determine the limit. For that we need the Kullback-Leibner information as a tool. This is a generalization of the well-known Shannon entropy for the case of probability measures with densities.

Definition 15.10 For two probability measures Q and P the *Kullback-Leibner information* $K(Q, P)$ is

$$K(Q, P) = \begin{cases} \int [\ln \frac{dQ}{dP}] \frac{dQ}{dP} dP & \text{if } Q \ll P, \\ \infty & \text{otherwise.} \end{cases} \tag{15.9}$$

Theorem 15.5 $K(Q, P) \geq 0$ *with equality if and only if* $Q = P$.

Proof $\ln(t) \leq t - 1$ with equality for $t = 1$. Thus

$$K(Q, P) \geq \int \frac{dQ}{dP} - 1 dP = \int dQ - \int dP = 0.$$

Since $Q \neq P$ on a set of positive P measure, $K(Q, P) > 0$ for $Q \neq P$. $\qquad\square$

Lemma 15.2 $\lim_{n\to\infty} L_n = 0$ *with probability* 1 *under* H_0.

Proof If P_0 is not absolutely continuous with respect to P_1 then there exists a set A with $P_0(A) > 0$ and $P_1(A) = 0$. In this case, with probability 1 there are infinitely many i with $X_i \in A$. Hence with probability one, $g(X_i) = 0$ infinitely often, i.e. $\lim_{n\to\infty} L_n = 0$ with probability 1.

Now assume that $P_0 \ll P_1$. Without loss of generality we can also assume $P_1 \ll \mu$, i.e. $g(\omega) \neq 0$.

Now assume $g(X_i) \neq 0$ for all i and take the logarithm

$$\lim_{n\to\infty} \sum_{i=1}^{n} \ln \frac{g(X_i)}{f(X_i)} = -\infty.$$

By the strong law of large numbers, under H_0 we get

$$\lim_{n\to\infty} \frac{1}{n} \sum_{i=1}^{n} \ln \frac{g(X_i)}{f(X_i)} = E_0 \left[\ln \frac{g(X_i)}{f(X_i)} \right]$$

$$= \int \left[\ln \frac{g(\omega)}{f(\omega)} \right] f(\omega) d\mu$$

$$= -\int \left[\ln \frac{f(\omega)}{g(\omega)} \right] \frac{f(\omega)}{g(\omega)} g(\omega) d\mu$$

$$= -\int \left[\ln \frac{dP_0}{dP_1} \right] \frac{dP_0}{dP_1} dP_1$$

$$= -K(P_0, P_1) < 0$$

and hence $\lim_{n\to\infty} \sum_{i=1}^{n} \ln \frac{g(X_i)}{f(X_i)} = \lim_{n\to\infty} -nK(P_0, P_1) = -\infty$. □

Interchanging the role of P_0 and P_1 we get

$$\lim_{n\to\infty} L_0 = \infty \quad \text{with probability 1 under } H_1.$$

This proves that for every $A < 1 < B$ the stopping time

$$\tau = \inf\{n \geq 1 | L_n \leq A \text{ or } L_n \geq B\}$$

is almost surely finite under H_0 and under H_1.

By the optimal sampling theorem (Theorem 15.3), the sequence $L_{\tau \wedge n}$, $n \geq 1$, is a martingale under H_0. In particular, $E_0[L_{\tau \wedge n}] = E_0[L_1] = 1$.

Furthermore, for each $F \in \mathcal{F}_\tau$

$$P_1\big(F \cap \{\tau < \infty\}\big) = \sum_{n=1}^{\infty} P_1\big(F \cap \{\tau = n\}\big).$$

We may write $F \cap \{\tau < n\} = \{\omega : (X_1, \ldots, X_n)(\omega) \in G_n\}$ for an appropriate Borel subset of $\mathbb{R}^n$. Since the observations $X_1, \ldots, X_n$ are independent

$$P_1(F \cap \{\tau = n\}) = \int_{G_n} g(x_1) \cdots g(x_n) \mu d(x_1) \cdots \mu d(x_n)$$

$$= \int_{G_n} \prod_{i=1}^{n} \frac{g(x_i)}{f(x_i)} f(x_1) \cdots f(x_n) \mu d(x_1) \cdots \mu d(x_n)$$

$$= \int_{F \cap \{\tau = n\}} L_\tau d P_0.$$

Hence

$$P_1(F \cap \{\tau < \infty\}) = \int_{F \cap \{\tau < \infty\}} L_\tau d P_0. \qquad (15.10)$$

We know already that τ is almost surely finite under H_0 and under H_1, so setting $F = \Omega$ in (15.10) we get

$$1 = \int L_\tau d P_0.$$

We are now in the position that we can analyze the test. We must determine A and B in such a way that the error probability of type-I is $\alpha = P_0(L_\tau \geq B)$ and the error probability of type-II is $\beta = P_1(L_\tau \leq A)$.

Then

$$\beta = P_1(L_\tau \leq A)$$

$$= \int_{L_\tau \leq A} L_\tau d P_0 \quad \text{by (15.10)}$$

$$\leq A P_0(L_\tau \leq A) = A(1 - \alpha).$$

Interchanging the role of P_0 and P_1 we get $\alpha \leq B^{-1}(1 - \beta)$.

Using the approximations $\beta \approx A(1 - \alpha)$ and $\alpha \approx B^{-1}(1 - \beta)$ we get the threshold values $A = \frac{\beta}{1-\alpha}$ and $B = \frac{1-\beta}{\alpha}$ given in Algorithm 15.1.

15.3.4 Brownian Motion

In classical statistics the central limit theorem is the justification that we can replace the often unknown or unwieldy exact distribution by the normal distribution. We would like to have something similar for sequential analysis. The appropriate tool is Brownian motion.

Definition 15.11 (Brownian motion) A *Brownian motion* or a *Wiener process* with drift μ and variance σ^2 is a stochastic process $[W_t : t \in R^{\geq 0}]$ on some probability space $(\Omega, \mathcal{F}, P)$ which satisfies the following three properties:

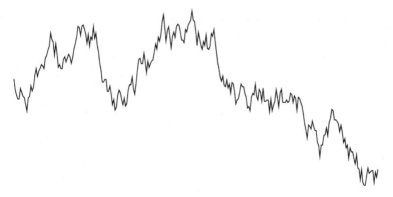

Fig. 15.1 A Brownian motion path

1. The process starts at 0, i.e. $P(W_0 = 0) = 1$.
2. The increments are independent, i.e. if $0 \leq t_1 \leq t_2 \leq \cdots \leq t_n$ then

$$P(W_{t_i} - W_{t_{i-1}} \in A_i, i \leq n) = \prod_{i \leq n} P(W_{t_i} - W_{t_{i-1}} \in A_i)$$

for all measurable sets A_i.
3. For $0 \leq s \leq t$ the increment $W_t - W_s$ is normally distributed with mean $\mu(t - s)$ and variance $\sigma^2(t - s)$.
4. W_t is a continuous function of t with probability 1.

Figure 15.1 shows a typical Brownian motion path.

It is rather difficult to prove that such a process exists (see, for example, Theorem 37.1 in [27]).

If not specified otherwise, by Brownian motion we always mean the standard Brownian motion with drift $\mu = 0$ and variance $\sigma^2 = 1$.

The next theorem collects all the properties of Brownian motion we need:

Theorem 15.6 *Let $[W_t : \in R^{\geq 0}]$ be a standard Brownian motion with drift $\mu = 0$ and variance $\sigma^2 = 1$. Then:*

(a) *For every $c > 0$ the process $[W'_t : t \in R^{\geq 0}]$ defined by*

$$W'_t(\omega) = c^{-1} W_{c^2 t}(\omega) \tag{15.11}$$

is a standard Brownian motion.
(b) *The process $[W''_t : t \in R^{\geq 0}]$ defined by*

$$W''_t(\omega) = \begin{cases} t W_{1/t}(\omega) & \text{for } t > 0, \\ 0 & \text{for } t = 0 \end{cases} \tag{15.12}$$

is a standard Brownian motion.

(c) *Let τ be any stopping time and let $\mathcal{F}_\tau$ be the σ-field of all measurable sets M*
with $M \cap \tau \le t \in \mathcal{F}_t = \sigma(W_s, s \le t)$.
Then $[W_t^* : t \in R^{\ge 0}]$ with

$$W_t^*(\omega) = W_{\tau(\omega)+t}(\omega) - W_{\tau(\omega)}(\omega) \tag{15.13}$$

is a standard Brownian motion and $[W_t^* : t \ge 0]$ is independent from $\mathcal{F}_\tau$.
This property is called the strong Markov property.

(d) *Let τ be a stopping time and let*

$$W_t'''(\omega) = \begin{cases} W_t(\omega) & \text{for } t \le \tau(\omega), \\ W_{\tau(\omega)}(\omega) - (W_t(\omega) - W_{\tau(\omega)}(\omega)) & \text{for } t \ge \tau(\omega). \end{cases} \tag{15.14}$$

Then $[W_t''' : t \in R^{\ge 0}]$ is a standard Brownian motion.
This property is called the reflection principle.

(e) *For every $a \ge 0$ we have*

$$P\left(\max_{0 \le t' \le t} W_{t'} \ge a\right) = 2P(W_t \ge a) = 2\Phi(-a/t). \tag{15.15}$$

(f) *For $a > 0$ and $y < a + bt$ the following equation holds:*

$$P\left(W_{t'} \ge a + bt' \text{ for some } t' \in [0, t] \mid W_t = y\right) = \exp\left[-\frac{2a(a + bt - y)}{t}\right]. \tag{15.16}$$

(g) *For $a > 0$ and $x < a + bt$ we have*

$$P\left(W_{t'} \ge a + bt' \text{ for some } t' \in [0, t], W_t \le x\right) = \exp(-2ab)\Phi\left(\frac{x - 2a}{\sqrt{t}}\right) \tag{15.17}$$

and in particular

$$P\left(W_{t'} \ge a + bt' \text{ for some } t' \in [0, t]\right)$$
$$= 1 - \Phi\left(\frac{a + bt}{\sqrt{t}}\right) + \exp(-2ab)\Phi\left(\frac{(a + bt) - 2a}{\sqrt{t}}\right). \tag{15.18}$$

Proof

(a) Since $t \mapsto c^2 t$ is a strictly increasing function, the process $[W_t' : t \in R^{\ge 0}]$ has
independent increments. Furthermore, $W_t' - W_s' = c^{-1}(W_{c^2 t} - W_{c^2 s})$ is normal
distributed with mean 0 and variance $c^{-2}(c^2 t - c^2 s) = t - s$. Since the paths
of $W'(\omega)$ are continuous and start at 0 we have proved that $[W_t' : t \ge 0]$ is a
standard Brownian motion.

(b) $t \mapsto 1/t$ is strictly decreasing and hence $[W_t'' : t \in R^{\geq 0}]$ has independent increments. As in (a) we check that the increments are normally distributed. The paths are continuous for $t > 0$. That $W''(\omega)$ is continuous at 0 with probability 1 is just the strong law of large numbers.

(c) See [27] p. 508.

(d) By the strong Markov property we know that $[W_t^* : t \geq 0]$ is a standard Brownian motion independent from $\mathcal{F}_\tau$. By (a) $[-W_t^* : t \in R^{\geq 0}]$ is also a standard Brownian motion independent from $\mathcal{F}_\tau$.

The Brownian motion $[W_t : t \in R^{\geq 0}]$ satisfies

$$W_t(\omega) = \begin{cases} W_t(\omega) & \text{for } \tau(\omega) < t, \\ W_{\tau(\omega)}(\omega) + W_{t-\tau(\omega)}^*(\omega) & \text{for } \tau(\omega) \geq t. \end{cases} \qquad (15.19)$$

We can replace the random variable $W_{t-\tau(\omega)}^*(\omega)$ in Eq. (15.19) by any other standard Brownian motion independent from $\mathcal{F}_\tau$, without changing the distribution. In particular, we get a Brownian motion if we replace $W_{t-\tau(\omega)}^*(\omega)$ by $-W_{t-\tau(\omega)}^*(\omega)$ in Eq. (15.19).

(e) Consider the stopping time $\tau = \inf\{s \mid W_s \geq a\}$ and the stopping time $\tau''' = \inf\{s \mid W_s''' \geq a\}$ with the process W''' defined in part (d). Then $\tau = \tau'''$ and

$$P\left(\max_{0 \leq t' \leq t} W_{t'} \geq a\right) = P(\tau \leq t)$$

$$= P(\tau \leq t, W_t \leq a) + P(\tau \leq t, W_t \geq a)$$

$$= P(\tau''' \leq t, W_t''' \leq a) + P(\tau \leq t, W_t \geq a)$$

$$= 2P(\tau \leq t, W_t \geq a)$$

$$= 2P(W_t \geq a).$$

(f) See [250] p. 375.

(g) Integrating (15.16) we get:

$$P\left(W_{t'} \geq a + bt' \text{ for some } t' \in [0,t], W_t \leq x\right)$$

$$= \int_{-\infty}^x P\left(W_{t'} \geq a + bt' \text{ for some } t' \in [0,t] \mid W_t = y\right) dP_t(y)$$

$$= \int_{-\infty}^x \exp\left[-\frac{2a(a+bt-y)}{t}\right] \frac{1}{\sqrt{2\pi t}} \exp\left[-\frac{y^2}{2t}\right] dy$$

$$= \exp(-2ab) \int_{-\infty}^x \frac{1}{\sqrt{2\pi t}} \exp\left[-\frac{(y-2a)^2}{2t}\right] dy$$

$$= \exp(-2ab) \Phi\left(\frac{x-2a}{\sqrt{t}}\right).$$

Furthermore,

$$P\left(W_{t'} \geq a + bt' \text{ for some } t' \in [0, t]\right)$$

$$= P\left(W_{t'} \geq a + bt' \text{ for some } t' \in [0, t], W_t \geq a + bt\right)$$

$$+ P\left(W_{t'} \geq a + bt' \text{ for some } t' \in [0, t], W_t \leq a + bt\right)$$

$$= P(W_t \geq a + bt) + \exp(-2ab)\Phi\left(\frac{(a + bt) - 2a}{\sqrt{t}}\right)$$

$$= 1 - \Phi\left(\frac{a + bt}{\sqrt{t}}\right) + \exp(-2ab)\Phi\left(\frac{(a + bt) - 2a}{\sqrt{t}}\right). \qquad \square$$

15.3.5 The Functional Central Limit Theorem

We are now ready to formulate the functional central limit theorem which is the justification that we can replace a sequence of accumulated independent variables by a Brownian motion.

Theorem 15.7 (Functional central limit theorem) *Suppose that $X_1, X_2, \ldots$ are iid. random variables with mean 0, variance σ^2 and finite fourth moment. Let*

$$Y_n(t) = \frac{1}{\sigma\sqrt{n}} \sum_{i=1}^{k} x_i \quad \text{for } \frac{k-1}{n} < t \leq \frac{k}{n}. \tag{15.20}$$

Then there exists for each n (on some probability space) a process $[Z_n(t) : 0 \leq t \leq 1]$ and $[W_n(t) : 0 \leq t \leq 1]$ such that $[Z_n(t) : 0 \leq t \leq 1]$ has the same finite dimensional distributions as $[Y_n(t) : 0 \leq t \leq 1]$ and such that $[W_n(t) : 0 \leq t \leq 1]$ is a standard Brownian motion, and

$$P\left(\sup_{0 \leq t \leq 1} |Z_n(t) - W_n(t)| \geq \epsilon\right) \to 0 \quad \text{for } \epsilon > 0. \tag{15.21}$$

Proof See [27] Theorem 37.8. $\qquad \square$

As an application, consider a sequence $X_1, X_2, \ldots$ of iid. random variables with mean 0 and variance σ^2. What is the limit distribution of

$$M_n = \max_{1 \leq i \leq n} \sum_{j=1}^{i} X_i \ ?$$

This is a typical question we have to answer when we deal with sequential tests. By the functional central limit theorem $\frac{M_n}{\sigma\sqrt{n}} = \sup_{0 \leq t \leq 1} Y_n$ has the same distribution as

$\sup_{0 \le t \le 1} Z_n$ and by (15.21)

$$P\left(\left| \sup_{0 \le t \le 1} Z_n(t) - \sup_{0 \le t \le 1} W_n(t) \right| \ge \epsilon \right) \to 0 \quad \text{for } \epsilon > 0.$$

By Theorem 15.6 (e) $P(\sup_{0 \le t \le 1}) W_n \ge x = 2\Phi(x)$ for $x \ge 0$. Therefore,

$$P\left[\frac{M_n}{\sigma \sqrt{n}} \ge x \right] \to 2\Phi(x)$$

for $x \ge 0$.

To answer similar questions for random variables with mean $\mu \ne 0$ we use part (g) of Theorem 15.6.

Chapter 16
Combinatorics

Combinatorics has several applications in cryptography. We use it a lot in the analysis of the RC4 key-scheduling (Sect. 9.3.1) where most of the results from this chapter find their application.

16.1 Asymptotic Calculations

This chapter collects the solution to some counting problems that arise during the analysis of RC4. We start with Euler's famous summation method.

First we need the *Bernoulli numbers* which are the coefficients in the following Taylor series:

$$\frac{x}{e^x - 1} = B_0 + B_1 x + \frac{B_2}{2!} x^2 + \frac{B_3}{3!} x^3 + \cdots . \tag{16.1}$$

Since

$$\frac{x}{e^x - 1} + \frac{x}{2} = \frac{x}{2} \cdot \frac{e^x + 1}{e^x - 1} = \frac{-x}{2} \cdot \frac{e^{-x} + 1}{e^{-x} - 1}$$

is an even function, we see that $B_{2k+1} = 0$ for $k \geq 1$.

We define the *Bernoulli polynomial* by

$$B_n(x) = \sum_{k=0}^{n} \binom{n}{k} B_k x^{n-k}. \tag{16.2}$$

A short calculation gives

$$B_n'(x) = \sum_{k=0}^{n} \binom{n}{k} B_k (n-k) x^{n-k-1}$$

$$= m \sum_{k=0}^{n-1} \binom{n-1}{k} B_k x^{(n-1)-k}$$

$$= m B_{n-1}(x).$$

A. Klein, *Stream Ciphers*, DOI 10.1007/978-1-4471-5079-4_16,
© Springer-Verlag London 2013

Fig. 16.1 Comparison of $\int_a^b f(x)dx$ and $1/2f(a) + f(a+1) + \cdots + f(b-1) + 1/2f(b)$

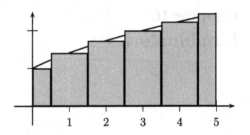

Another important identity arises if we multiply the defining equation (16.1) of the Bernoulli numbers by $e^x - 1 = \sum_{n=1}^{\infty} \frac{1}{n!} x^n$ and compare the coefficients of x^n in the corresponding power series. This gives

$$\sum_{k=0}^{n} \frac{B_k}{k!} \frac{1}{(n-k)!} = \delta_{n1}$$

or, after some simple transformations,

$$\sum_{k=0}^{n} \binom{n}{k} B_k = B_n + \delta_{n1}. \tag{16.3}$$

For the Bernoulli polynomials Eq. (16.3) implies

$$B_n(1) = B_n = B_n(0) \text{ for } n > 1. \tag{16.4}$$

As Fig. 16.1 shows, $\int_a^b f(x)dx$ is approximately $1/2f(a) + f(a+1) + \cdots + f(b-1) + 1/2f(b)$. Euler's formula gives precise error terms. One can view it either as a method to evaluate sums by integration or as a tool for numerical integration.

Theorem 16.1 (Euler's summation formula) *Let* $f : \mathbb{R} \to \mathbb{R}$ *be* m *times differentiable. Then for* $a, b \in \mathbb{Z}$

$$\sum_{k=a}^{b} f(k) = \int_a^b f(x)dx + \frac{f(a) + f(b)}{2} + \sum_{k=2}^{m} \frac{B_k}{k!} f^{(k-1)}(x) \Big|_a^b + R_m \tag{16.5}$$

where

$$R_m = (-1)^{m+1} \int_a^b \frac{B_m(\{x\})}{m!} f^m(x)dx. \tag{16.6}$$

($\{x\}$ *denotes the fractional part of* x.)

Proof We begin with the following identity:

$$\int_k^{k+1} \left(\{x\} - \frac{1}{2} \right) f'(x)dx = \left(x - k - \frac{1}{2} \right) f(x) \Big|_k^{k+1} - \int_k^{k+1} f(x)dx$$

$$= \frac{1}{2}\big(f(k+1) + f(k)\big) - \int_k^{k+1} f(x)dx \qquad (16.7)$$

which is just integration by parts. Summation from a to b gives

$$\int_a^b \left(\{x\} - \frac{1}{2} \right) f'(x)dx = \frac{1}{2}f(a) + f(a+1) + \cdots + f(b-1) + \frac{1}{2}f(b) - \int_a^b f(x)dx$$

which is Eq. (16.5) for $m = 1$.

Integrating by parts gives

$$\frac{1}{m!} \int_a^b B_m\big(\{x\}\big) f^{(m)}(x)dx = \frac{1}{(m+1)!} B_{m+1}\big(\{x\}\big) f^{(m)}(x)\Big|_a^b$$

$$- \frac{1}{(m+1)!} \int_a^b B_{m+1}\big(\{x\}\big) f^{(m+1)}(x)dx. \qquad (16.8)$$

Here we use that $B(\{x\})$ is continuous by Eq. (16.4).

Applying (16.8) repeatedly to (16.1) proves the theorem. □

As an example of how to apply Euler's summation formula we derive an asymptotic expansion for the *harmonic numbers* $H_n = \sum_{i=1}^n \frac{1}{i}$.

Theorem 16.2 *There exists a constant γ (Euler's constant) such that*

$$H_{n-1} = \ln n + \gamma + \sum_{k=1}^m \frac{(-1)^{k-1} B_k}{kn^k} + \int_n^\infty \frac{B_m(x)}{x^{m+1}} dx$$

$$= \ln n + \gamma + \sum_{k=1}^m \frac{(-1)^{k-1} B_k}{kn^k} + O\left(\frac{1}{n^{m+1}} \right).$$

$(\gamma \approx 0.57721\ldots)$

Proof By Euler's summation formula, we get

$$H_{n-1} = \ln n + \sum_{k=1}^m \frac{B_k}{k!} \frac{(-1)^{k-1}(k-1)!}{x^k} \Big|_1^n$$

$$+ (-1)^{m+1} \int_1^n \frac{B_m(\{x\})}{m!} \cdot \frac{(-1)^m m!}{x^{m+1}} dx. \qquad (16.9)$$

Taking the limit we get

$$\gamma = H_{n-1} - \ln n = \sum_{k=1}^{n} \frac{B_k}{k}(-1)^k - \int_1^{\infty} \frac{B_m(x)}{x^{m+1}}. \qquad (16.10)$$

Since $B_m(x)$ is bounded in the interval $[0, 1]$, the limit exists.
   Putting (16.9) and (16.10) together proves the theorem.   □

   More examples of the use of Euler's formula can be found in [119, 151].

16.2 Permutations

The bijective functions from $\{1, \ldots, n\}$ to $\{1, \ldots, n\}$ form the symmetric group S_n. S_n has $n! = n \cdot (n-1) \cdots 2 \cdot 1$ elements. The elements of S_n are called *permutations*.
   For the permutations of S_n we use the famous cycle notation, i.e. we write $(123)(45)$ for $(1 \mapsto 2, 2 \mapsto 3, 3 \mapsto 1, 4 \mapsto 5, 5 \mapsto 4)$. When we multiply permutations we read the terms from right to left (think of function composition), i.e. $(12)(23) = (231)$.
   An *involution* is a permutation of order 2. In our analysis of RC4 (see Theorem 9.7) we need the number of involutions in S_n, which will be determined in this section.
   Let t_n be the number of involution in S_n. The element n may either be a fixed point or exchanged with any of the other elements. This leads to the recursion

$$t_n = t_{n-1} - (n-1)t_{n-2},$$

which in turn leads to an easy way to tabulate t_n for small n.
   An alternative is to count the number $t_{n,k}$ of permutations with k two-cycles. To generate such a permutation write down any permutation of n elements. Then pair the first $2k$ elements into k two-cycles and declare the last $n - 2k$ elements as fixed points. By this procedure, one counts every permutation of $t_{n,k}$ exactly $k!2^k(n-2k)!$ times ($k!$ ways to order k two-cycles, each two-cycle can be written in two ways as (ab) or (ba), and the $n - 2k$ fixed points can be in any order). Thus

$$t_{n,k} = \frac{n!}{k!2^k(n-2k)!}.$$

No explicit formula for $t_n = \sum_{k=0}^{\lfloor n/2 \rfloor} t_{n,k}$ is known, but we can determine its asymptotic behavior. Our derivation of the asymptotic behavior follows [153].

Theorem 16.3 *The number t_n of involutions in S_n satisfies*

$$t_n = \frac{1}{\sqrt{2}} n^{n/2} e^{-n/2+\sqrt{n}-1/4} \left(1 + \frac{7}{24} n^{-1/2} + O\left(n^{-3/4}\right) \right).$$

Proof First note that

$$\frac{t_{n,k+1}}{t_{n,k}} = \frac{(n-2k)(n-2k-1)}{2(k+1)}$$

so $t_{n,k}$ increases until k reaches approximately $\frac{1}{2}(n - \sqrt{n})$.

The transformation $k = \frac{1}{2}(n - \sqrt{n}) + x$ brings the large parts near the origin. To get rid of the factorials we apply Stirling's formula. If $|x| \geq n^{\epsilon+1/4}$, Stirling's formula gives the bound

$$t_{n,k} \leq e^{-2n^{2\epsilon}} \exp\left(\frac{1}{2}n \ln n - \frac{1}{2}n + \sqrt{n} - \frac{1}{4}\ln n - \frac{1}{4} - \frac{1}{2}\ln \pi + O\left(n^{3\epsilon-1/4}\right)\right).$$

Because of the term $e^{-2n^{2\epsilon}}$, this is much smaller than any polynomial in n and, as we will see, these terms do not contribute to the order of t_n.

So we may restrict x to the interval $[-n^{\epsilon+1/4}, n^{\epsilon+1/4}]$ for some ϵ. For x in that interval we get by Stirling's formula the expansion

$$t_{n,k} = \exp\left(\frac{1}{2}n \ln n - \frac{1}{2}n + \sqrt{n} - \frac{1}{4}\ln n - \frac{2x^2}{\sqrt{n}} - \frac{1}{4} - \frac{1}{2}\ln \pi \right.$$
$$\left. - \frac{4x^3}{3n} + \frac{2x}{\sqrt{n}} + \frac{1}{3\sqrt{n}} - \frac{4x^4}{3n\sqrt{n}} + O\left(n^{5\epsilon-3/4}\right)\right). \tag{16.11}$$

The common factor

$$\exp\left(\frac{1}{2}n \ln n - \frac{1}{2}n + \sqrt{n} - \frac{1}{4}\ln n - \frac{1}{4} - \frac{1}{2}\ln \pi + \frac{1}{3\sqrt{n}}\right)$$

in (16.11) causes no difficulty. So we must sum

$$\exp\left(-\frac{2x^2}{\sqrt{n}} - \frac{4x^3}{3n} + \frac{2x}{\sqrt{n}} - \frac{4x^4}{3n\sqrt{n}} + O\left(n^{5\epsilon-3/4}\right)\right)$$

$$= \exp\left(-\frac{2x^2}{\sqrt{n}}\right)\left(1 - \frac{4}{3}\frac{x^3}{n} + \frac{8}{9}\frac{x^6}{n^2}\right)\left(1 + 2\frac{x}{\sqrt{n}} + 2\frac{x^2}{n}\right)$$

$$\times \left(1 - \frac{4}{3}\frac{x^4}{n\sqrt{n}}\right)\left(1 + O\left(n^{9\epsilon-3/4}\right)\right) \tag{16.12}$$

from α to β where α and β are integers near $\pm n^{\epsilon+1/4}$.

Multiplying out we must evaluate the sum

$$\sum_{x=\alpha}^{\beta} x^t e^{-x^2}$$

for integers t. As e^{-x^2} approaches 0 quickly for $|x| \to \infty$, we can replace the limits α and β by $-\infty$ and $+\infty$ without changing the asymptotic behavior.

Applying Euler's summation formula we get

$$\sum_{x=-\infty}^{+\infty} f(x) = \int_{-\infty}^{+\infty} f(x)dx - \frac{1}{2}f(x)\Big|_{-\infty}^{+\infty} + \frac{B_2}{2}\frac{f'(x)}{1!}\Big|_{-\infty}^{+\infty} + \cdots$$

$$+ \frac{B_m}{m+1}\frac{f^{(m)}(x)}{m!}\Big|_{-\infty}^{+\infty} + R_{m+1}.$$

Thus for $f(x) = x^t e^{-x^2}$ we get $f^m(x)|_{-\infty}^{+\infty} = 0$ and $R_{m+1} = O(n^{(t+1-m)/4})$, so only the integral gives a significant contribution.

The integral $\int_{-\infty}^{+\infty} x^t e^{-x^2}dx$ is not difficult to evaluate. For odd t a simple substitution $x^2 = y$ does the job. For even t we use partial integration to reduce the problem to $\int_{-\infty}^{\infty} e^{-x^2}dx$. This integral is solved by

$$\left(\int_{-\infty}^{\infty} e^{-x^2}\right)^2 dx = \int_{-\infty}^{+\infty}\int_{-\infty}^{+\infty} e^{-x^2-y^2}dxdy$$

then transforming Cartesian coordinates to polar coordinates

$$= \int_0^{2\pi}\int_0^{\infty} re^{-r^2}drd\phi$$

$$= \pi.$$

Plugging everything together gives the asymptotic expansion of t_n. With a bit of extra work the error term could be improved to $O(n^{-k})$ for any k, instead of $O(n^{-3/4})$.   □

16.3 Trees

A tree is a graph without cycles and a labeled tree with n vertices is a tree whose vertices are numbered from 1 to n.

We encountered labeled trees in Sect. 9.3.1. There we cited a famous theorem of Cayley on the number of finite trees, which will be proved in this section.

Theorem 16.4 (Cayley [47]) *There are n^{n-2} labeled trees with n vertices.*

Proof A very elegant proof of Cayley's theorem is due to Prüfer [215].

Let T be a labeled tree with n vertices. We define a sequence starting with $T_1 = T$. To obtain T_{i+1} from T_i delete the monovalent vertex of T_i with the smallest label. Let x_i be the label of that vertex and let y_i be the adjacent vertex in T_i. The *Prüfer code* of T is $(y_1, \ldots, y_{n-2})$.

Fig. 16.2 A labeled tree

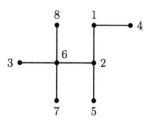

For example the tree of Fig. 16.2 has the Prüfer code $(6, 1, 2, 2, 6, 6)$. Note that in the example $y_7 = 8$ as for all trees with 8 vertices and the sequence x_i is $(3, 4, 1, 5, 2, 7, 6)$.

We claim that the Prüfer code is a bijection from the set labeled trees to the set $\{1, \ldots, n\}^{n-2}$. To prove this we must reconstruct the tree from the Prüfer code.

Note the following simple facts:

- Since every tree has at least two monovalent vertices, we never delete the vertex n, i.e. $y_{n-1} = n$.
- The edges of T_k are $(x_k, y_k), \ldots, (x_{n-1}, y_{n-1})$.
- The number of occurrences of a vertex v in the sequence $y_k, \ldots, y_{n-2}$ is $(\deg_{T_k} v) - 1$. (v occurs $\deg_{T_k} v$ times in the list of edges $(x_k, y_k), \ldots, (x_{n-1}, y_{n-1})$ and exactly once in $x_k, \ldots, x_{n-1}, y_{n-1}$.)
- x_k is the smallest number missing in $\{x_1, \ldots, x_{k-1}, y_k, \ldots, y_{n-1}\}$.

Thus one can reconstruct the sequence $x_1, \ldots, x_{n-1}$ uniquely from the Prüfer code $(y_1, \ldots, y_{n-2})$ and hence the Prüfer code determines the unique tree with edges $(x_1, y_1), \ldots, (x_{n-1}, y_{n-1})$. □

Part IV
Exercises with Solutions

Chapter 17
Exercises

The exercises are sorted roughly according to the corresponding chapters. The number inside the parentheses indicates the difficulty ranging from 1 (easy) to 5 (hard). The letter P indicates that the exercise involves programming and the letter M indicates that the exercise needs a bit more mathematics than the other exercises.

17.1 (1) What are the differences between the different operation modes of a block cipher? In addition to the three stream-oriented modes OFB, CFB and CTR (see Fig. 1.2) also consider the block-oriented modes ECB (electronic code book) and CBC (cipher block chaining) defined by $c_i = E(m_i, k)$ and $c_i = E(c_{i-1} \oplus m_i, k)$, respectively.

Complete the following table:

Mode	ECB	CBC	OFB	CFB	CTR
parallel encoding?	Yes				
parallel decoding?	Yes				
The equation $c_i = c_j$ implies	$m_i = m_j$				
an input bit affects	only the current block				
a flipped bit in the ciphertext affects	a whole block				

17.2 (1) What are the differences between synchronous and self-synchronizing stream ciphers with respect to bit flipping and deletion errors?

17.3 (2) Break the following Vigenère cipher

```
OVSTL ZOHKI LLUZL HALKM VYZVT LOVBY
ZPUZP SLUJL DPAOO PZSVU NAOPU IHJRJ
BYCLK VCLYH JOLTP JHSCL ZZLSP UDOPJ
OOLDH ZIYLD PUNHW HYAPJ BSHYS FTHSV
KVYVB ZWYVK BJAOP ZOLHK DHZZB URBWV
UOPZI YLHZA HUKOL SVVRL KMYVT TFWVP
```

A. Klein, *Stream Ciphers*, DOI 10.1007/978-1-4471-5079-4_17,
© Springer-Verlag London 2013

UAVMC PLDSP RLHZA YHUNL SHURI PYKDP
AOKBS SNYLF WSBTH NLHUK HISHJ RAVWR
UVA

Hint: The keyword has length five characters.

17.4 (3) Break the following auto key cipher:

LSHSF LYTRO REETS ZSFEG PFQLL IOSUL
LBMOM LUOMY NQLRO GUUAR XHVRG LMBUD
RTMOU ROLBW ORWXJ FHHBJ MVZRZ SYQVP
FNWST DHXML PTOVJ XEMON OMGG

Hint: The message contains the word "secret".

17.5 (1) Explain why it is an error to use a binary additive stream cipher two times with the same keystream for encryption. What is the consequence for the minimal length of an initialization vector? (Assume for simplicity that the initialization vector is chosen at random.)

17.6 (2) There exist six m-sequences of order 5. Compute for all pairs of them the cross-correlation function and verify in these special cases the results of Sect. 2.3.3.

17.7 (P3) Algorithm 17.1 is a variation of Algorithm 2.6. Prove that it computes the sideway addition mod 2.

Algorithm 17.1 Sideway addition mod 2

Require: $x = (x_{31} \dots x_0)$ is 32-bit word
Ensure: $y = SADD(x) \mod 2$
 1: $y \leftarrow x \oplus (x \gg 16)$
 2: $y \leftarrow y \oplus (y \gg 8)$
 3: $y \leftarrow y \oplus (y \gg 4)$
 4: $y \leftarrow 0x6996 \gg (y \,\&\, 0xF)$ {$0x6996$ and $0xF$ are hexadecimal constants}
 5: $y \leftarrow y \,\&\, 1$ {$y \leftarrow y \mod 2$}

17.8 (2) Enumerate all binary sequences $a_0, \dots, a_{2n}$ which have a perfect linear complexity profile.

17.9 (P5) Develop an algorithm that chooses a random de Bruijn sequence of order n. All $2^{2^{n-1}}$ possibilities should have equal probability.

17.10 (2) Consider a Geffe generator built from three LFSRs with feedback polynomials $f_1 = x^5 + x^2 + 1$, $f_2 = x^2 + x + 1$ and $f_3 = x^4 + x + 1$. Compute the linear complexity of the generator and the feedback polynomial of the minimal LFSR that generates the same sequence.

17.11 (M2) Let $b_1, \ldots, b_n \in \{0, 1\}$. Let $s = b_1 + \cdots + b_n$ (summation is done in the domain of integers). Determine the Boolean function $f_{j,n}$ for the jth bit of s. What is the degree of f_j?

17.12 (M2) In a correlation attack you observe a sequence $z_1, z_2 \ldots$ and want to compare it with a possible internal sequence $x_1, x_2, \ldots$.
   Test the hypothesis

- H_0: The observed sequence is correlated to the internal sequence by $P(z_i = x_i) = \frac{1}{2} + \varepsilon$.

against the alternative

- H_1: both sequences are uncorrelated.

   How must you choose the stopping conditions to bound the error probability for an error of the first type by 1 % and for an error of the second type by 0.1 %? Let $\varepsilon = 0.1$. Assuming that the hypothesis is false, what is the expected number of observed bits before rejection?

17.13 (P3) Prove: If $f = x ? f_h : f_l$, $g = x ? g_h : g_l$ and $h = x ? h_h : h_l$ then $f \wedge g \wedge h = x ?(f_h \wedge g_h \wedge h_h) : (f_l \wedge g_l \wedge h_l)$.
   Turn this into an algorithm with computes the BDD of $f \wedge g \wedge h$ directly. Find functions f, g and h for which the ternary algorithm is faster than computing $f \wedge g \wedge h$ in any order by a binary algorithm.
   Also find functions f, g and h for which the binary computation $(f \wedge g) \wedge h$ is faster than the ternary algorithm.

17.14 (P4) Sometimes we are not interested in the complete Boolean function f, but only in the values that f takes on some set $S \subseteq \{0, 1\}^n$. In these cases we can try to simplify f by changing its values outside S.
   Define $f \downarrow g$ (read: f is constrained with respect to g) by $f \downarrow g = 0$ if $g = 0$ and $(f \downarrow g)(y) = f(y')$ where y' is the first element in the series $y, y \oplus 1, y \oplus 2, \ldots$ for which $g(y') = 1$ otherwise.
   Prove that $f \downarrow g$ satisfies:

(a) $0 \downarrow g = 0$.
(b) $f \downarrow 1 = f$.
(c) $1 \downarrow g = g \downarrow g = 1$ for $g \neq 0$.
(d) If $f = (x_0 ? f_h : f_l) \neq 0$ and $g = (x_0 ? g_h : g_l) \neq 0$ then

$$f \downarrow g = \begin{cases} f_l \downarrow g_l & \text{if } g_h = 0, \\ f_h \downarrow g_h & \text{if } g_l = 0, \\ x_0 ?(f_h \downarrow g_h) : (f_l \downarrow g_l) & \text{otherwise.} \end{cases}$$

   Turn this equation into a top down algorithm for computing the BDD of $f \downarrow g$ similar to the one described in the text for Boolean operations.

17.15 (P5) The *self-shrinking generator* is constructed from an LFSR by applying the following transformation to the output. Group the output bits in pairs, delete the pairs 00 and 01 and replace the pairs 10 and 11 by 0 and 1, respectively. For example 11 01 10 00 10 is mapped to 1 0 0.

Implement a BDD-based attack against the self-shrinking generator. Estimate the complexity of the attack.

17.16 (M3) Let $f, g \in R[x, y]$. Prove that if $(\hat{x}, \hat{y})$ is a solution of $f(x, y) = g(x, y) = 0$ then $\hat{x}$ is a zero of $\mathrm{Res}_y(f, g)$.

17.17 (2) Let $\prec$ satisfy (1) and (2) of Definition 6.1. Show that the graded version of $\prec$ defined by $\alpha \prec_{gr} \beta \iff (|\alpha| < |\beta|) \vee (|\alpha| = |\beta| \wedge \alpha \prec \beta)$ is a monomial order.

17.18 (1) Consider a shrinking generator in which the control and the output sequence are generated by LFSRs with the same feedback polynomial. Why is it a terrible idea to also use the same initial state for both LFSRs?

17.19 (3) Transfer the ideas of Sect. 7.3.3 to the self-shrinking generator (described in Exercise 17.15). Develop a correlation attack against the self-shrinking generator.

17.20 (2) Prove that if RC4 starts in the state $j = i + 1$, $S[j] = 1$, then it runs through a cycle of length n.

Why can this cycle never cause problems?

17.21 (5) Consider the following variation of RC4 which replaces the additions mod 256 by bitwise XOR (see Algorithm 17.2).

Algorithm 17.2 A weak variation of the RC4 pseudo-random generator
---
1: {initialization}
2: $i \leftarrow 0$
3: $j \leftarrow 0$
4: {generate pseudo-random sequence}
5: **loop**
6:     $i \leftarrow (i + 1)265$
7:     $j \leftarrow j \oplus S[i]$
8:     Swap $S[i]$ and $S[j]$
9:     $k \leftarrow S[i] \oplus S[j])$
10:    **output** $S[k]$
11: **end loop**

Give a reason why Algorithm 17.2 is weaker than the original RC4.

17.22 (2) Bring the Blum-Blum-Shub generator into the formal form of Algorithm 11.1, so that you can apply Theorem 11.1 directly.

17.23 (M2) Prove that the number $\text{SADD}(x)$ of set bits in 32-bit integer (standard two's complement representation) satisfies:

$$\text{SADD}(x) = -\sum_{i=0}^{31}(x \lll i). \tag{17.1}$$

(Here $x \lll i$ denotes the rotation by i bits.)
Does this lead to an efficient way to compute the sideway addition?

17.24 (P3) Show that $x \leftarrow x \,\&\, (x - 1)$ removes the leftmost bit of x. Turn this into a method that computes $\text{SADD}(x)$ with $O(\text{SADD}(x))$ operations.

17.25 (M3) Show that Algorithm 13.1 works correctly. Divide the task into the following steps:

(a) The Jacobi symbol is multiplicative in both arguments, i.e. $\left(\frac{aa'}{b}\right) = \left(\frac{a}{b}\right)\left(\frac{a'}{b}\right)$ and $\left(\frac{a}{bb'}\right) = \left(\frac{a}{b}\right)\left(\frac{a}{b'}\right)$.
(b) If $b \geq 3$ is odd then

$$\left(\frac{a}{b}\right) = \left(\frac{a \bmod b}{b}\right).$$

(c) If $a, b \geq 3$ are odd then

$$\left(\frac{a}{b}\right)\left(\frac{b}{a}\right) = (-1)^{(a-1)(b-1)/4}.$$

(d) $\left(\frac{2}{b}\right) = 1$ if $b \equiv \pm 1 \bmod 8$ and $\left(\frac{2}{b}\right) = -1$ if $b \equiv \pm 3 \bmod 48$.
(e) Prove that Algorithm 13.1 works correctly.

17.26 (M5) The basic variant of the knapsack-based public key cryptosystem [190] works as follows:

Choose a sequence $(b_i)_{i=0,\ldots,k-1}$ with $b_{i+1} \geq 2b_i$ and $m > \sum_{i=0}^{k-1} b_i$. Choose c with $\gcd(c, m) = 1$. Publish the values $a_i = b_i c \bmod m$.

To encrypt a binary number $x = \sum_{i=0}^{k-1} x_i 2^i$ ($x_i \in \{0, 1\}$) one computes $s = \sum_{i=0}^{k-1} a_i x_i$.

You can easily compute $s' = sc^{-1} \bmod m$ and solve the easy knapsack problem $s' = \sum_{i=0}^{k-1} b_i x_i$.

The (general?) knapsack problem, where $s = \sum_{i=0}^{k-1} a_i x_i$, had been assumed to be secure because it is $\mathcal{NP}$-complete (average case complexity). However, the weights a_i have a very special form which is far from the average case.

Break the cipher by writing the knapsack problem as a lattice problem and using the LLL-algorithm.

Try your attack on the following toy example. The pairs AA, AB, ..., ZZ of letters are identified with the binary numbers from $0, \ldots, 26^2 - 1 = 675$.

The weights a_i are given by the following table.

i	0	1	2	3	4	5	6	7	8	9
a_i	4881	2662	1774	4879	4285	5020	2940	3661	3772	3698

The message is

$$12938, 14196, 16432, 18620$$

No rigorous proof is needed as long as you recover the plaintext.

Bonus question:

Find at least one protocol failure in the toy example, besides using a broken cipher and it being too small.

17.27 (M3) Let $n = pq$ be the product of two primes. Show that if $|p - q| \le \sqrt[4]{n}$ then the integer n can be factored in polynomial time.

17.28 (M3) Use Zsigmondy's theorem to prove Wedderburn's theorem: Every finite skew field is a (commutative) field.

Hint:

Write down the class formula for the conjugation classes. Use Zsigmondy's theorem to obtain a contradiction.

17.29 (M2) Compute the variation distance of an $\mathcal{N}(0, 1)$ and an $\mathcal{N}(0, 2)$ distribution.

17.1 Proposals for Programming Projects

Every algorithm described in the book provides a programming project. The following proposals should help you to find interesting starting points. All the proposals are open-ended problems.

Project 1 (correlation attacks) Implement a correlation attack. The program can benefit a lot from clever data structures and memory layout (Sect. 4.1.3 gives some hints). You may want to optimize the bitset handling (Sect. 12.1.2). There is also plenty of opportunity for parallel programming. The Fourier transform (Sect. 4.1.6.2) is suitable for GPU computing (perhaps a good reason to learn Opencl). The pre-processing phase can be implemented as distributed programs.

To test your correlation attack I recommend a simple non-linear combiner (for example $2n + 1$ LFSRs with the majority function as combiner).

Project 2 (correlation attack) Adopt your correlation attack program to a real cipher, for example the shrinking generator (Sect. 7.3.3) or A5/1 (Sect. 8.3.3).

In comparison to the theory from Chap. 4, some modifications will be necessary since the correlations are now less regular.

What is the biggest size of the shrinking generator you can break?

Project 3 (algebraic attack) Implement the XL-Algorithm (Sect. 6.1.2.3). Play around with the parameters and try to find good heuristics. Your implementation can benefit a lot from fast linear algebra (Sect. 12.5).

Project 4 (RC4) The attacks based on Golic's correlation (Sect. 9.5) are better than the FMS-attack (Sect. 9.4) because they can use more information.

Try to integrate all the information you can get into one algorithm (Sect. 9.5.2 gives some ideas).

This can become a programming contest. The winner is the one who needs the fewest sessions to break RC4.

Project 5 (RC4) Implement a state recovering attack against RC4 (Sect. 9.6). Try to optimize the program as much as possible (d-order, g-generative patterns or fortuitous states can be of help).

What is the largest n for which you can break RC4 working on $\mathbb{Z}/n\mathbb{Z}$? This is another possible programming contest.

Project 6 (side channel attacks) Choose any cipher (for example an irregular shift register or the Blum-Blum-Shub generator) and try to secure it against side channel attacks (Sect. 7.4). The RC4 implementation I put on the homepage of the book shows some ideas.

Perhaps you can do this project together with a friend. After the implementation you can exchange your programs and try to hack the other's implementation.

Project 7 (BDD-based attack) Implement a BDD-based attack against your favorite cipher (see also Exercise 17.15). Try time-memory trade-off techniques (Sect. 5.2.2).

Can you make use of BDD-variants (like zero-suppressed BDDs) or of advanced BDDs techniques (like variable reordering)?

This project benefits from a lot of clever memory layout (Sect. 12.2 gives some ideas).

Chapter 18
Solutions

17.1 The complete table is:

Table 18.1 Comparison of block cipher modes

Mode	ECB	CBC	OFB
parallel encoding?	Yes	No	NO
parallel decoding?	Yes	Yes	NO
The equation $c_i = c_j$ implies	$m_i = m_j$	$m_i \oplus m_j = c_{i-1} \oplus c_{j-1}$	$m_i \neq m_j$
an input bit affects	only the current block	all following blocks	only the corresponding ciphertext bit
a flipped bit in the ciphertext affects	a whole block	a whole block and one bit in the following block	only one bit

Mode	CFB	CTR
parallel encoding	No	Yes
parallel decoding?	Yes	Yes
The equation $c_i = c_j$ implies	$m_i \neq m_j$	$m_i \neq m_j$
an input bit affects	all following blocks	only the corresponding ciphertext bit
a flipped bit in the ciphertext affects	a bit and the next block	only one bit

Note that in the block-oriented modes two identical cipher blocks ($c_i = c_j$) leak one equation (128 bit information on a standard block cipher like AES), but the stream-oriented modes leak almost no information. This is one reason why the stream-oriented modes are considered superior.

A. Klein, *Stream Ciphers*, DOI 10.1007/978-1-4471-5079-4_18,
© Springer-Verlag London 2013

Note also that the two modes OFB and CTR which use the clock cipher as synchronous stream cipher have no error propagation. So they must be secured against active attacks by some other means (see Algorithm 1.1).

17.2 In a synchronous stream cipher, bit flipping affects only the current bit. So a synchronous cipher provides absolutely no resistance against an active attack. One always needs an extra cryptographic primitive to secure it against an active attack (see Algorithm 1.1). In a self-synchronizing stream cipher with delay k, bit flipping affects the k following bits, providing at least a minimal security against active attacks.

In a self-synchronizing stream cipher with delay k, a deletion error affects only the k following bits. This is what the name self-synchronizing means.

In a synchronous stream cipher a deletion error affects all following bits (the synchronization is lost and will not be established automatically). You must ensure the synchronization by other parts of the protocol.

17.3 Since we know already that the key word has length 5, we must break just 5 simple Caesar ciphers.

Since the length of the cipher text is relative short, we cannot expect that 'E' is always the most frequent letter. (Only three of the five Caesar ciphers are solved by searching for the most frequent letter.) However, if we look at the whole letter distribution it is not difficult to find the key word DANCE and to decrypt the cipher as:

"Holmes had been seated for some hours in silence with his long, thin back curved over a chemical vessel in which he was brewing a particularly malodorous product. His head was sunk upon his breast, and he looked from my point of view like a strange, lank bird, with dull grey plumage and a black top-knot."

The message is the beginning of the novel [81].

17.4 This problem shows that the auto-key cipher provides no security against a known plaintext attack. If we know or suspect that the plaintext contains the word "secret" we can try each possible position of the key-phrase "secret" and test if we get a plausible plain text. The best way to do this manually is to shift the cipher text by S, E, C, R, E and T respectively and write these texts in shifted rows as shown below.

Then we must search for columns which contain meaningful text.

```
     TAPANTGB   ...   TUJCLZBUWCZWTJEWZEFRNPPJRUDHZHAGY   ...
     HODOBHUPN  ...   IXQZNPIKQNKHXSKNSTFBDDXFIRVNVOUMR   ...
     JQFQDJWRPM ...   ZSBPRKMSPMJZUMPUVHDFFZHKTXPXQWOTN   ...
     UBQBOUHCAXA ...  DMACVXDAXUKFXAFGSOQQKSVEIAIBHZEYO   ...
  HODOBHUPNKNA  ...   ZNPIKQNKHXSKNSTFBDDXFIRVNVOUMRLBJ   ...
SZOZMSFAYVYLL   ...   YATVBYVSIDVYDEQMOOIQTCGYGZFXCWMUD   ...
```

We find two such columns. From the distance between the two columns we can conclude that the key has 7 letters and that the plaintext contains the phrase ... ETHODO?SECRET?RITING .... A plausible possibility for the two missing letters is ... ETHODOFSECRETWRITING ....

Decoding the whole cipher under this assumption reveals the message:

"Few persons can be made to believe that it is not quite an easy thing to invent a method of secret writing which shall baffle investigation."

The key is GOLDBUG.

The message is a part of Edgar Allan Poe's essay "A Few Words on Secret Writing" [212]. The key was of course chosen in honor of Poe's novel "The Gold-Bug" [213].

More techniques for attacking classical ciphers can be found in [98].

17.5 Let $c_i = m_i \oplus z_i$ and $c'_i = m'_i \oplus z_i$ be two cipher texts that belong to different messages, but to the same keystream.

Then the attacker can compute $d = c_i \oplus c'_i = m_i \oplus m'_i$. Now the messages m_i and m'_i can be reconstructed with Friedman's attack (see also Problem 17.4).

As a practical consequence we must require that an initialization vector is never used twice. If we want to use 2^k sessions the birthday paradox forces us to work with an initialization vector of length 2^{2k} (see also the protocol failure in WEP described in 9.2.1).

Normally we want to have initialization vectors of size 80 bits or more.

17.6 The multiplicative group $\mathbb{F}_{2^5}^{\times}$ is cyclic of order 31. As 31 is prime all elements of $\mathbb{F}_{2^5} \backslash \{0, 1\}$ are primitive elements. Hence there exist $30/5 = 6$ primitive polynomials of degree 5. The polynomial are $f_1 = x^5 + x^2 + 1$, $f_2 = x^5 + x^4 + x^3 + x^2 + 1$, $f_3 = x^5 + x^4 + x^2 + x + 1$, $f_4 = x^5 + x^3 + x^2 + x + 1$, $f_5 = x^5 + x^4 + x^3 + x + 1$ and $f_6 = x^5 + x^3 + 1$.

The corresponding LFSR sequences are

$$L_1 = 0010110011110001101110101000001$$

$$L_2 = 1111011100010101101000011001001$$

$$L_3 = 0010110011110001101110101000001$$

$$L_4 = 0010011000010110101000111011111$$

$$L_5 = 1101100111000011010100100010111$$

$$L_6 = 0000101011101100011111001101001$$

The cross-correlation function of L_1 and L_6 take 5 values ($5 \times -1, 1 \times 11, 10 \times 3, 5 \times -5, 5 \times -9$). The cross-correlation of L_2 and L_3 and the cross-correlation of L_4 and L_5 have the same spectrum.

The spectrum of all the other cross-correlation functions is ($15 \times -1, 10 \times 7, 6 \times -9$).

The spectrum of the auto-correlation functions is ($1 \times 31, 30 \times -1$), as we proved in Theorem 2.3.

17.7 After line 3 the last four bits y_0, y_1, y_2 and y_3 of y satisfy

$$y_i = \left(\sum_{j=0}^{7} x_{i+4j} \right) \bmod 2.$$

Algorithm 2.6 adds two further steps to compute $y_0 + y_1 + y_2 + y_3 \bmod 2$. In Algorithm 17.1 this computation is replaced by a table look up.

There are only 16 possible inputs and a table of 16 bits fits in the register. What happens in line 4 is that the ith bit of the constant 0x6996 is moved to the left where $i = (y_3, \ldots, y_0)_2$. The constant 0x6996 is chosen in such a way that this bit has the right value.

In comparison to Algorithm 2.6 this "in register table look-up" saves two instructions (at least if loading the constants does not count).

This trick was found by Falk Hueffner [277].

17.8 The simplest way to enumerate the sequences with a perfect linear complexity profile is to use Theorem 2.19.

Theorem 2.19 allows us to choose $a_1, a_3, \ldots$ arbitrarily, while $a_0, a_2, \ldots$ are given by the recursion $a_{2i} = a_{2i-1} + a_i$.

So all we have to is to enumerate the n-tuples $a_1, \ldots, a_{2n-1}$. There are many ways to do this: lexicographic, Gray codes, ... see [157].

Note that we can determine in advance which a_i we have to change if we toggle a_{2j-1} (namely all a_i with $i \in \{2^k j, 2^k j - 2^{k-1} \mid k \in \mathbb{N}\}$).

So the enumeration must be no slower than the simple enumeration of all n-tuples.

17.9 By Theorem 3.3 we get a random Eulerian cycle of the de Bruijn graph starting with the edge $0 \to 0$ in the de Bruijn graph D_n by choosing a random oriented spanning subtree with root 0.

To do this we turn the counting procedure of Theorem 3.4 into an algorithm (see Algorithm 18.1).

A small example makes clear what happens in Algorithm 18.1. We want a random de Bruijn sequence of order 2.

The Laplacian matrix of D_2 is

$$L = \begin{pmatrix} 1 & -1 & 0 & 0 \\ 0 & 2 & -1 & -1 \\ -1 & -1 & 2 & 0 \\ 0 & 0 & -1 & 1 \end{pmatrix}.$$

Deleting the 0th column and row of L we get

$$L_0 = \begin{pmatrix} 2 & -1 & -1 \\ -1 & 2 & 0 \\ 0 & -1 & 1 \end{pmatrix}.$$

Now we choose a vertex $v = 10$ and an edge $e = 10 \to 01$. Our matrix splits into

$$L' = \begin{pmatrix} 2 & -1 & -1 \\ 0 & 1 & 0 \\ 0 & -1 & 1 \end{pmatrix} \quad \text{and} \quad L'' = \begin{pmatrix} 2 & -1 & -1 \\ -1 & 1 & 0 \\ 0 & -1 & 1 \end{pmatrix}.$$

Algorithm 18.1 Choosing a random de Bruijn sequence

1: {initialization}
2: Let L be the Laplacian matrix
3: $L \leftarrow L_0$ {see Theorem 3.4}
4: Set the current subtree to the empty set.
5: **repeat**
6:     Choose a vertex v different from 0 for which you don't have an edge in the current subtree.
7:     Choose an edge e starting at v.
8:     Write $L = L' + L''$ where L' is matrix for the graph with the edge e removed and L'' is the graph with all edges starting at v except with e removed.
9:     Add e to your current subtree with probability $p = \frac{\det L''}{\det L}$.
10:    If you add e, set $L \leftarrow L''$, otherwise set $L \leftarrow L'$.
11: **until** you have an oriented spanning subtree.
12: Turn the oriented spanning subtree into an Eulerian cycle as described in Theorem 3.3.
13: Choose a random starting point in the Eulerian cycle and output the corresponding de Bruijn sequence.

Computing the determinants we get $\det L_0 = 2$ (as proved in Sect. 3.1), $\det L' = 2$ and $\det L'' = 0$. Hence $p = 0$, which means that e cannot be part of any oriented spanning subtree. So we must put the $e = 10 \to 00$ into the tree and continue with $L = L'$.

In the next step we choose $v = 10$ and $e = 10 \to 01$. Now we get

$$L' = \begin{pmatrix} 1 & 0 & -1 \\ 0 & 1 & 0 \\ 0 & -1 & 1 \end{pmatrix} \quad \text{and} \quad L'' = \begin{pmatrix} 1 & -1 & 0 \\ 0 & 1 & 0 \\ 0 & -1 & 1 \end{pmatrix}.$$

We get $\det L' = \det L'' = 1$ and hence $p = 1/2$. So we have to choose with probability $1/2$ either $e = 10 \to 01$ or $10 \to 11$ as part of our oriented spanning subtree.

After this choice, the remaining part of the oriented spanning subtree is uniquely determined and we have made a random choice between the two possible Eulerian cycles.

17.10 The non-linear combination function of the Geffe generator is $f(a, b, c) = ab + bc + c$.

A feedback polynomial of the sequence $L_1 L_2$ is, by Theorem 3.7:

$$f_{12}(z) = \text{Res}_x \left(f_1(x), f_2^*(zx) \right)$$
$$= z^{10} + z^8 + z^6 + z^5 + 1.$$

Fig. 18.1 Binomial
coefficients modulo 2

Similarly we get a feedback polynomial of the sequence $L_2 L_3$:

$$f_{23}(z) = \text{Res}_x \left(f_3(x), f_2^*(zx) \right)$$
$$= z^8 + z^7 + z^6 + z^4 + 1.$$

Finally we get a feedback polynomial of the Geffe generator, by Theorem 2.9, as

$$f(z) = f_{12}(z) f_{23}(z) f_3(z)$$
$$= z^{22} + z^{21} + z^{18} + z^{17} + z^{15} + z^{14} + z^{10} + z^7 + z^6 + z + 1.$$

Since the feedback polynomials f_1, f_2 and f_3 are all primitive and of pairwise different degree, Theorem 3.8 shows that the Geffe generator has linear complexity $5 \cdot 2 + 2 \cdot 4 + 4 = 22$ and hence f is indeed the minimal feedback polynomial of the Geffe generator.

17.11 For any prime p, let $n = \sum_i n_i p^i$ and $k_i = \sum_i k_i p^i$, then

$$\binom{n}{k} \equiv \prod_i \binom{n_i}{k_i} \quad \text{mod } p. \tag{18.1}$$

Specializing for $p = 2$ and $k = 2^j$ we get $\binom{n}{2^j} \equiv 1 \mod 2$ if and only if the jth position in the binary representation of n is a 1.

Hence

$$f_{j,n}(b_1, \ldots, b_n) = \sum_{S \in \binom{\{1,\ldots,n\}}{2^j}} \prod_{i \in S} x_i$$

where the sum runs over all subsets of $\{1, \ldots, n\}$ of size 2^j.

One can avoid the use of Eq. (18.1) and use a more elementary approach. Just draw the binomial coefficients modulo 2 (Fig. 18.1). The pattern is obvious and can easily be proved.

17.12 Use Wald's sequential test (Algorithm 15.1).

Let

$$l_i = \begin{cases} 1 + 2\varepsilon & \text{if } z_i = x_i, \\ 1 - 2\varepsilon & \text{if } z_i \neq x_i. \end{cases}$$

The test statistic is $L_n = \prod_{i=1}^{n} l_i$. We choose the alternative H_1 if the test statistic becomes $\leq A$ and we choose the hypothesis H_0 if the test statistic becomes $\geq B$.
The thresholds are $A = \frac{0.001}{1-0.01} \approx 10^{-3}$ and $B = \frac{1-0.0001}{0.01} \approx 10^2$.
Now let $\varepsilon = 0.1$. Then $E_{H_1}[\ln l_i] = \frac{1}{2}(\ln 1.2 + \ln 0.8) = -0.04$.
Let τ be the stopping time for our test. We have

$$E_{H_1}[\ln L_\tau] \approx 0.999 \ln A + 0.001 \ln B = -6.88.$$

(The $\approx$ sign would be an $=$ if we could be sure that the bounds A and B are met exactly.)
By the optimal sampling theorem (Theorem 15.3) we have

$$E_{H_1}[\ln L_\tau] = E_{H_1}[\tau] E_{H_1}[\ln l_i]$$

and hence $E_{H_1}[\tau] = 172.15$, i.e. we expect that we must observe 172.15 bits before we can safely reject the hypothesis.

17.13 By definition $f = (x \wedge f_h) \vee (\bar{x} \wedge f_l)$, and so on. The distributive law gives $f \wedge g \wedge h = (x \wedge f_h \wedge g_h \wedge h_h) \vee (\bar{x} \wedge f_l \wedge g_l \wedge h_l)$, which proves the recursion formula. One can turn the recursion formula directly into a top down algorithm by adding the starting point $0 \wedge g \wedge h = 0$, and so on. All function calls must be memoized to avoid duplicated work. In the simple Common Lisp implementation that I prepared for this book, the real code looks almost like pseudo-code (see Algorithm 18.2).

Algorithm 18.2 Binary Decision Diagrams: The ternary-and operator

```
(defun BDD-ternary-and (x y z)
  (cond ((eql x T) (BDD-and y z))
        ((eql x nil) nil)
        ((eql y T) (BDD-and x z))
        ((eql y nil) nil)
        ((eql z T) (BDD-and x y))
        ((eql z nil) nil)
        ((eql x y) (BDD-and x z))
        ((eql x z) (BDD-and y z))
        ((eql y z) (BDD-and x y))
        (t (let ((m (min (V x) (V y) (V z))))
             (BDD-make-node
              m
              (BDD-ternary-and (left x m) (left y m) (left z m))
              (BDD-ternary-and (right x m) (right y m) (left z m))))))))

(memoize 'BDD-ternary-and)
```

Let $f = (x_1 \wedge x_2 \wedge f_1) \vee (x_1 \wedge \overline{x_2} \wedge f_2)$, $g = (x_1 \wedge x_2 \wedge g_1) \vee (\overline{x_1} \wedge x_2 \wedge g_2)$ and $h = (x_1 \wedge \overline{x_2} \wedge h_1) \vee (\overline{x_1} \wedge x_2 \wedge h_2)$ then $f \wedge g = (x_1 \wedge x_2 \wedge f_1 \wedge g_1)$, $f \wedge h =$

$(x_1 \wedge \overline{x_2} \wedge f_2 \wedge h_1)$ and $g \wedge h = (\overline{x_1} \wedge x_2 \wedge g_2 \wedge h_2)$. Thus for general f_1, f_2, g_1, g_2, h_1 and h_2 the size of the BDD for the binary-and will grow quadratically with the size of the input BDDs. But the ternary-and will reach, after two recursion steps, the endpoints of the recursion that simplifies to 0.

On the other hand, let f', g' and h be three functions for which the size BDD of $f' \wedge g' \wedge h$ is cubic in comparison to the input BDDs. Let $f = f' \wedge x_n$ and $g = g' \wedge \overline{x_n}$. Then computing $f \wedge g$ needs at most quadratic time and space. Since $f \wedge g = 0$, the time needed for the computation of $(f \wedge g) \wedge h$ is essentially the time needed to compute $f \wedge g$. Hence, in this case, two applications of the binary-and algorithm are faster than the ternary-and.

17.14

(a) If $f = 0$ then $f(y') = 0$, no matter which value y' has.
(b) If $g = 1$ then by definition $y' = y$ and hence $(f \downarrow 1)(y) = f(y)$ for all y.
(c) If $g \neq 0$ then $(1 \downarrow g)(y) = 1(y') = 1$ for all y and $(g \downarrow g)(y) = g(y') = 1$ (the last equality follows from the definition of y').
(d) If $g_h = 0$ then for $x = 0x_2x_3 \cdots x_n$ and $y = 1x_2x_3 \cdots x_n$ we get $x' = y' = 0x'_2 \cdots x'_n$ and hence $f \downarrow g(x) = f \downarrow g(y) = f(x') = f_l(x'_2, \ldots, x'_n) = f_l \downarrow g_l(x_2, \ldots, x_n)$.
    Similarly we get $f \downarrow g = f_h \downarrow g_h$ if $g_l = 0$.
    If $g_l \neq 0$ and $g_h \neq 0$ then the first coordinate of y' is always identical to the first coordinate of y and hence

$$f \downarrow g(y) = f(y')$$
$$= y'_1 ? f_h(y'_2, \ldots, y'_n) : f_l(y'_2, \ldots, y'_n)$$
$$= y_1 ? f_h \downarrow g_h(y_2, \ldots, y_n) : f_l \downarrow g_l(y_2, \ldots, y_n).$$

Turning this into a recursive algorithm is straightforward. The conditions (a)–(c) are the endpoints of the recursion. Again we need memoization to make the algorithm effective.

Algorithm 18.3 shows example code in Common Lisp which is almost as simple as pseudo-code.

The constraint operator was introduced in [61].

17.15 As variables we choose the outputs $x_1, x_2, x_3, \ldots$ of the LFSR in their natural order. We need a BDD that describes the feedback function of the LFSR. Here we can use essentially the same BDD as in the attack against E_0, see Fig. 5.9 (a).

In addition, we need a BDD that describes the compression function. We need two nodes to describe the next step for the case where the first $2k$ bits of the LFSR sequences produce j output bits (see Fig. 18.2). Whether the high or low pointer of the node label x_i is connected to $\bot$ depends on the value z_{j+1} of the $(j + 1)$-th observed bit. Figure 18.2 shows the drawing for the case $z_{j+1} = 1$.

Algorithm 18.3 Binary Decision Diagrams: The constrain operator

```
(defun BDD-constrain (f g)
  (cond ((eql g nil) nil)    ; note that we must check for g=0 first
        ((eql g T) f)
        ((eql f nil) nil)
        ((eql f T) T)
        ((eql f g) g)
        (T (let ((m (min (V f) (V g)))
                 (fl (left f m))
                 (fh (right f m))
                 (gl (left f m))
                 (gh (right f m)))
             (cond ((eql gl nil) (BDD-constrain fh gh))
                   ((eql gh nil) (BDD-constrain fl gl))
                   (T (BDD-make-node
                        m
                        (BDD-constrain fl gl)
                        (BDD-constrain fh gh)))))))))

(memoize 'BDD-constrain)
```

Fig. 18.2 Basic BDD for attacking the self-shrinking generator

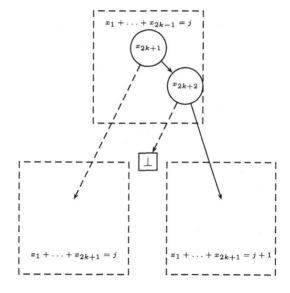

Instead of building the BDD directly from Fig. 18.2 we can use melting to construct it from more elementary BDDs. All we need is a BDD for the predicate

$$P_{j,k}(x_1, \ldots, x_k) = \begin{cases} 1 & \text{if exactly } j \text{ of the } k \text{ variables } x_1, \ldots, x_k \text{ are true,} \\ 0 & \text{otherwise.} \end{cases}$$

The BDD for $P_{j,k}$ is very simple and I have added it as an elementary function to my BDD C-library. Once we have the predicate $P_{j,k}$ we can build the BDD for the expression

$$\left[P_{j,k}(x_1, \ldots, x_{2k-1}) \wedge (x_{2k+1} = 1)\right] \Rightarrow (x_{2k+2} = z_{j+1})$$

by melting. The advantage of this approach is that it is not necessary to decide how many output bits we want to use in advance. We can extend the BDD step by step.

Let us now estimate the size of the BDDs. When we use $2k$ bits, the size of the BDD sketched in Fig. 18.2 is $2\binom{k}{2}$, which is quite small. The size of the BDD describing the LFSR grows as 2^{2k} (as long as $2k < n$). So the size of the BDD that describes the self-shrinking generator is bounded by $k^2 2^{2k}$.

On the other hand, there are 2^{2k} possible bit sequences of length $2k$, but only 3^k of them are mapped by the self-shrinking compression to the observed output sequence. So we estimate that only $2^n \frac{3^k}{2^{2k}}$ of the 2^n possible initial values of the LFSR (the size of the LFSR is n) will lead to a sequence for which the first k steps are compatible with the observed output. This bounds the size of the BDD by $(2k)2^n \frac{3^k}{2^{2k}}$.

These two bounds meet for $k \approx \frac{n}{4 - \log_2 3}$. At this point the BDD reaches its largest size, which is $\approx 2^{0.82n}$.

The self-shrinking generator was introduced in [188]. A BDD-attack against the self-shrinking generator was first described in [166].

17.16 $f(\hat{x}, \hat{y}) = g(\hat{x}, \hat{y}) = 0$ means that the polynomials $f(\hat{x}, y)$ and $g(\hat{x}, y)$ have a common zero $\hat{y}$. Hence $\text{Res}_y(f(\hat{x}, y), g(\hat{x}, y)) = 0$. Thus $\text{Res}_y(f, g)$, which is a polynomial in x, has the zero $\hat{x}$.

Thus by resultants one can reduce the problem of solving a system of non-linear equations to the problem of finding zeros of a univariate polynomial.

Solving non-linear systems of equations by resultants is quite successful in computational algebraic geometry, but in cryptography Gröbner bases or linearization are normally more effective. Anyway, resultants are tool worth knowing about.

17.17 For every $\alpha \in \mathbb{N}^n$ the number of elements $\beta \in \mathbb{N}^n$ with $|\beta| \le |\alpha|$ is finite (to be exact the number is $\binom{n+|\alpha|-1}{|\alpha|}$). Hence every descending chain in a graded order must become stationary.

17.18 In this case the control and the output sequence are identical. Hence all zeros will be deleted from the output sequence. The output of the shrinking generator will be constant.

17.19 Let $(s_i)_{i \in \mathbb{N}}$ be the sequence generated by the LFSR of the self-shrinking generator. The self-shrinking generator can be interpreted as a normal shrinking generator with the control sequence defined by $c_i = s_{2i-1}$ and the generation sequence defined by $x_i = s_{2i}$.

Golic's correlation attack (see Sect. 7.3.3) makes no assumption about the form of the control and the generation sequence. So we can use it to reconstruct the sequence $x_i = s_{2i}$. Once the attacker knows every second bit of the LFSR sequence it is not difficult to invert the LFSR and reconstruct the seed.

As shown in Sect. 7.3.3, the correlation decreases as $\frac{1}{\sqrt[4]{i}}$, so one can use only the first 10000 to 100000 bits before the loss of synchronization becomes relevant.

If the LFSR is large enough (128 bits or better 256 bits) the self-shrinking generator is still secure.

17.20 If $j = i + 1$ and $S[j] = 1$ then the following happens during one step of the RC4 generator: i is increased by i. Now $S[i] = S[j] = 1$ and hence j is increased by one. $S[i]$ and $S[j]$ are swapped. After the swap $S[j] = 1$.

Thus the condition $j = i + 1$ and $S[j] = 1$ is again satisfied and the only thing that the RC4 generator does is to increase i and j by 1. Thus after n steps we will be back at the same point where we started.

Since RC4 is invertible we cannot reach this cycle if we are not in it right from the beginning. However, RC4 is initialized with $i = j = 0$, i.e. at the beginning we are not in the cycle and hence it can never be reached.

This cycle was first noted by Finny [91].

17.21 This is an open-ended problem. Here is just one possible answer:

The many subgroups of $(\mathbb{Z}/2\mathbb{Z})^{256}$ allow a lot of fortuitous states. For example, if the first 8 elements of S form a permutation of $\{0, \ldots, 7\}$, this partial S-box is a fortuitous state of length 8. This already gives 8! fortuitous states of length 8. Moving the region around further increases the number of fortuitous states. A high number of fortuitous states increases the probability that an adversary can mount an attack on them.

Another reason why Algorithm 17.2 is weak is that the more regular jumps of j reduce the number of necessary guessing steps in a state recovering attack. Hence state recovering attacks against Algorithm 17.2 will be faster.

17.22 Let D_n be the set of integers x in the range $[1, \ldots, \frac{n-1}{2}]$ with $(\frac{x}{n}) = 1$. Let $f : D_n \to D_n$ be the permutation that is defined by

$$f(x) = \begin{cases} x^2 \bmod n & \text{if } x^2 \bmod n \leq \frac{n-1}{2}, \\ -x^2 \bmod n & \text{if } \frac{n+1}{2} \leq x^2 \bmod n \leq n - 1. \end{cases}$$

The predicate $B_n : D_n \to \{0, 1\}$ is

$$B_n(x) = \begin{cases} x \bmod 2 & \text{if } x \text{ is a quadratic residue modulo } n, \\ x + 1 \bmod 2 & \text{if } x \text{ is a quadratic non-residue modulo } n. \end{cases}$$

With these functions Algorithm 11.1 computes the Blum-Blum-Shub sequence.

B_n is hard, since it is the decision of the quadratic residue problem. You can easily choose a random element in D_n, so the predicate is accessible. Thus all conditions of Theorem 11.1 are satisfied.

17.23 Consider a 1-bit of x, by the rotations it is moved to every position, thus the contribution of this bit to the whole sum is $\sum_{i=0}^{31} = 2^i = 2^{32} - 1$. In the two's complement representation $2^{32} - 1$ is identified with -1. So the sum $\sum_{i=0}^{31}(x \lll i)$ adds -1 for every 1-bit of x, which is exactly the statement of Eq. (17.1).

Computing SADD(x) by Eq. (17.1) needs 64 operations which is much slower than any of the algorithms presented in Sect. 12.1.2.

17.24 Here is an example of what the single operations do:

x	1001101000
$x - 1$	1001100111
$x \,\&\, (x - 1)$	1001100000

$x - 1$ differs from x by changing the tailing zeros into ones and the preceding 1 into a 0. Hence $x \leftarrow x \,\&\, (x - 1)$ removes the leftmost bit.

One can turn this into an algorithm for computing SADD(x) by successively removing bits and counting how often this is possible.

Algorithm 18.4 Sideway addition for sparse words

1: $s \leftarrow 0$
2: **while** $x \neq 0$ **do**
3:    $x \leftarrow x \,\&\, (x - 1)$
4:    $s \leftarrow s + 1$
5: **end while**
6: **return** s

Algorithm 18.4 needs 4 operations per loop (including the comparison with 0) and hence $4 \, \text{SADD}(x)$ in total. It beats the algorithm from Sect. 12.1.2 if SADD(x) is at most 3 or 4.

This method was developed by Wegner [279].

17.25

(a) $\left(\frac{a}{bb'}\right) = \left(\frac{a}{b}\right)\left(\frac{a}{b'}\right)$ is part of the definition of the Jacobi symbol. $\left(\frac{aa'}{b}\right) = \left(\frac{a}{b}\right)\left(\frac{a'}{b}\right)$ follows immediately from Corollary 13.11 and the definition of the Jacobi symbol.

(b) Let $b = p_1^{e_1} \cdots p_k^{e_k}$. Since for each prime factor p_i of b the two numbers a and $a \bmod b$ are in the same equivalence class modulo p_i and $\left(\frac{a}{p_i}\right) = \left(\frac{a \bmod b}{p_i}\right)$ by the definition of the Legendre symbol. Hence by definition of the Jacobi symbol

$$\left(\frac{a}{b}\right) = \left(\frac{a}{p_1}\right)^{e_1} \cdots \left(\frac{a}{p_n}\right)^{e_n}$$
$$= \left(\frac{a \bmod b}{p_1}\right)^{e_1} \cdots \left(\frac{a}{p_n}\right)^{e_n}$$
$$= \left(\frac{a \bmod b}{b}\right).$$

(c) We prove the result by induction on a and b. If a and b are primes this is the statement of the quadratic reciprocity law (Theorem 13.12).

If $a \equiv a'a''$, with $a', a'' \geq 3$ then

$$\left(\frac{a}{b}\right) = \left(\frac{a'}{b}\right)\left(\frac{a''}{b}\right) \quad \text{(by (a))}$$

$$= (-1)^{(a'-1)(b-1)/4}\left(\frac{b}{a'}\right)(-1)^{(a''-1)(b-1)/4}\left(\frac{b}{a''}\right) \quad \text{(by induction)}$$

$$= (-1)^{(a-1)(b-1)/4}\left(\frac{b}{a}\right)$$

$$\big(\text{by (a) and checking all 8 cases for } a', a'', b \text{ modulo 4.}\big)$$

Exactly the same argument handles the case $b = b'b''$.

(d) Let $b = p_1^{e_1} \cdots p_k^{e_k} q_1^{d_1} \cdots q_j^{d_j}$ be the factorization of b with $p_i \pm 1 \mod 8$ for $i = 1, \ldots, k$ and $q_i \equiv \pm 3 \mod 8$ for $i = 1, \ldots, j$.

Then by definition of the Jacobi symbol and Theorem 13.12 we obtain:

$$\left(\frac{2}{b}\right) = \left(\frac{2}{p_1}\right)^{e_1} \cdots \left(\frac{2}{p_k}\right)^{e_k}\left(\frac{2}{q_1}\right)^{d_1} \cdots \left(\frac{2}{q_j}\right)^{d_j}$$

$$= 1^{e_1} \cdots 1^{e_k}(-1)^{d_1} \cdots (-1)^{d_j}$$

$$= (-1)^{d_1+\cdots+d_k}.$$

But $b \equiv \pm 3 \mod 8$ if and only if it has an odd number of factors of the form $\pm 3 \mod 8$, which proves (d).

(e) By definition $\left(\frac{1}{b}\right) = 1$ for all b. This is what Algorithm 13.1 does in line 2. If $a = 2^k a'$ with a' odd then by (a), we obtain

$$\left(\frac{a}{b}\right) = \left(\frac{2}{b}\right)^k \left(\frac{a'}{b}\right).$$

By (d) $\left(\frac{2}{b}\right) = -1$ if and only if $b \equiv \pm 3 \mod 8$. Thus the condition in line 6 holds if $\left(\frac{2}{b}\right)^k = -1$.

If a is odd, Algorithm 13.1 uses (c) to express $\left(\frac{a}{b}\right)$ in terms of $\left(\frac{b}{a}\right)$. $\left(\frac{b}{a}\right)$ is immediately reduced to $\left(\frac{b \mod a}{a}\right)$ by (b). There are cases where $b \mod a = 0$ and hence $\left(\frac{b}{a}\right) = \left(\frac{0}{a}\right) = 0$. These cases are caught by lines 13–15. If $b \equiv a \equiv 3 \mod 4$ then $(-1)^{(a-1)(b-1)/4} = -1$ and by (c) $\left(\frac{a}{b}\right) = -\left(\frac{b}{a}\right)$ (line 17) or finally $(-1)^{(a-1)(b-1)/4} = 1$ and by (c) $\left(\frac{a}{b}\right) = \left(\frac{b}{a}\right)$ (line 19).

In the recursive step Algorithm 13.1 reduces the size of a and b. So the algorithm must finally stop at line 2 if $\gcd(a, b) = 1$ or at line 14 if $\gcd(a, b) > 1$.

As in the Euclidean algorithm, the worst case is if a and b are consecutive Fibonacci numbers. Thus the runtime is bounded by $\log(a)/\log((\sqrt{5}+1)/2)$.

17.26 Consider the lattice with basis

$$\begin{pmatrix} 1 & & 0 & 0 \\ & \ddots & & \vdots \\ 0 & & 1 & 0 \\ a_0 & \cdots & a_{k-1} & s \end{pmatrix}.$$

If $s = \sum_{i=0}^{k-1} a_i x_i$ then $v = (x_0, \ldots, x_{k-1}, 0)^t$ is an element of the lattice. The length of v is at most $\sqrt{k}$ which is very short in comparison to the basis vectors used to define the lattice, which are all of length about 2^k.

Thus v is (most likely) the first element of the LLL-base.

Applying the LLL algorithm to the given weights a_i and $s = 12938$ in the exercise yields $v_{min} = (1, 0, 0, 0, 1, 0, 0, 0, 1, 0, 0)^t$ as first vector in the LLL-basis, i.e. $x = 273$ or KN.

The second number in the message $s = 14196$ gives the vector $v_{min} = (1, 1, 1, 1, 0, 0, 0, 0, 0, 0, 0)^t$

The third number $s = 16432$ is a bit tricky. We obtain

$$v_{min} = (1, 0, 0, -1, 0, 0, 0, 0, 0, 0, 2)^t$$

as the minimal element in the LLL-basis, but the second element in the basis is $v = (0, 0, 1, 0, 1, 0, 1, 0, 1, 1, 1, 0, 0)^t$, which is the desired solution of the knapsack problem. This is a heuristic approach, there is no guarantee that it will always work.

Finally $s = 18620$ gives us $v = (0, 1, 1, 1, 1, 1, 0, 0, 0, 0, 0)^t$ as the second element in the LLL-basis.

The message is KN AP SA CK.

This attack is only the beginning. More elaborate versions (see [200, 243]) use the LLL-algorithm to construct a c' and m' such that the sequence $b'_i = a_i c^{-1}$ mod m' satisfies $b'_{i+1} \geq 2b'_i$. This allows the attacker to decode the ciphertext without always running the relatively slow LLL-algorithm. Nevertheless, our basic attack shows the danger that the LLL-algorithm provides for most knapsack-based cryptosystems.

Answer to the bonus question:

The toy example fails against a guess and check attack. If the attacker is able to guess a part of the plaintext, he can easily check his guess was correct. This is normally something you want to prevent.

A way to deal with this problem is to add some random bits as padding, but the padding expands the cipher's ciphertext which is often unacceptable. This is another reason why asymmetric ciphers play almost no role in data encryption.

17.27 Let $m = \lceil \sqrt{n} \rceil$. Then $p = m + \epsilon_1$ and $q = m - \epsilon_2$ with $0 \leq \epsilon_1, \epsilon_2 \leq \sqrt[4]{n}$. The equation $n = pq$ is equivalent to

$$n = m^2 + m(\epsilon_1 - \epsilon_2) - \epsilon_1 \epsilon_2$$

or

$$\epsilon_1 - \epsilon_2 = \frac{n - m^2 + \epsilon_1 \epsilon_2}{m} = \left\lceil \frac{n - m^2}{m} \right\rceil$$

where the last equation holds since $\epsilon_1 \epsilon_2 \le m$.

Once we know $\delta = \epsilon_1 - \epsilon_2$ we just solve the quadric equation $n = (m + \epsilon_1)(m - \epsilon_1 + \delta)$ to obtain ϵ_1 and factor n.

(Note: This exercise is a special case of Coppersmith's method to factor $n = pq$ if at least the upper half bits of p are known [57].)

17.28 Let F be a skew field over $\mathbb{F}_p$ of order n and let C be the centralizer of F. The class formula states that

$$|F| = |C| + \sum_x |C(x)|$$

where the sum runs over all classes. The size of the class $C(x)$ is $\frac{p^n - 1}{p^{n_x} - 1}$ where p^{n_x} is the size of the stabilizer of x. Hence we have:

$$p^n - 1 = p^c - 1 + \sum_x \frac{p^n - 1}{p^{n_x} - 1}. \tag{18.2}$$

By Zsigmondy's theorem either $n = 2$ or $n = 6$ and $p = 2$ or there exists a prime q that divides $p^n - 1$, but not $p^k - 1$ for $k < n$. In the small cases we can check directly that F must be a commutative field.

So assume that there exist a Zsigmondy prime q such that q divides the left-hand side of Eq. (18.2) and all terms in the sum on the right-hand side. Then it must also divide the last term $p^c - 1$. By the property of a Zsigmondy prime we conclude that $c = n$ and hence F is a commutative field.

This is one of Wedderburn's original proofs.

17.29 This is just a calculus problem related to Theorem 15.1.

Both distributions have a Lebesgue density. The density of $\mathcal{N}(0, 1)$ is $\varphi_1(x) = \frac{1}{\sqrt{2\pi}} e^{-x^2/2}$ and the density of $\mathcal{N}(0, 2)$ is $\varphi_2(x) = \frac{1}{2\sqrt{\pi}} e^{-x^2/4}$ (see Fig. 18.3).

The graphs of φ_1 and φ_2 intersect at $x_0 = \sqrt{4 \ln \sqrt{2}} \approx 1.17$ and $-x_0$. Hence by Theorem 15.1

$$\|\mathcal{N}(0, 1) - \mathcal{N}(0, 2)\| = \frac{1}{2} \int_{-\infty}^{\infty} |\varphi_1(x) - \varphi_2(x)| dx$$

$$= \frac{1}{2} \int_{-x_0}^{x_0} \varphi_1(x) - \varphi_2(x) dx + \int_{-\infty}^{x_0} \varphi_2(x) - \varphi_1(x) dx$$

$$= \int_0^{x_0} \varphi_1(x) - \varphi_2(x) dx + \int_{-\infty}^{x_0} \varphi_2(x) - \varphi_1(x) dx.$$

Since the antiderivative of e^{-x^2} has no closed form the integrals must be evaluated numerically. We reduce the problem to calculating the quantiles of the standard

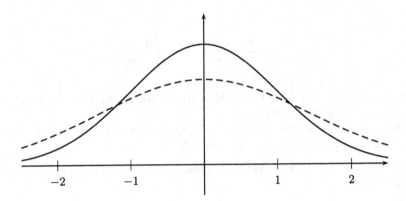

Fig. 18.3 The densities of $\mathcal{N}(0, 1)$ and $\mathcal{N}(0, 2)$

normal distribution

$$\Phi(x) = \int_{-\infty}^{x} \frac{1}{\sqrt{2\pi}} e^{x^2/2}.$$

The function Φ is tabulated.

$$\|\mathcal{N}(0, 1) - \mathcal{N}(0, 2)\| = \Phi(x_0) - \Phi(x_0/\sqrt{2}) + \Phi(-x_0/\sqrt{2}) - \Phi(-x_0)$$
$$= 2\Phi(X_0) - 2\Phi(x_0/\sqrt{2})$$
$$\approx 0.1646.$$

The maximal success probability for distinguishing an $\mathcal{N}(0, 1)$ and an $\mathcal{N}(0, 2)$ distribution is therefore 0.1646.

Part V
Programs

Chapter 19
An Overview of the Programs

For the reader's convenience I have implemented most of the algorithms described in this book. Most programs are in C++, but I also use Common Lisp and C where it seems appropriate. I do not claim that these programming languages are the best for the given problem. I just like C and C++ as near-machine languages and Common Lisp is my favorite language for high level abstraction.

You can download the programs for free at http://cage.ugent.be/~klein/streamcipher. The reader may use and modify the programs as he wishes. You may use small code snippets without giving a reference, but if you distribute modified versions of the programs you should cite this book and allow the user to modify your modifications.

The programs are planned for research experiments and not for use in high security settings, so most of the work was focused on clear and simple code; I did not strengthen the implementation against side channel attacks.

Special attention was given to extensive documentation. For the interface of the C++ libraries I use the tool Doxygen [125]. It can generate interface documentation in different formats (HTML, LaTeX (PDF), Unix-Manpages). Figure 19.1 shows as an example a part of the documentation of the LFSR library in an HTML-browser.

For the documentation of the implementation of the algorithms I developed a new literate programming tool which is described in Chap. 20. Figure 19.2 shows as an example a page from the implementation of the Berlekamp-Massey algorithm.

Here follows a list of all programs. Unless otherwise stated, the programs are written in C++ and run on all platforms.

Chapter 2

- *LFSR.hpp* and *LFSR.web*: A library for LFSR. It is optimized for convenience not for speed.
- *linear_complexity.hpp* and *linear_complexity.web*: An implementation of the Berlekamp-Massey algorithm and the asymptotic fast variant described in Sect. 2.4.2.
- *LFSR-test.web*: A C-Program which implements the different algorithms for simulating LFSRs in software, which are described in Sect. 2.6.2. This program con-

A. Klein, *Stream Ciphers*, DOI 10.1007/978-1-4471-5079-4_19,
© Springer-Verlag London 2013

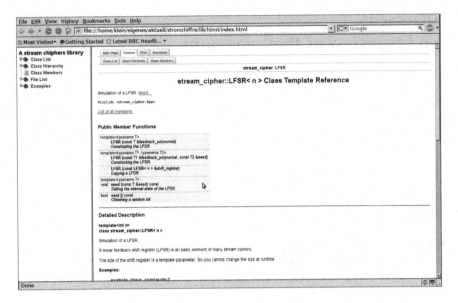

Fig. 19.1 An example of the Doxygen documentation

Fig. 19.2 An example of the pweb documentation

The definitions for all functions are in the header file and the interface is documented by Doxygen.

@c
#include⟨linear_complexity.hpp⟩

2 The Berlekamp-Massey Algorithm

In 1968 E. R. Berlekamp presented an efficient algorithm for decoding BCH-Codes. (BCH-Codes are an important class of cyclic error-correcting codes.) One year later Massey noticed that the decoding problem is in its essential parts equivalent to the determination of the shortest LFSR that generates a given sequence.

For a detailed description of the algorithm see Section 2.4.2 of the book.

2.1 Read a bit

One advantage of the Berlekamp-Massey Algorithm is its iterative nature. It always investigates one bit at a time and compares the new bit with its prediction. If the prediction is wrong, it has to update the guess for the feedback polynomial.

Our implementation reflects this as follows. The function `read_next_bit` has a bool as input an returns `true` if the prediction of Berlekamp-Massey Algorithm was correct.

The linear complexity and the feedback polynomial are stored in internal data structures and can be accessed via `get_linear_complexity` and `get_feedback`.

@c

```
bool stream_cipher::BM_algorithm::read_next_bit(bool bit) {
    X.push_back(bit);
    bool d ← false;

    for(int j←0; j≤Li; j++)
        if (f[j] ∧ X[i+1-f.size()+j])
            d ← ¬d;
    if(¬d) {
        i++;
        return true; // do nothing
    } else {
        ⟨UPDATE POLY, 2.2⟩
        i++;
        return false;
    }
}
```

tains a small piece of assembler code (see Algorithm 2.4). Thus it needs an ix86 platform and the GNU gcc.

Chapter 3

• *Geffe.hpp* and *Geffe.web*: An implementation of the Geffe generator.

Chapter 4

• *preprocessing2.web*: There are different programs for searching relations (see Sect. 4.1.3). The reason is that, even if many parts of the programs are equal, each possible weight has its specialities. This program is specialized for relations of weight 2. It can also search for extended relations as described in Sect. 4.1.4.

  All pre-processing programs have no input routines. You must enter the information about the LFSR that should be attacked into the source code.
• *preprocessing3.web*: This is the program for relations of weight 3. It implements the trick described in Sect. 4.1.3 to speed up the program by a factor of 3 in comparison to the generic algorithm.
• *preprocessing4.web*: This is the program for relations of weight 4. As explained in Sect. 4.1.3, for this case it is better to use a sorted list instead of a hash table as the main data structure.

  It implements the trick described in Sect. 4.1.3 to speed up the program by a factor of 2 in comparison to the generic algorithm.
• *preprocessing5.web*: This is the program for relations of weight 5. It implements the trick described in Sect. 4.1.3 to speed up the program by a factor of about 8 in comparison to the generic algorithm.
• *text2bin.web* and *bin2text.web*: To keep the output platform independent, the preprocessing program writes the relation set in a simple text format. If you want you can convert the relation set to a binary format. The binary format is faster and needs less space on the hard disk, but it is no longer portable between different architectures.
• *attack-CJS-simple-count.web* and *attack-CJS-Fourier.web*: These programs implement the basic version of the CJS-attack. They differ in the method used to evaluate the relations. The first program uses the sideway addition in arrays (see Sect. 4.1.6.1) and the second program uses the fast Fourier transform (see Sect. 4.1.6.2).

  Compare the speed of the two programs. In general the first is better for cases with few relations of low weight and the second is better for many relations of high weight.
• *attack-CJS-extended-rel.web*: A version of the CJS attack that uses extended relations (see Sect. 4.1.4).
• *attack-Viterbi.web*: An attack based on convolutional codes and Viterbi decoding (see Sect. 4.2.3.1).
• *attack-sparse.web*: An implementation of the attack of W. Meier and O. Staffelbach against an LFSR with a sparse feedback polynomial (see Sect. 4.3).

Chapter 5

- A strange feature of the programming language C is that it has many functions for dealing with floating point numbers, but almost no functions for dealing with integers. Even the function max with integer arguments is missing. Since I work a lot with integers, I decided to solve the problem once and for all.

  inttools.h and *inttools.c* provide a full featured C library for dealing with integers.

- *BDD.h* and *BDD.web*: A full BDD library written in C. It is optimized for compact memory and efficient use of the cache.

  To test the library I have prepared the following programs.

  queen.web: A program solving the n-queen problem. Note that BDDs are not the optimal way to solve the n-queen problem, but it is a good stress test and can be used to compare different BDD libraries.

  BDDcalc.web A reverse Polish notation style calculator for BDDs.

- *simpleBDD.web*: A simple BDD library written in Common Lisp. It has almost no memory management and it is intended to be used to quickly test new ideas.

- *E0.hpp* and *E0.web*: A simplified implementation of the cipher E_0 as described in Sect. 5.2.1.

- *E0-attack.web*: A BDD-based attack against E_0, as described in Sect. 5.2.2.

Chapter 6

- *LILI.sing*: An input file for the computer algebra system Singular, which does the degree reduction for LILI-128 as described in Sect. 6.3.1.

- *E0.sing*: An input file for Singular, which does the pre-processing for the algebraic attack against E_0 as described in Sect. 6.3.2.

Chapter 7

- *StopAndGo.hpp* and *StopAndGo.web*: An implementation of the stop-and-go generator (see Fig. 7.1).

- *StepOnceTwice.hpp* and *StepOnceTwice.web*: An implementation of the step-once-twice generator (see Sect. 7.1).

- *AlternatingStep.hpp* and *AlternatingStep.web*: An implementation of the alternating step generator (Algorithm 7.1).

- *Shrinking.hpp* and *Shrinking.web*: A simple implementation of the shrinking generator (see Sect. 7.3.1). It takes no precaution to avoid side channel attacks. The program *timing_attack.web* demonstrates a simple timing attack on this example. For measurement it uses the ix86 instruction RDTSC. Thus this program is not platform independent.

Chapter 8

- *A5-1.web* and *A5-2.web*: Simulation of the A5 algorithms in software.

- *A5-1-special.web*: Generating special states of A5/1 as described by Algorithm 8.3.

- *A5-1-invert.web*: The fact that A5/1 can be effectively inverted is crucial for the attack described in Sect. 8.3.2. This program demonstrates this part of the attack.

Chapter 9

- *RC4.hpp* and *RC4.web*: An implementation of RC4. The implementation contains a lot of checks against fault attacks and trades speed against security.
- *digraph-simple.web*: A direct translation of the pseudo-code of Algorithm 9.10 into Common Lisp.
- *fortuitous-state.web*: A highly optimized program to search fortuitous states that was used to create Table 9.3.
- *RC4-FMS.web*: A demonstration of the FMS-attack (see Sect. 9.4).
- *RC4-attack.web*: A demonstration of the attack based on Golić's relation (see Sect. 9.5).

Chapter 10

- *trivium4.hpp* and *trivium.web*: An implementation of the cipher Trivium. The program favors clarity over speed.
- *rabbit.hpp* and *rabbit.web*: An implementation of the cipher Rabbit. The program favors clarity over speed.
- *moustique.hpp* and *moustique.web* the cipher Moustique.

Chapter 12

- *popcnt.web*: A C implementation of different algorithms for the sideway addition (population count) problem.
- *EK-bitset.tar.gz*: An optimized C++ bitset library similar to the standard bitset class.
- *matrix-multiplication.web*: The matrix multiplication algorithms described in Sect. 12.5.1.

Chapter 14

- *irrpoly.gap*: A GAP program that generates a random irreducible polynomial with Theorem 14.2.
- *primelt.gap*: A GAP program that generates a random primitive with Algorithm 14.1.

Chapter 20

- *pweb.tar.gz*: The pweb system used to document the programs described in Chap. 20.

To make use of all programs you must install the following software on your system. The makefiles assumes that everything is stored in standard places. If you have installed the software in unusual places you will have to add appropriate path information.

- A working make (for example [174]).
- A working LaTeX environment (for example [264]).
- A implementation of the M4 macro processor (for example [238]).
- The pweb system (Chap. 20).
- The optimized C++ bitset library EK-bitset (see above).

- A C-Compiler that supports (at least partially) the C99 standard. The GNU C Compiler will work [102].
- A C++ compiler (for example [102]).
- The Boost C++ Libraries [35].
- The curses library for text based interfaces (for example [65]).
- The Doxygen code documentation tool [125].
- V. Shoup's NTL: A Library for doing Number Theory [244].
- The GNU Multiple Precision Arithmetic Library (needed by NTL) [105].
- A Common Lisp environment. I have tested all Lisp programs with clisp [54] and sbcl [233], but other Common Lisp implementations should work without any difficulties.
- The computer algebra system Singular [253].
- The computer algebra system GAP [99].

For all these components there exist freely available implementations for all relevant platforms.

Chapter 20
Literate Programming

20.1 Introduction to Literate Programming

Most people working in mathematics know D.E. Knuth's TEX typesetting system. Less well-known is that as a companion to TEX Knuth developed a new programming style called literate programming (see [158]). The idea is that a program should read like an article.

A literate programming environment, also called a web-system, consists of two programs called `tangle` and `weave`. Figure 20.1 illustrates the work flow in a literate programming environment.

The tangle program extracts the source code of the web-file and generates an input file for the compiler (interpreter) of our programming language. It must resort the code from the human readable form into the machine readable form.

The weave program generates the input for a text processor. It must apply syntax highlighting to the code and generate an index, etc.

In principle you can use any text processing tool and any computer language as a basis for a literate programming environment. In practice the weave and tangle program need some information about the destination language, so a specific literate programming environment supports only a few languages.

20.2 Pweb Design Goals

When I started this project, I found that no literate programming tool satisfies my exact needs. So I began with the development of a new literate programming tool. I called it `pweb`, since its main part comprises a bunch of Perl scripts.

I designed `pweb` with the following goals in mind:

- It should be easy to add support for additional programming languages. To guarantee this, pweb should only use regular expressions and not a complete parser. I was willing to scarify advanced formatting functions, such as auto-detection of variables, for this goal.

A. Klein, *Stream Ciphers*, DOI 10.1007/978-1-4471-5079-4_20,
© Springer-Verlag London 2013

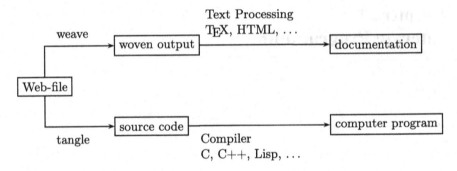

Fig. 20.1 The literate programming environment

- Not every language has such a good pre-processor as LISP. Therefore pweb should provide an extra pre-processor. I decided to use the macro processor M4 for that purpose.
- It should be possible to have (in the documentation) variable names like α, x', $\hat{x}$.

20.3 Pweb Manual

20.3.1 Structure of a WEB-Document

All web commands start with an @ symbol. The first line of a pweb-document must be of the form `@<programing language>2<formating language>`. For example a C++ Program that is documented in LaTeX starts with `@cpp2latex`.

The next lines up to the first `@*` command form the limbo section. The contents of the limbo section are copied directly into the header of the documentation (the part between `\documentclass{article}` and `\begin{document}` if you use LaTeX or the part between `<head>` and `</head>` if you use HTML).

If you use LaTeX as a formatting language the limbo section must define the title and the author.

The remaining part of the web-document splits into text sections, code sections, named modules and macros. A text section must precede each section of another type.

20.3.2 Text Sections

The first line of a text section has the form

`@*n <title of the section>`.

where $n \in \mathbb{N}$. This defines a section of depth n with the given title. pweave transforms it to the LaTeX commands `\section`, `\subsection` and so on or the

HTML tags <H1>...</H1>. It depends on the chosen formatting language how many substructures exist.

@* is a short form for the highest level @*0.

Inside a text section you can use all commands from the chosen formatting language (like \emph{...} in LaTeX or ... in HTML). Of course, once you start using these commands you cannot change the formatting language.

20.3.3 Code Sections and Modules

A code section starts either with @a or @c and ends at the next text section. Both commands are equivalent and the only reason for having two commands is the smooth transition from other literate programming environments (cweb uses @c and fweb uses @a). I recommend to use only @c in new pweb projects.

The effect of a code section is that ptangle copies that code directly into the source code of the program.

In addition to code sections fweb uses modules. A module starts with the line @<Module name@>= and ends at the next text section.

You can insert a module in any code section or any other module by @<Module name@>.

It is not necessary to define a module before its first use. You can define it afterwards.

A typical use of a module is an initialization code like:

```
@* Initialize variables.
Explain what has to be done
@<init@>=
   set some variables ...
@* The main Program.
Explain the program.
@c
int main() {
   @<init@>

   Do something ...
}
```

The presence of modules changes the programming style. It is no longer necessary to use functions for structuring the program. You should only define a function if it really is a function; for structuring the code, modules do a better job.

In addition you can split the code into smaller parts than is normally possible in a programming language. D.E. Knuth [150] recommends that a single section should contain no more than 12 lines of code. If you ever find yourself writing larger code sections, you should consider defining a new module to structure the code.

20.3.4 Macros

Macros are a useful technique for doing computations at compile time, but the support of macros is very different between programming languages. For example, Common Lisp has a perfect macro support, where macros can be written in almost the same syntax as the rest of the programming language. C has limited macro support (no recursion in macros) and the macro language differs completely from the programming language. Java has no macros at all.

pweb offers an interface to the M4 macro processor that can be used in addition to the macros provided by the chosen programming language.

A simple macro has the form

```
@define MACRO-NAME SUBSTITUTION
```

and the effect is that in the program MACRO-NAME is replaced by SUBSTITUTION. This corresponds to the C-Macro

```
#define MACRO-NAME SUBSTITUTION
```

More complex macros must be defined in a special section.

The macro section must start with @m and end at the beginning of the next text section. Inside a macro section you can use the entire syntax of the M4 macro language. In particular, you have conditions and recursion. Here is a small example of how to define a macro for the factorial function.

```
@m
define('FACTORIAL','ifelse($1,1,1,
          eval($1*'FACTORIAL(eval($1-1)')))')
```

A complete description of M4 can be found in [238].

A side-effect of using macros is that a single line of the fweb document can expand to multiple lines in the programming language. This tweaks the line numbering and makes it difficult to interpret a later error message which refers to the wrong line number. To help you fweb contains a special command @l which resets the line numbering.

You should always use long macros like this:

```
@c
some commands
A_BIG_LONG_MACRO @l
some more commands
```

The @l command only works with programming languages that support redirection of error messages like C and C++ do with #line macro. In other languages the @l command will be ignored.

20.3.5 Special Variable Names

Most programming languages require that a variable name must start with a letter and allow only letters, digits and the underscore sign (_) afterwards. However, in mathematics we often use variables like α, x' or $\hat{x}$. Fweb pays attention to this by providing special commands for such variables.

With the command `@alpha` you define a variable (or to be more precise, a code fragment) named `alpha` in the source code and printed α in the documentation. All Greek letters are supported by this facility.

With `x@'` you define a variable x'. It becomes `x_prime` or `xprime` in the programming language. Similarly you can define variables $\hat{x}$, $\tilde{x}$ by `x@hat` and `x@tilde`, respectively.

Combinations like `@beta@'` for β' are possible.

20.3.6 Include Files

You can use `@include <filename>` to include other pweb documents in your document. The include mechanism is similar to the one used by C and C++.

- If the include-file has a preamble it will be ignored.
- Each include-file must start with a text section and it can contain only complete sections.
- The `@include` command must stand alone in a line.

20.3.7 Conditional Compilation

In addition to the conditional compilation features that may be provided by the underlying programming languages, `pweb` provides an extra mechanism.

Code between `@tangle off` and `@tangle on` is ignored by `ptangle`. The commands `@tangle off` and `@tangle on` must stand alone in a line and are *not* nestable. Typically one uses them to give a short example code in the documentation.

```
@* How to use the new functions.
Some explanation ...
@tangle off
@c
some lines of example code ...
@tangle on
```

Similarly `pweave` ignores the code between `@weave off` and `@weave on`. One can use these commands to hide some definitions in the documentation. Since the purpose of literate programming is to provide full documentation, one should only rarely use @weave off and @weave on. An exception is that it can be a good idea to use code like

```
@weave off
@include somefile.web
@weave on
```

and run pweave directly on somefile.web. This gives two separate articles, one for the main program and one for the include-file.

In addition, fweb has comments. Everything from @% until the end of a line will be ignored by fweb.

20.3.8 More pweb Commands

Inside a text section you can insert small code examples by |code example|.

Since @ and | have a special meaning in pweb documents you must escape them if you want use them in the text. Write @@ for a literal @ and @| for a literal |.

20.3.9 Compatibility Features

To improve the compatibility with other literate programming systems pweb ignores the following commands:

- @; used by fweb as an invisible semicolon.
- @\ used by fweb as a forced line break.
- @~ used by fweb to inhibit a line break.

You should not use any of these commands in a new pweb project.

20.3.10 Common Errors

Typical errors when typing a pweb document are:

- The title of a text section must end with a dot! If you forget the dot at the end of the line the pweave formation goes wrong.
- If you forget to define a module ptangle will wait on the command line until you type the module. No error message is displayed! Stop the program by typing Control-d and add the module definition to your file.

20.3.11 Editing pweb Documents

You can use any text editor to create your pweb documents, but the correct syntax
highlighting can be problem. If you use Emacs for editing you can use the mmm-
mode from M.A. Shulman [246] to patch together the syntax highlighting for LaTeX,
C and M4. Install the MMM Mode and add the following code to your .emacs.

```
(mmm-add-group
 'web
 '((web1
    :submode c-mode
    :face mmm-code-submode-face
    :delimiter-mode nil
    :front "@<[^>]*@>="
    :back  "@[* ]"
    )
   (web2
    :submode c-mode
    :face mmm-code-submode-face
    :delimiter-mode nil
    :front "@[ac]"
    :back "@[* ]" )
   (web3
    :submode m4-mode
    :face mmm-code-submode-face
    :delimiter-mode nil
    :front "@[m]"
    :back "@[* ]" )
   ))


(setq mmm-global-mode 'maybe)
(mmm-add-mode-ext-class 'latex-mode "\\.web\\'" 'web)
(add-to-list 'auto-mode-alist '("\\.web\\'" . latex-mode))
```

20.3.12 Extending pweb

The formatting of pweb is driven by a bunch of Perl scripts. For each programming
language lang and each formatting language text you need the following files
listed in Table 20.1.

Table 20.1 Files needed by pweb

`begin_lang2text.pl`	Prints the text at the beginning of the formatted document (like `\documentclass{article}` if you use LaTeX).
`middle_lang2text.pl`	Prints whatever comes after the limbo section (if you use HTML it would most likely be `</head><body>`).
`end_lang2text.pl`	Prints the text that closes the formatted document (like `\end{document}` if you use LaTeX).
`formater_lang2text`	Converts the source code of the programming language in printable code. A typical entry of `formater_cpp2tex` is `s/(?<![a-zA-Z0-9_])\b(asm|auto|bool|` `...|while)\b/\{\\rm \\bf $1\}/g;` This regular expression prints the keywords in boldface.
`formater_m4lang2text`	Like `formater_lang2text` but treats the macro sections.
`codestart_lang2text`	Contains things that have to be printed at the start of a code section.
`codeend_lang2tex`	Same as before but for the end of the code section.
`codestart_m4lang2text`	Contains things that have to be printed at the start of a macro section.
`codeend_m4lang2text`	End of a macro section.

In principle you can add any combination of programming languages and formatting languages.

Notations

$x \equiv y \bmod n$	$x \equiv y \bmod m \iff m \mid (x - y)$ (mod as equivalence relation)
$x = y \bmod n$	x is the unique integer in the interval $[0, m-1]$ with $x \equiv y \bmod m$ (mod as function)
$\mathbb{N}$	the set of non-negative integers
$\mathbb{N}^+$	the set of positive integers
$\mathbb{Z}$	the set of integers
$\mathbb{R}$	the set of real numbers
$\mathbb{C}$	the set of complex numbers
$\Re(x)$	the real part of the complex number x
$\mathbb{F}_q$	the finite field with q elements (q is a prime power)
$F[z]$	the polynomial ring over the field F in the variable z
$F[[z]]$	the ring of formal power series over the field F
$F((z))$	the Laurent series over F
$\mathrm{Tr}_{F'/F}$	the trace function in the field extension F'/F
S_n	the symmetric group on n elements
$\prec$	a monomial ordering
$\prec_{\mathrm{lex}}$	the lexicographic order
$\prec_{\mathrm{grlex}}$	the graded lexicographic order
$\prec_{\mathrm{grevlex}}$	the graded reverse lexicographic order
$\prec_M$	the matrix order defined by M
$a \wedge b$	logic AND
$a \vee b$	logic OR
$\neg a, \bar{a}$	logic NOT
$a \& b$	bitwise AND operation $((1011)_2 \& (1110)_2 = (1010)_2)$
$a \mid b$	bitwise OR operation $((1001)_2 \mid (1100)_2 = (1101)_2)$
$a \oplus b$	bitwise XOR operation $((1001)_2 \mid (1100)_2 = (0101)_2)$
$\bar{a}$	bitwise complement, $a = (1001)_2 \implies \bar{a} = (0110)_2$
$a \gg k$	right shift of k bits $(11010100)_2 \gg 3 = (00011010)_2$
$a \ll k$	left shift of k bits $(11010100)_2 \ll 3 = (10100000)_2$
$a \ggg k$ or $RotR(a, k)$	right rotation of k bits $(11000101)_2 \ggg 1 = (11100010)_2$
$a \lll k$ or $RotL(a, k)$	left rotation of k bits $(11000101)_2 \lll 1 = (10001011)_2$

A. Klein, *Stream Ciphers*, DOI 10.1007/978-1-4471-5079-4,
© Springer-Verlag London 2013

μ_k	the infinite 2-adic fraction $-1/(2^{2^k}+1)$
SADD(w)	sideway addition SADD(w) = $w_0 + \cdots + w_n$
$\lfloor x \rfloor$	the greatest integer smaller than or equal to $a \in \mathbb{R}$
$\lceil x \rceil$	the smallest integer greater than or equal to $a \in \mathbb{R}$
$\binom{x}{m}$	the binomial coefficient $\frac{x(x-1)\cdots(x-m+1)}{m!}$, $x \in \mathbb{R}$, $m \in \mathbb{N}$
$\gcd(a,\dots,a_k)$	the greatest common divisor of $a_1,\dots,a_k$
$\mathrm{lcm}(a_1,\dots,a_k)$	the least common multiple of $a_1,\dots,a_k$
$\left(\frac{a}{b}\right)$	the Jacobi symbol (see Definition 13.3)
$\varphi(n)$	Eulerian φ-function, $\varphi(n) = |\{a \mid \gcd(a,n)=1, 1 \le a \le n\}|$
$\lambda(n)$	the Carmichael function $\lambda(n) = \min\{e|a^e \equiv 1 \bmod n$ for all e with $\gcd(e,n)=1\}$
$O(g(n))$	$f(n) = O(g(n)) \iff \limsup_{n\to\infty} \frac{f(n)}{g(n)} < \infty$
$O^\sim(g(n))$	$f(n) = O^\sim(g(n)) \iff f(n) = O(g(n)\log^k g(n))$ for some $k \in \mathbb{N}$
$o(g(n))$	$f(n) = o(g(n)) \iff \lim_{n\to\infty} \frac{f(n)}{g(n)} = 0$
$\Omega(g(n))$	$f(n) = \Omega(g(n)) \iff \limsup_{n\to\infty} \frac{g(n)}{f(n)} = 0$
$\omega(g(n))$	$f(n) = \omega(g(n)) \iff \lim_{n\to\infty} \frac{g(n)}{f(n)} = 0$
$\Theta(g(n))$	$f(n) = \Theta(g(n)) \iff f(n) = O(g(n))$ and $f(n) = \Omega(g(n))$
$f(n) \sim g(n)$	$f(n) = \Theta(g(n)) \iff \lim_{n\to\infty} \frac{f(n)}{g(n)} = 1$
$(\Omega, \mathcal{F}, P)$	a probability space on Ω with σ-algebra $\mathcal{F}$ and probability measure P
iid.	identically and independently distributed (random variables)
$\mathcal{N}(\mu, \sigma^2)$	the normal distribution with mean μ and variance σ^2
$\Phi(x)$	the distribution function of the standard normal distribution, $\Phi(x) = \int_{-\infty}^x \frac{1}{\sqrt{2\pi}} e^{-u^2/2} du$
$[W_t : t \ge 0]$	a Brownian motion (Wiener process)
$\nu \ll \mu$	the measure ν is absolutely continuous with respect to μ, $\mu(A) = 0 \implies \nu(A) = 0$

References

1. van Aardenne-Ehrenfest, T., de Bruijn, N.G.: Circuits and trees in oriented linear graphs. Bull. Belg. Math. Soc. Simon Stevin **28**, 203–217 (1951) (p. 61)
2. Aciiçmetz, O., Koç, Ç.K., Seifert, J.P.: Predicting secret keys via branch prediction. In: Abe, M. (ed.) CT-RSA. LNCS, vol. 4377, pp. 225–242 (2007) (p. 246)
3. Agrawal, M., Kayal, N., Saxena, N.: PRIMES is in P. Ann. Math. **2**, 781–793 (2002) (p. 251)
4. Aircrack-ng a toolsuite. http://www.aircrack-ng.org (pp. 185, 207)
5. Alexi, W., Chor, B., Goldreich, O., Schnorr, C.P.: RSA and Rabin functions: certain parts are as hard as the whole. SIAM J. Comput. **17**, 194–200 (1988) (p. 251)
6. Amrhein, B., Gloor, O., Küchlin, W.: On the walk. Theor. Comput. Sci. **187**, 179–202 (1997) (pp. 140, 142)
7. Arbaugh, W.A.: An inductive chosen plaintext attack against WEP/WEP2 (2001). http://www.cs.umd.edu/~waa/attack/v3dcmnt.htm (p. 189)
8. Armknecht, F., Krause, M.: Algebraic attacks on combiners with memory. In: Proceedings of Crypto 2003. LNCS, vol. 2729, pp. 162–176. Springer, Berlin (2003) (pp. 149, 151)
9. Ars, G., Faugère, J.C., Imai, H., Kawazoe, M., Sugita, M.: Comparison between XL and Gröbner basis algorithms. In: Advances in Cryptology—ASIACRYPT 2004. LNCS, vol. 3329, pp. 338–353. Springer, Berlin (2004) (p. 147)
10. Atkin, A.O.L., Bernstein, D.J.: Prime sieves using binary quadratic forms. Math. Comput. **73**, 1023–1030 (2004) (p. 251)
11. Aumasson, J.P.: On a bias of Rabbit. http://www.ecrypt.eu.org/stream/papersdir/2006/058.pdf (p. 235)
12. Bach, E.: Improved asymptotic formulas for counting correlation-immune Boolean functions. Technical report 1616, Computer Science Dept., University of Wisconsin (2007) (p. 81)
13. Bailey, D.H., Lee, K., Simon, H.D.: Using Strassen's algorithm to accelerate the solution of linear systems. J. Supercomput. **4**(4), 357–371 (1990) (p. 281)
14. Barkan, E., Biham, E.: Conditional estimators: an effective attack on A5/1. In: Selected Areas in Cryptography. Lecture Notes in Comput. Sci., vol. 3897, pp. 1–19. Springer, Berlin (2006) (p. 176)
15. Barkan, E., Biham, E., Keller, N.: Instant ciphertext-only cryptanalysis of GSM encrypted communication. In: Advances in Cryptology—CRYPTO 2003. LNCS, vol. 2729, pp. 600–616. Springer, Berlin (2003) (p. 172)
16. Bauer, F.L.: Kryptologie, Methoden und Maximen, 2 Auflage. Springer, Berlin (1994) (p. 8)
17. Baumert, L.D.: Cyclic Difference Sets. LNM, vol. 182. Springer, Berlin (1971) (p. 29)
18. Baur, W., Strassen, V.: On the complexity of partial derivatives. Theor. Comput. Sci. **22**, 317–330 (1983) (p. 291)

A. Klein, *Stream Ciphers*, DOI 10.1007/978-1-4471-5079-4,
© Springer-Verlag London 2013

19. Berbain, C., Gilbert, H., Patarin, J.: Quad: a practical stream cipher with provable security. In: Vaudenay, S. (ed.) EUROCRYPT. Lecture Notes in Computer Science, vol. 4004, pp. 109–128. Springer, Berlin (2006) (p. 255)

20. Berlekamp, E.R.: Algebraic Coding Theory. McGraw-Hill, New York (1968) (p. 33)

21. Bernasconi, J., Günter, C.G.: Analysis of nonlinear feedback logic for binary sequence generators. In: Advances in Cryptology—Eurocrypt '85. LNCS, vol. 219, pp. 161–166 (1986) (p. 75)

22. Bernstein, D.J.: Cache-timing attacks on AES. http://cr.yp.to/antiforgery/cachetiming-20050414.pdf (p. 165)

23. Bernstein, D.J.: Why haven't cube attacks broken anything. http://cr.yp.to/cubeattacks.html (p. 231)

24. Bernstein, D.J.: Related-key attacks: who cares? eStream discussion forum. http://www.ecrypt.eu.org/stream/phorum, 22 June 2005 (p. 239)

25. Beth, T., Piper, F.: The stop-and-go generator. In: Advances in Cryptology—Eurocrypt'84. LNCS, vol. 209, pp. 88–92 (1985) (p. 155)

26. Biham, E., Shamir, A.: Differential cryptanalysis of DES-like cryptosystems. J. Cryptol. **4**(1), 3–72 (1991) (p. 30)

27. Billingsley, P.: Probability and measure. In: Wiley Series in Probability and Mathematical Statistics, 3rd edn. Wiley-Interscience, New York (1995) (pp. 316, 317, 319, 323, 325, 326)

28. Biryukov, A., Shamir, A., Wagner, D.: Real time cryptanalysis of A5/1 on a PC. In: Fast Software Encryption (2000). http://cryptome.org/a5.ps (pp. 176, 178, 179)

29. Blake, I.F., Gao, S., Lambert, R.: Constructive problems for irreducible polynomials over finite fields. In: Information Theory and Applications, Rockland, ON, 1993. Lecture Notes in Comput. Sci., vol. 793, pp. 1–23. Springer, Berlin (1994) (p. 309)

30. Blake, I.F., Gao, S., Lambert, R.J.: Construction and distribution problems for irreducible trinomials over finite fields. In: Applications of Finite Fields, Egham, 1994. Inst. Math. Appl. Conf. Ser. New Ser., vol. 59, pp. 19–32. Oxford Univ. Press, New York (1996) (p. 309)

31. Bluetooth special interest group: bluetooth specification, vol. 2, part H, November 2003. Available online http://bluetooth.org/foundry/adopters/document/Bluetooth_Core_Specification_v1.2 (p. 126)

32. Blum, L., Blum, M., Shub, M.: A simple unpredictable pseudorandom number generator. SIAM J. Comput. **15**(2), 364–383 (1986) (p. 244)

33. Blum, M., Micali, S.: How to generate cryptographically strong sequences of pseudo-random bits. SIAM J. Comput. **13**, 850–864 (1984) (pp. 241, 242)

34. Boesgaard, M., Vesterager, M., Peterson, T., Christiansen, J., Scavenius, O.: Rabbit: a new high-performance stream cipher. In: Johansson, T. (ed.) Proceedings of Fast Software Encryption 2003. LNCS, vol. 2887, pp. 307–329. Springer, Berlin (2003) (p. 232)

35. Boost C++ libraries. http://www.boost.org/ (p. 370)

36. Briceno, M., Goldberg, I., Wagner, D.: An implementation of GSM A3A8 algorithm (1998). http://www.iol.ie/~kooltex/a3a8.txt (p. 169)

37. Briceno, M., Goldberg, I., Wagner, D.: GSM cloning (1998). http://www.isaac.cs.berkeley.edu/isaac/gsm-faq.html (p. 169)

38. Briceno, M., Goldberg, I., Wagner, D.: A pedagogical implementation of the GSM A5/1 and A5/2 "voice privacy" encryption algorithms (1999). http://cryptome.org/gsm-a512.htm (p. 169)

39. Buchberger, B.: Gröbner bases: an algorithmic method in polynomial ideal theory. In: Bose, N.K., Reidel, D. (eds.) Multidimensional Systems Theory, pp. 184–232. Reidel, Dordrecht (1985) (p. 140)

40. Bügisser, P., Clausen, M., Shokrollahi, M.A.: Algebraic Complexity Theory. Grundlehren der mathematischen Wissenschaften, vol. 315. Springer, Berlin (1997) (p. 281)

41. Camion, P., Carlet, C., Charpin, P., Sendrier, N.: On correlation-immune functions. In: Feigenbaum, J. (ed.) Advances in Cryptology—CRYPTO'91. Lecture Notes in Computer Science, vol. 576, pp. 86–100. Springer, Berlin (1992) (p. 76)

42. Camion, P., Canteaut, A.: Generalization of Siegenthaler inequality and Schnorr-Vaudenay multipermutations. In: Advances in Cryptology—CRYPTO '96 (Santa Barbara, CA). Lecture Notes in Comput. Sci., vol. 1109, pp. 372–386. Springer, Berlin (1996) (pp. 78, 80)

43. Canfield, E.R., Gao, Z., Greenhill, C., McKay, B.D., Robinson, R.W.: Asymtotic enumeration of correlation-immune Boolean functions. Cryptogr. Commun. **2**, 111–126 (2010). arXiv:0909.3321 (p. 81)

44. Carlet, C., Gouget, A.: An upper bound on the number of m-resilient Boolean functions. In: ASIACRYPT 2002. LNCS, vol. 2501, pp. 484–496. Springer, Berlin (2002) (p. 81)

45. Carlet, C., Klapper, A.: Upper bounds on the number of reslient functions and of bent functions. In: Lecture Notes Dedicated to Philippe Desarte. Springer Verlag, to appear. A shorter version has appeared in the Proceedings of the 23rd Symposium on Information Theory in the Benelux, Louvain-La-Neuve, Belgian, 2002 (p. 81)

46. Casti, J.L.: Dynamical Systems and Their Applications: Linear Theory. Academic Press, San Diego (1977) (p. 33)

47. Cayley, A.: A theorem on trees. Q. J. Math. **23**, 376–378 (1889) (p. 334)

48. Chabaud, F., Vaudenay, S.: Links between differential and linear cryptanalysis. In: Santis, A.D. (ed.) Advances in Cryptology EUROCRYPT 94. LNCS, vol. 950, pp. 356–365. Springer, New York (1995) (p. 30)

49. Chan, A.H., Games, R.A., Key, E.L.: On the complexity of de Bruijn sequences. J. Comb. Theory, Ser. A **33**, 233–246 (1982) (p. 157)

50. Cheng, Q.: On the construction of finite field elements of large order. Finite Fields Appl. **11**, 358–366 (2005) (p. 307)

51. Cheng, U., Golomb, S.W.: On the characterisation of PN sequences. IEEE Trans. Inf. Theory **29**, 600 (1983) (p. 29)

52. Chepyzhov, V., Johansson, T., Smeets, B.: A simple algorithm for fast correlation attacks on stream ciphers. In: Proceedings of the 7th International Workshop on Fast Software Encryption. LNCS, vol. 1978, pp. 181–195 (2001) (pp. 91, 94)

53. Chose, P., Joux, A., Mitton, M.: Fast correlation attacks: an algorithmic point of view. In: Knudsen, L.R. (ed.) EUROCRYPT 2002. LNCS, vol. 2332, pp. 209–221. Springer, Berlin (2002) (pp. 96, 105)

54. GNU CLISP—an ANSI Common Lisp Implementation. http://clisp.cons.org/ (p. 370)

55. Collart, S., Kalkbrener, M., Mall, D.: Converting bases with the Gröbner walk. J. Symb. Comput. **24**, 465–469 (1997) (pp. 140, 142)

56. Cook, S.A.: On the minimum computation time of functions. PhD thesis, Harvard University (1966) (p. 249)

57. Coppersmith, D.: Finding a small root of a bivariate equation; factoring with high bits known. In: Maurer, U.M. (ed.) EUROCRYPT. LNCS, vol. 1070, pp. 178–189. Springer, Berlin (1996) (p. 361)

58. Coppersmith, D., Krawczyk, H., Mansour, Y.: The shrinking generator. In: Advances in Cryptology—CRYPTO '93, Santa Barbara, CA, 1993. LNCS, vol. 773, pp. 22–39. Springer, Berlin (1994). doi:10.1007/3-540-48329-2_3 (p. 159)

59. Coppersmith, D., Winograd, S.: Matrix multiplications via arithmetic progressions. J. Symb. Comput. **9**, 251–280 (1990) (pp. 281, 289)

60. Coppersmith, D.: Small solutions to polynomial equations and low exponent vulnerabilities. J. Cryptol. **10**(4), 223–260 (1997) (p. 303)

61. Coudert, O., Berthet, C., Madre, J.C.: Verification of synchronous sequential machines based on symbolic execution. In: Automatic Verification Methods for Finite State Systems. LNCS, vol. 407, pp. 365–373. Springer, Berlin (1989) (p. 354)

62. Courtois, N.: Fast algebraic attacks on stream ciphers with linear feedback. In: Proceedings of Crypto 2003. LNCS, vol. 2729, pp. 177–194. Springer, Berlin (2003) (p. 149)

63. Courtois, N., Klimov, A., Patarin, J., Shamir, A.: Efficient algorithms for solving overdefined systems of multivariate polynomial equations. In: Advances in Cryptology—EUROCRYPT 2000. LNCS, vol. 1807, pp. 392–407. Springer, Berlin (2000) (p. 145)

64. Courtois, N., Meier, W.: Algebraic attacks on stream ciphers with linear feedback. In: Proceedings of Eurocrypt 2003. LNCS, vol. 2656, pp. 345–359. Springer, Berlin (2003). An extended version is available at http://www.cryptosystem.net/stream/ (pp. 148, 149, 153)

65. ncurses. http://www.gnu.org/software/ncurses/ncurses.html (p. 370)

66. Daemen, J.: Cipher and hash function design strategies based on linear and differential cryptanalysis. Doctoral dissertation. K.U. Leuven, March 1995 (p. 239)

67. Daemen, J., Govaert, R., Vandewalle, J.: On the design of high speed self-synchonizing stream ciphers. In: Kam, P.Y., Hirota, O. (eds.) Singapore ICCS/ISITA '92 Conference Proceedings, pp. 183–279. IEEE, New York (1992) (p. 239)

68. Daemen, J., Kitsos, P.: The self-synchonizing stream cipher MOSQUITO: eSTREAM documentation, version 2 (2005). http://www.ecrypt.eu.org/stream/p3ciphers/mosquito/mosquito. pdf (p. 235)

69. Daemen, J., Kitsos, P.: The self-synchonizing stream cipher MOUSTIQUE (2006). http://www.ecrypt.eu.org/stream/p3ciphers/mosquito/mosquito_p3.pdf (p. 235)

70. Daemen, J., Lano, J., Preneel, B.: Chosen ciphertext attack on SSS (2005). http://www.ecrypt.eu.org/stream/papersdir/044.pdf (p. 235)

71. Dai, Z.d.: Proof of Rueppel's linear complexity conjecture. IEEE Trans. Inf. Theory **32**, 440–443 (1986) (p. 50)

72. Dawson, E., Clark, A., Golić, J., Millan, W., Penna, L., Simpson, L.: The LILI-128 keystream generator. In: Proc. of First NESSIE Workshop (2001) (p. 151)

73. de Bruijn, N.G.: A combinatorial problem. Nedel. Akad. Wetensch. Proc. **49**, 758–764 (1946). other name: Indag. Math. **8**, 461–467 (1946) (pp. 59, 60)

74. de Bruijn, N.G.: Acknowledgment of priority to C. Fyle Sainte-Marie on the counting of circular arrangements of 2^n zeroes and ones that show each n-letter word exactly once. Technical Report TH-Report 75-WSK-06, Technolocical University Eidhoven, June 1975 (p. 60)

75. De Cannière, C., Preneel, B.: Trivium. http://www.ecrypt.eu.org/stream/triviump3.html (p. 229)

76. Denisov, O.V.: An asymptotic formula for the number of correlation-immune boolean functions of order k. Discrete Appl. Math. **2**(4), 407–426 (1992). English Translation from Diskr. Math. **3**(2), 25–46 (1991) (p. 81)

77. Denisov, O.V.: A local limit theorem for the distribution of a part of the spectrum of a random binary function. Discrete Appl. Math. **10**, 87–101 (2000). English Translation from Diskr. Math. **12**, 1 (2000) (p. 81)

78. Dholakia, A.: Introduction to convolutional codes with applications. In: The Kluwer International Series in Engineering and Computer Science. Kluwer Academic, Boston (1994) (pp. 95, 106)

79. Diaconis, A.: In: Group Representations in Probability and Statistics. Lecture Notes-Monographs Series, vol. 11. IMS, Hayward (1988) (p. 197)

80. Dinur, I., Shamir, A.: Cube attacks on tweakable black box polynomials. In: Joux, A. (ed.) EUROCRYPT. LNCS, vol. 5479, pp. 278–299. Springer, Berlin (2009). Also available as Cryptology ePrint Archive, Report 2008/385 http://eprint.iacr.org/ (pp. 231, 232)

81. Doyle, A.C.: The Return of Sherlock Holmes, chapter The Adventure of the Dancing Men. Georges Newnes, Ltd (1905). Originally published 1903, Available online http://en.wikisource.org/wiki/The_Adventure_of_the_Dancing_Men (pp. 2, 348)

82. Edel, Y., Klein, A.: Computational aspects of fast correlation attacks. Submitted, available online http://cage.ugent.be/~klein/corr-comp.pdf (pp. 113, 263, 264)

83. Edel, Y., Klein, A.: Population count in arrays. Submitted, available online http://cage.ugent.be/~klein/popc.html (pp. 95, 96, 97, 113)

84. Ekdahl, P., Johansson, T.: Another attack on A5/1. IEEE Trans. Inf. Theory **49**(1), 284–289 (2003) (pp. 179, 180)

85. The estream project. http://www.ecrypt.eu.org/stream/ (pp. 12, 229)

86. Fano, R.M.: A heuristic discussion of probabilistic decoding. IEEE Trans. Inf. Theory **IT-9**, 64–74 (1963) (pp. 109, 110)

87. Faugère, J.C.: A new efficient algorithm for computing Gröbner bases (F4). J. Pure Appl. Algebra **139**(1), 61–88 (1999). Available online http://fgbrs.lip6.fr/@papers/F99a.pdf (p. 140)
88. Faugère, J.C.: A new efficient algorithm for computing Gröbner bases without reduction to zero (F5). In: Proceedings of the 2002 International Symposium on Symbolic and Algebraic Computation (ISSAC), pp. 75–83. ACM, New York (2002). Available online http://fgbrs.lip6.fr/@papers/F02a.pdf (p. 140)
89. Faugère, J.C., Gianni, P., Lazard, D., Mora, T.: Efficient computation of zero-dimensional Gröbner bases by change of ordering. J. Symb. Comput. **16**(4), 329–344 (1993) (p. 143)
90. Ferguson, N., Schneier, B.: Practical Cryptography. Wiley, New York (2003) (p. 5)
91. Finny: A RC4 cycle that can't happen. Posting to sci.crypt, September 1994 (p. 357)
92. Fisher, J.B., Stern, J.: An efficient pseudo-random generator provably as secure as syndrome decoding. In: Advances in Cryptology—EUROCRYPT '96. LNCS, vol. 1070, pp. 245–255 (1996) (p. 255)
93. Fluhrer, S., Mantin, I., Shamir, A.: Weaknesses in the key scheduling algorithm of RC4. In: Selected Areas in Cryptography. LNCS, vol. 2259, pp. 1–24. Springer, Berlin (2001) (pp. 196, 199, 200, 202)
94. Fluhrer, S.R., Lukes, S.: Analysis of the E_0 encryption system. In: Proc. 8th Workshop on Selected Areas in Cryptography. LNCS, vol. 2259. Springer, Berlin (2001) (p. 127)
95. Fluhrer, S.R., McGrew, D.A.: Statistical analysis of the alleged RC4 keystream generator. In: Proceedings of the 7th International Workshop on Fast Software Encryption. LNCS, vol. 1978, pp. 19–20. Springer, Berlin (2000) (pp. 213, 214, 218)
96. Forney, G.D. Jr.: The Viterbi algorithm. Proc. IEEE **61**(3), 268–278 (1973) (p. 107)
97. Friedlander, J.B., Pomerance, C., Shparlinski, I.E.: Period of the power generator and small values of the Carmicael's function. Math. Comput. **70**, 1591–1605 (2001) (pp. 254, 294)
98. Gaines, H.F.: Cryptanalysis. Dover, New York (1956) (p. 349)
99. Gap—groups, algorithms, programming—a system for computational discrete algebra. http://www.gap-system.org/ (p. 370)
100. von zur Gathen, J., Gerhard, J.: Modern Computer Algebra. Cambridge University Press, Cambridge (1999) (pp. 146, 153, 247, 292, 302)
101. von zur Gathen, J., Shparlinski, I.: Predicting subset sum pseudorandom generators. In: Selected Areas in Cryptography. LNCS, vol. 3357, pp. 241–251. Springer, Berlin (2005) (p. 255)
102. GCC, the GNU Compiler Collection. http://gcc.gnu.org/ (p. 370)
103. Gebauer, R., Möller, H.M.: On an installation of Buchberger's algorithm. In: Robbiano, L. (ed.) Computational Aspects of Communicative Algebra, pp. 141–152. Academic Press, New York (1988) (p. 140)
104. Geffe, P.R.: How to protect data with ciphers that are really hard to break. Electronics **4**, 129–156 (1973) (p. 64)
105. The GNU Multiple Precision Arithmetic Library. http://gmplib.org/ (pp. 249, 370)
106. Gold, R.: Maximal recursive sequences with 3-valued cross-correlation functions. IEEE Trans. Inf. Theory **14**, 154–156 (1968) (p. 30)
107. Goldberg, I., Wegner, D., Green, L.: The (real-time) cryptanalysis of A5/2. Presented at the Rump Session of Crypto '99 (1999). Slides available at www.cs.berley.edu/~daw/tmp/a52-sliedes.ps (p. 175)
108. Goldstein, D., Moews, D.: The identity is the most likely exchange shuffle for large n. Aequ. Math. **65**(1–2), 3–30 (2003) (p. 195)
109. Golić, J.Dj.: Cryptanalysis of the alleged A5 stream cipher. In: Advances in Cryptology—EUROCRYPTO '97. LNCS, vol. 1233, pp. 239–255. Springer, Berlin (1997) (p. 65)
110. Golić, J.Dj.: Linear statistical weakness of alleged RC4 keystream generator. In: Advances in Cryptology—EUROCRYPTO '97. LNCS, vol. 1233, pp. 226–238. Springer, Berlin (1997) (p. 213)
111. Golić, J.Dj.: Linear models for a time-variant-permutation generator. IEEE Trans. Inf. Theory **45**(7), 2374–2382 (1999) (p. 213)

112. Golić, J.Dj.: Iterative probabilistic cryptanalysis of RC4 keystream generator. In: ACISP 2000, pp. 220–233 (2000) (p. 202)
113. Golić, J.Dj.: Correlation analysis of the shrinking generator. In: Advances in Cryptology—CRYPTO 2001 (Santa Barbara, CA). Lecture Notes in Comput. Sci., vol. 2139, pp. 440–457. Springer, Berlin (2001) (pp. 161, 162, 163)
114. Golomb, S.W.: On the classification of balanced binary sequences of period $2^n - 1$. IEEE Trans. Inf. Theory **26**, 730–732 (1980) (p. 27)
115. Golomb, S.W.: Shift Register Sequences. Aegean Park, Laguna Hills, revised edition (1982) (pp. 17, 24, 25, 27)
116. Gong, G., Gupta, K.C., Hell, M., Nawaz, Y.: Towards a general RC4-like keystream generator. In: Information Security and Cryptology. Lecture Notes in Comput. Sci., vol. 3822, pp. 162–174. Springer, Berlin (2005) (pp. 222, 223)
117. Good, I.J.: Normal recurring decimals. J. Lond. Math. Soc. **21**(3), 169–172 (1946) (p. 59)
118. Gopalakrishnan, K., Stinson, D.R.: Three characterisations of non-binary correlation-immune and resilient functions. Des. Codes Cryptogr. **5**, 241–251 (1995) (p. 76)
119. Graham, R.L., Knuth, D.E., Patashnik, O.: Concrete Mathematics, 2nd edn. Addison-Wesley, Reading (1994) (pp. 19, 332)
120. Günther, C.G.: Alternating step generators controlled by De Bruijn sequences. In: Advances in Cryptography, Eurocrypt '87. LNCS, vol. 304, pp. 5–14 (1988) (pp. 157, 158)
121. Gupta, K., Gong, G., Nawaz, Y.: A 32-bit RC4-like keystream generator. Technical Report CACR 2005-21, Center for Applied Cryptographic Research, University of Waterloo, 2005. http://www.cacr.math.uwaterloo.ca/tech_reports.html (p. 224)
122. Hardy, G.H., Wright, E.M.: The Theory of Numbers, 4th edn. Oxford University Press, London (1960) (pp. 295, 296, 300)
123. Hawkes, P., Paddon, M., Rose, G.G., de Vries, M.W.: Primitive specification for SSS (2005). http://www.ecrypt.eu.org/stream/ciphers/sss/sss.pdf (p. 235)
124. Hawkes, P., Rose, G.G.: Rewriting variables: the complexity of fast algebraic attacks on stream ciphers. In: Advances in Cryptology—CRYPTO 2004. Lecture Notes in Comput. Sci., vol. 3152, pp. 390–406. Springer, Berlin (2004) (p. 149)
125. van Heesch, D.: Doxygen, source code documentation generator tool. http://www.stack.nl/~dimitri/doxygen/ (pp. 365, 370)
126. Helleseth, T., Kumar, P.V.: Sequences with low correlation. In: Pless, V.S., Huffman, W.C. (eds.) Handbook of Coding Theory, vol. II, pp. 1765–1853. Elsevier, Amsterdam (1998). Chap. 21 (p. 29)
127. Higham, N.J.: Exploiting fast matrix multiplication within the level 3 BLAS. ACM Trans. Math. Softw. **16**(4), 352–368 (1990) (p. 281)
128. Hinek, M.J.: Cryptanalysis of RSA and its variants. In: Cryptography and Network Security. CRC Press, Boca Raton (2010) (p. 247)
129. Hoch, J., Shamir, A.: Fault analysis of stream ciphers. In: Joye, M., Quisquater, J. (eds.) Cryptographic Hardware and Embedded Systems—CHES 2004. LNCS, vol. 3156, pp. 240–253. Springer, Berlin (2004) (p. 165)
130. Hopcraft, J.E., Ullman, J.D.: Introduction to Automata Theory, Languages and Computation, 1st edn. Addision-Wesley, Reading (1979) (p. 273)
131. Howgrave-Graham, N.: Finding small roots of univariate modular equations revisited. In: Proceeding of Cryptography and Coding. LNCS, vol. 1355, pp. 45–50. Springer, Berlin (1997) (p. 303)
132. Huang, X., Huang, W., Liu, X., Wang, C., Wang, Z.J., Wang, T.: Reconstructing the non-linear filter function of LILI-128 stream cipher based on complexity (2007). http://arxiv.org/abs/cs.CR/0702128 (p. 151)
133. Hulton, D.: Practical exploration of RC4 weaknesses in WEP environments. Presented at HiverCon (2002) (p. 201)
134. Jakobsson, M., Wetzel, S.: Security weaknesses in bluetooth. In: Proc. RSA Security Conf.—Cryptographer's Track. LNCS, vol. 2020, pp. 176–191. Springer, Berlin (2001) (p. 127)

135. Jelinek, F.: Sequential decoding algorithm using a stack. IBM J. Res. Dev. **13**, 675–678 (1969) (p. 109)
136. Johansson, T., Jönsson, J.J.: Improved fast correlation attacks on stream ciphers via convolutional codes. In: Advances in Cryptology—EUROCRYPT '99. LNCS, vol. 1592, pp. 347–362. Springer, Berlin (1999) (p. 112)
137. Johansson, T., Jönsson, J.J.: Theoretical analysis of a correlation attack based on convolutional codes. IEEE Trans. Inf. Theory **48**(8) (2002) (p. 112)
138. Joux, A.: Algorithmic Cryptanalysis. CRC Press, Boca Raton (2009) (pp. 97, 247)
139. Joux, A., Muller, F.: Chosen-ciphertext attacks against MOSQUITO. In: Robshaw, M. (ed.) Fast Software Encryption 2006. LNCS, vol. 4047, pp. 390–404. Springer, Berlin (2006) (p. 235)
140. Jungnickel, D.: Finite Fields: Structure and Arithmetics. BI Wissenschaftsverlag, Mannheim (1993) (pp. 297, 305)
141. Kahn, D.: The Codebreakers. MacMillan, New York (1967) (pp. 1, 2, 3)
142. Kailath, T., Sayed, A.H.: Displacement structure: theory and applications. SIAM Rev. **35**, 297–386 (1995) (pp. 31, 37)
143. Karatsuba, A., Ofman, Yu.: Multiplication of multidigit numbers on automata. Sov. Phys. Dokl. **7**, 595–596 (1963). Original in: Dokl. Akad. Nauk SSSR **145**, 293–394 (1963) (p. 247)
144. Kasami, T.: Weight distribution formula for some class of cyclic codes. Technical Report R-285, Coordinated Science Laboratory, University of Illinois, Urbana, April 1966 (p. 30)
145. Käsper, E., Rijmen, V., Bjørstad, T.E., Rechberger, C., Robshaw, M.J.B., Sekar, G.: Correlated keystreams in Moustique. In: AFRICACRYPT, pp. 246–257 (2008) (pp. 235, 238, 239)
146. Kerckhoffs, A.: La cryptographie militaire. Journal des sciences militaires, 9th series, 1883. (January 1883) pp. 5–83, (Feburary 1883) pp. 161–191 (p. 1)
147. Kipnis, A., Shamir, A.: Cryptanalysis of the HFE public key cryptosystem. In: Proceedings of CRYPTO '99. Springer, Berlin (1999) (p. 144)
148. Klein, A.: Attacks against the RC4 stream cipher. Des. Codes Cryptogr. **48**, 269–286 (2008) (pp. 202, 203, 205, 209, 224, 227)
149. Knudsen, L.R., Meier, W., Preneel, B., Rijmen, V., Verdoolaege, S.: Analysis methods for (alleged) RC4. In: Ohta, K., Pei, D. (eds.) Advances in Cryptology—ASIACRYPT'98. Lecture Notes in Computer Science, vol. 1998, pp. 327–341. Springer, Berlin (1998) (pp. 210, 211)
150. Knuth, D.E.: Literate programming. Comput. J., **27**, 97–111 (1985). Reprinted with corrections in Knuth, D.E.: Literate Programming, Number, 27, CSLI Lecture Notes. Center for the Study of Language and Information, 1992, Stanford, California (p. 373)
151. Knuth, D.E.: The Art of Computer Programming, vol. 1. Fundamental Algorithms, 3rd edn. Addison-Wesley, Reading (1998) (p. 332)
152. Knuth, D.E.: The Art of Computer Programming, vol. 2. Seminumerical Algorithms, 3rd edn. Addison-Wesley, Reading (1998) (pp. 191, 247, 249)
153. Knuth, D.E.: The Art of Computer Programming, vol. 3. Sorting and Searching, 3rd edn. Addison-Wesley, Reading (1998) (p. 332)
154. Knuth, D.E.: MMIXware: A RISC Computer for the Third Millennium. Lecture Notes in Computer Science, vol. 1750. Springer, Berlin (1999) (p. 288)
155. Knuth, D.E.: MMIX, the Art of Computer Programming. Fasc. 1. Addison-Wesley, Upper Saddle River (2005) (p. 288)
156. Knuth, D.E.: The Art of Computer Programming, vol. 4. Bitwise Tricks and Techniques, Binary Decision Diagrams. Addison-Wesley, Upper Saddle River (2009) (pp. 123, 125, 261, 272)
157. Knuth, D.E.: The Art of Computer Programming, vol. 4. Generating All Tuples and Permutations. Addison-Wesley, Upper Saddle River (2005) (p. 350)
158. Knuth, D.E.: Literate Programming. Number 27 in CSLI Lecture Notes. Center for the Study of Language and Information, Stanford, California (1992) (p. 371)

159. Koblitz, N., Menezes, A.: Another look at "provable security". Journal of Cryptology **20** (2004). See also Cryptology ePrint Archive, Report 2004/152 http://eprint.iacr.org/ (p. 247)

160. Koç, Ç.K. (ed.): Cryptographic Engineering Springer, Berlin (2009) (p. 246)

161. Kocher, P.: Timing attacks on implementations of Diffi-Hellman, RSA, DSS and other systems. In: Kobliz, M. (ed.) CRYPTO '96. LNCS, vol. 1109, pp. 104–113 (1996) (p. 246)

162. Kocher, P., Jaffe, J., Jun, B.: Differential power analysis. In: Wiener, M. (ed.) Advances in Cryptology—CRYPTO 1999. LNCS, vol. 1666, pp. 288–297. Springer, Berlin (1999) (p. 164)

163. Kohno, T., Viega, J., Whiting, D.: Cwc: A high-performance conventional authenticated encryption mode. Cryptology ePrint Archive, Report 2003/106 (2003). http://eprint.iacr.org/ (p. 239)

164. KoreK: chopchop (experimental WEP attacks) (2004). http://www.netstumbler.org/ showthread.php?t=12489 (p. 189)

165. KoreK: Next generation of WEP attacks? (2004). http://www.netstumbler.org/showthread. php?p=93942&postcount=35 (p. 201)

166. Krause, M.: BDD-based attacks of keystream generators. In: Knudson, L. (ed.) Advances in Cryptology—EUROCRYPT '02. LNCS, vol. 1462, pp. 222–237. Springer, Berlin (2002) (pp. 117, 356)

167. Lano, J.: Cryptanalysis and design of synchronous stream ciphers. PhD thesis, Katholieke Universiteit Leuven, Faculteit Ingenierswtenschappen, Departement Elektrotechniek-ESAT, Kasteelpark Arenberg 10, 3001 Leuven-Heverlee. Juni 2006 (p. 164)

168. Lenstra, A.K., Lenstra, H.W., Lovász, L.: Factoring polynomials with rational coefficients. Math. Ann. **261**, 515–572 (1982) (p. 302)

169. Lidl, R., Niederreiter, H.: Introduction to Finite Fields and Their Applications. Cambridge University Press, Cambridge (1986) (p. 25)

170. van Lint, J.H.: Introduction to Coding Theory, 3rd edn. Springer, Berlin (1998) (p. 94)

171. Lu, P., Huang, L.: A new correlation attack of LFSR sequences. In: Feng, K., Niederreiter, H., Xing, C. (eds.) Coding, Cryptography and Combinatorics. Progress in Computer Science and Applied Logic, vol. 23, pp. 67–84. Birkhäuser, Basel (2004) (pp. 94, 95, 99)

172. Lu, Y., Vaudenary, S., Meier, W.: The conditional correlation attack: a practical attack on bluetooth encryption. In: Crypto 2005. LNCS, vol. 3621, pp. 97–117 (2005). Available online http://www.terminodes.org/micsPublicationsDetail.php?pubno=1216 (p. 127)

173. Lüneburg, H.: Ein einfacher Beweis für den Satz von Zsigmondy über primitive Primteiler von $A^N - 1$. In: Aigner, M., Jungnickel, D. (eds.) Geometries and Codes. Lecture Notes in Mathematics, vol. 893, pp. 219–222. Springer, Berlin (1981) (p. 297)

174. GNU make. http://www.gnu.org/software/make/ (p. 369)

175. Mantin, I.: A practical attack against RC4 in the WEP mode. In: Roy, B.K. (ed.) ASIACRYPT. LNCS, vol. 3788, pp. 395–411. Springer, Berlin (2005) (p. 203)

176. Mantin, I.: Predicting and distinguishing attacks on RC4 keystream generator. In: Cramer, R. (ed.) Advances in Cryptology—EUROCRYPT 2005. LNCS, vol. 3494, pp. 491–506. Springer, Berlin (2005) (p. 216)

177. Mantin, I., Shamir, A.: A practical attack on broadcast RC4. In: Matsui, M. (ed.) Revised Papers from the 8th International Workshop on Fast Software Encryption. LNCS, vol. 2355, pp. 152–164. Springer, London (2001) (p. 222)

178. Massey, J.L.: Shift-register synthesis and BCH-decoding. IEEE Trans. Inf. Theory **15**, 122–127 (1969) (p. 33)

179. Matsui, M.: Linear cryptanalysis method for DES cipher. In: Desmedt, Y. (ed.) Advances in Cryptology, Eurocrypt '93. LNCS, vol. 839, pp. 1–11. Springer, Berlin (1994) (p. 30)

180. Maurer, U., Massey, J.L.: Perfect local randomness in pseudo-random sequences. J. Cryptol. **4**, 135–149 (1993) (p. 256)

181. Maximov, A., Johansson, T., Babbage, S.: An improved correlation attack on A5/1. In: Selected Areas in Cryptography. Lecture Notes in Comput. Sci., vol. 3357, pp. 1–18. Springer, Berlin (2005) (pp. 179, 180)

182. Maximov, A., Khovratovich, D.: New State Recovering Attack on RC4. Technical report, Laboratory of Algorithmics, Cryptology and Security, University of Luxembourg (2008). http://eprint.iacr.org/2008/017 (pp. 210, 212)

183. May, A.: Using LLL-Reduction for solving RSA and Factorization Problems: A Survey. Available online http://citeseerx.ist.edu/viewdoc/summary?doi=10.1.1.86.9908 (p. 247)

184. McEliece, R.J.: The algebraic theory of convolutional codes. In: Pless, V.S., Huffman, W.C. (eds.) Handbook of Coding Theory, vol. I, pp. 1065–1138. Elsevier, Amsterdam (1998). Chap. 12 (p. 106)

185. McGuire, G., Calerbank, A.R.: Proof of a conjecture of Sarwarte and Pursley regarding pairs of binary m-sequences. IEEE Trans. Inf. Theory 41, 1153–1155 (1995) (p. 30)

186. Meidel, W., Niederreiter, H.: Linear complexity, k-error linear complexity, and the discrete Fourier transform. J. Complex. 18, 87–103 (2002) (p. 44)

187. Meier, W., Staffelbach, O.: Fast correlation attacks on stream ciphers. J. Cryptol. 1, 159–176 (1989) (p. 114)

188. Meier, W., Staffelbach, O.: The self-shrinking generator. In: Proceedings of EUROCRYPT '94. LNCS, vol. 950, pp. 205–214 (1994) (p. 356)

189. Meinel, C., Theobald, T.: Algorithmen und Datenstrukturen im VLSI-Design: OBDD – Grundlagen und Anwendungen. Springer, Berlin (1997) (p. 125)

190. Merkel, R.C., Hellman, M.E.: Hiding information and signatures in trapdoor knapsack. IEEE Trans. Inf. Theory IT-24(5), 525–530 (1978) (pp. 254, 343)

191. Miller, G.L.: Riemann's hypothesis and tests for primality. J. Comput. Syst. Sci. 13, 300–317 (1976) (p. 251)

192. Mironov, I.: (Not so) random shuffles of RC4. In: Advances in Cryptology—CRYPTO 2002. LNCS, vol. 2442, pp. 304–319. Springer, Berlin (2002) (pp. 196, 198)

193. Mitchell, C.: Enumerating Boolean functions of cryptographic significance. J. Cryptol. 2, 155–170 (1990) (p. 81)

194. Moen, V., Raddum, H., Hole, K.J.: Weakness in the temporal key hash of WPA. Mob. Comput. Commun. Rev. 8(2), 76–83 (2004) (p. 187)

195. Montgomery, P.L.: Modular multiplication without trial division. Math. Commun. 44, 519–521 (1985) (p. 250)

196. Montgomery, P.L.: A survey of modern integer factoring algorithms. Quart. - Cent. Wiskd. Inform. 7, 337–366 (1994) (p. 247)

197. Niederreiter, H.: Sequences with almost perfect linear complexity profile. In: Chaum, D., Price, W.L. (eds.) Advances in Cryptology, Eurocrypt '87. LNCS, vol. 304, pp. 37–51. Springer, Berlin (1988) (pp. 47, 48)

198. Niederreiter, H.: Keystream sequncence with a good linear complexity profile for every starting point. In: Advances in Cryptology—Eurocrypt '89. Lecture Notes in Computer Science, vol. 434, pp. 523–532 (1990) (pp. 46, 48)

199. The OCB authenticated-encryption algorithm. http://datatracker.ietf.org/doc/draft-krovetz-ocb/?include_text=1 (p. 239)

200. Odlyzko, A.M.: The rise and fall of the knapsack cryptosystem. In: Pomerance, C. (ed.) Cryptology and Computational Number Theory. Proceeding of Symposia in Applied Mathematics, vol. 42, pp. 75–88. American Mathematical Society, Providence (1990) (pp. 254, 277, 278, 360)

201. The openSSL library. http://www.openssl.org (p. 251)

202. Palmer, E.M., Read, R.C., Robonson, R.W.: Balancing the n-cube: a census of colorings. J. Algebr. Comb. 1, 257–273 (1992) (p. 81)

203. Pan, V.: Strassen's algorithm is not optimal. Trilinear technique of aggregating, uniting and canceling for constructing fast algorithms for matrix muliplication. In: Proc. Nineteenth Ann. Symp. on Foundations of Computer Science, pp. 28–38 (1978) (pp. 281, 286)

204. Pan, V.: How to Multiply Matrices Faster. Lecture Notes in Computer Science., vol. 179. Springer, Berlin (1984) (pp. 281, 282)

205. Parker, M.G., Kemp, A.H., Shepherd, S.J.: Fast Blum-Blum-Shub sequence generation using Montgomery multiplication. IEEE Proc. Comput. Digit. Techn. 147, 252–254 (2000) (p. 251)

206. Patarin, J.: Hidden field equations (HFE) and isomorphisms of polynomials (IP): two new families of asymmetric algorithms. In: Eurocrypt '96, pp. 33–48. Springer, Berlin (1996). An extended version can be found at http://www.minrank.org/courtois/hfe.ps (p. 254)

207. Paul, G., Maitra, S.: RC4 Stream Cipher and Its Variants. Discrete Mathematics and Its Applications. CRC Press, Boca Raton (2011) (p. 183)

208. Paul, G.K.: Analysis and design of RC4 and its variants. PhD thesis, Department of Computer Science & Engineering, Jadavpur University, Kolkata, India (2008) (p. 227)

209. Paul, S., Preneel, B.: A new weakness in the RC4 keystream generator and an approach to improve the security of the cipher. In: FSE 2004. LNCS, vol. 3017, pp. 245–259 (2004) (pp. 222, 224, 226)

210. Peikari, C., Chuvakin, A.: Security Warrior. O'Reilly (2004) (p. 246)

211. Perron, O.: In: Die Lehre von den Kettenbrüchen. Elementare Kettenbrüche. Band 1. 3 Auflage. Teubner, Stuttgart (1954) (p. 47)

212. Poe, E.A.: A Few Words on Secret Writing. Graham's Magazine. July 1841 (p. 349)

213. Poe, E.A.: The Gold-Bug. The Dollar Newspaper (Philadelphia, PA), vol. I, no. 23, pp. 1 and 4, June 28 1843. Available online http://www.eapoe.org/works/tales/goldbga2.htm (pp. 2, 349)

214. Pritchard, P.: A sublinear additive sieve for finding prime numbers. Commun. ACM **24**(1), 18–23 (1981) (p. 251)

215. Prüfer, H.: Neuer Beweis eines Satzes über Permutationen. Arch. Math. Phys. **27**, 742–744 (1918) (p. 334)

216. Pyshkin, A., Tews, E., Weinmann, R.P.: Breaking 104 bit WEP in less than 60 seconds. In: WISA. LNCS, vol. 4867, pp. 188–202 (2007). http://eprint.iacr.org/2007/120.pdf (pp. 185, 207)

217. Rabin, M.O.: Probabilistic algorithms for testing primality. J. Number Theory **12**, 128–138 (1980) (p. 251)

218. Ranjan, R.K., Gosti, W., Brayton, R.K., Sangiovanni-Vincentelli, A.: Dynamic reordering in a breadth-first manipulation based BDD package: challenges and solutions. In: International Conference on Computer Design IEEE, pp. 344–351, October (1997) (p. 268)

219. Reischuk, K.R.: Komplexitätstheorie. Grundlagen, Band 1. B.G. Teubner, Stuttgart, Leipzig (1999) (p. 273)

220. Rejewski, M.: An application of the theory of permutations in breaking the Enigma cipher. Appl. Math. **16**(4), 543–559 (1980) (p. 10)

221. Rivest, R.: RSA: Security response to weaknesses in key scheduling algorithm of RC4. Technical report, RSA Security, Inc. (2001). http://www.rsasecurity.com/rsalabs/technotes/wep.html (p. 183)

222. Rivest, R.L., Silverman, R.D.: Are 'strong' primes needed for RSA? Technical report, The RSA Laboratories Seminar Series (1997) (p. 251)

223. de Riviére, A.: Question 48. l'Intermédiare des Mathématiciens **1**, 19–20 (1894) (p. 60)

224. Robbiano, L.: Term orderings on the polynomial ring. In: EUROCAL'85. LNCS, vol. 204, 513–517 (1985) (p. 133)

225. Robbins, D., Bolker, E.: The bias of three pseudo-random shuffles. Aecquationes Mathematicae **22**, 268–292 (1981) (pp. 190, 192, 195)

226. Robshaw, M., Billet, O. (eds.): New Stream Cipher Designs, the ESTREAM Finalists. Lecture Notes in Computer Science, Security and Cryptology, vol. 4986. Springer, Berlin (2008) (p. 229)

227. Rudell, R.: Dynamic variable ordering for binary decision diagrams. In: Proc. Intl. Conf. on Computer-Aided Design, pp. 42–47, November 1993 (pp. 267, 268)

228. Rueppel, R.A.: Analysis and Design of Stream Chiphers. Springer, Berlin (1986) (pp. 42, 43, 44, 75)

229. Rueppel, R.A., Massey, J.L.: Knapsack as nonlinear function. In: IEEE Intern. Symp. of Inform. Theory. IEEE Press, New York (1985) (p. 255)

230. Rueppel, R.A., Staffelbach, O.J.: Products of linear recurring sequences with maximum complexity. IEEE Trans. Inf. Theory **33**, 124–131 (1987) (p. 70)

231. Sainte-Marie, Fly C.: Solution to question 48. l'Intermédiare des Mathématiciens **1**, 107–110 (1894) (p. 60)
232. Sarkar, P., Maitra, S.: Nonlinearity bounds and construction of resilient Boolean functions. In: Advances in Cryptology—CRYPTO 2000. LNCS, vol. 1880, pp. 515–532. Springer, Berlin (2000) (p. 152)
233. Steel Bank Common Lisp. http://www.sbcl.org/ (p. 370)
234. Schmidt, F., Simion, R.: Card shuffling and a transformation on S_n. Aequations Mathematicae **44**, 11–34 (1992) (p. 193)
235. Schneider, M.: A Note on the Construction and Upper Bounds of Correlation-Immune Functions. LNCS, vol. 1355, 295–306 (1997) (p. 81)
236. Schönhage, A., Strassen, V.: Schnelle Multiplikation großer Zahlen. Computing **7**, 281–292 (1971) (p. 249)
237. Seal, D.: Newsgroup comp.arch.arithmetic, 13 May 1997 (p. 263)
238. Seindal, R., Pinard, F., Vaughan, G.V., Blake, E.: GNU M4, version 1.4.11 (2008). Available online http://www.gnu.org/software/m4/manual/index.html (pp. 369, 374)
239. Shaked, Y., Wool, A.: Cryptanalysis of the bluetooth E_0 cipher using OBDDs. In: Information Security. LNCS, vol. 4176, pp. 187–202. Springer, Berlin (2006) (p. 127)
240. Shamir, A.: On the generation of cryptographically strong pseudo-random sequences. In: 8th International Colloquium on Automata Languages and Programming. LNCS, vol. 62 (1981) (p. 241)
241. Shamir, A., Kipnis, A.: Cryptanalysis of the HFE public key cryptosystem. In: CRYPTO '99 (1990). Available online http://www.minrank.org/courtois/hfesubreg.ps (p. 254)
242. Shamir, A., Tsaban, B.: Guaranteeing the diversity of number generators. Inf. Comput. **171**(2), 350–363 (2001) (p. 234)
243. Shamir, A., Zippel, R.E.: On the security of the Merkel-Hellman cryptographic scheme. IEEE Trans. Inf. Theory **IT-26**(3), 339–340 (1980) (pp. 254, 360)
244. Shoup, V.: NTL: A library for doing number theory. http://www.shoup.net/ntl/ (pp. 38, 370)
245. Shparlinski, I.: Cryptographic Applications of Analytic Number Theory: Complexity, Lower Bounds and Pseudorandomness. Progress in Computer Science and Applied Logic. Birkhäuser, Basel (2003) (p. 254)
246. Shulman, M.A.: MMM Mode for Emacs, version 0.4.8 edition (2004). http://mmm-mode.sourceforge.net/ (p. 377)
247. Sidorenko, A., Schoenmakers, B.: Concrete security of the Blum-Blum-Shub pseudorandom generator. In: Smart, N.P. (ed.) Cryptography and Coding 2005. LNCS, vol. 3796, pp. 355–375. Springer, Berlin (2005) (pp. 245, 246, 253)
248. Siegenthaler, T.: Correlation-immunity of nonlinear combining functions for cryptographic applications. IEEE Trans. Inf. Theory **30**, 776–780 (1984) (p. 78)
249. Siegmund, D.: Sequential Analysis (Tests and Confidence Intervals). Springer Series in Statistics. Springer, New York (1985) (pp. 113, 316)
250. Siegmund, D.: Boundary crossing probabilities and statistical application. Ann. Stat. **14**, 361–404 (1986) (p. 325)
251. Sieling, D., Wegener, I.: Reduction of OBDDs in linear time. Inf. Process. Lett. **48**, 139–144 (1993) (p. 119)
252. Silverman, R.D.: Fast generation of random, strong RSA primes. Technical report, RSA CrypoBytes, volume 3, No 2, 1997. Available online http://www.rsa.com/rsalabs/node.asp?id-2149 (p. 251)
253. The SINGULAR computer algebra system. http://www.singular.uni-kl.de/ (pp. 140, 370)
254. Solovay, R., Strassen, V.: A fast Monte Carlo test for primality. SIAM J. Comput. **6**, 84–85 (1977) (p. 251)
255. St Denis, T.: Cryptography for Developers. Syngress (2007) (p. 53)
256. Stanley, R.P.: Enumerative Combinatorics. Cambridge Studies in Advanced Mathematics, vol. 49. Cambridge University Press, Cambridge (1997) (p. 221)

257. Stern, J.: A method for finding codewords of small weight. In: Cohen, G.D., Wolfmann, J. (eds.) Coding Theory and Applications. Lecture Notes in Computer Science, vol. 388, pp. 106–113. Springer, Berlin (1989) (p. 254)

258. Sterndark, D.: RC4 algorithm revealed. Usenet posting sternCVKL4B.Hyy@netcom.com. September 1994 (p. 183)

259. Stinson, D.R.: Cryptography, Theory, Practice, Discrete Mathematics and Its Applications, 3rd edn. Chapman & Hall/CRC, London (2006) (p. 3)

260. Strassen, V.: Gaussian elimination is not optimal. Numer. Math. **13**, 354–356 (1969) (p. 280)

261. Swan, R.C.: Factorisation of polynomials over finite fields. Pac. J. Math. **12**, 1099–1106 (1962) (p. 308)

262. Tews, E.: Attacks on the WEP protocol. Master's thesis, TU Darmstadt, Fachgebiet Theoretische Informatik (CDC) (2007) (p. 185)

263. Tews, E., Klein, A.: Attacks on Wireless LANs: About the security of IEEE 802.11 based wireless networks. Vdm Verlag Dr. Müller (2008) (p. 185)

264. $\mathrm{T_EX}$ Live. http://www.tug.org/texlive/ (p. 369)

265. Toom, A.L.: The complexity of a scheme of functional elements realising the multiplication of integers. J. Sov. Math. **3**, 714–716 (1963). Original in: Dokl. Akad. Nauk SSSR **150**, 496–498 (1963) (p. 249)

266. Traverso, C.: Hilbert functions and Buchberger's algorithm. J. Symb. Comput. **22**, 355–376 (1997) (p. 142)

267. Tutte, W.T.: The dissection of equilateral triangles into equilateral triangles. Proc. Camb. Philos. Soc. **44**, 463–482 (1948) (p. 62)

268. Ulbricht, H.: Die Chriffriermaschine ENIGMA Trügerische Sicherheit, Ein Beitrag zur Geschichte der Nachrichtendienste. PhD thesis, Fachbereich Mathematik und Informatik, Technische Universität Carolo-Wilhelmina zu Braunschweig (2005) (p. 10)

269. Unwin, S.: The Probability of God: A Simple Calculation Proves the Ultimate Truth. Crown Forum, New York (2003) (p. 315)

270. Vazirani, U., Vazirani, V.: Efficient and secure pseudorandom number generation. In: Proceedings of the 25th Annual Symposium on the Foundations of Computer Science, pp. 458–463. IEEE Press, New York (1984) (pp. 245, 251, 252, 253)

271. Viterbi, A.J.: Error bounds for convolutional codes and an asymptotically optimum decoding algorithm. IEEE Trans. Inf. Theory **IT-13**(2), 260–269 (1967) (p. 107)

272. Vogel, R.: On the linear complexity of cascaded sequences. In: Advances in Cryptology—Eurocrypt '84. LNCS, vol. 209, pp. 99–109 (1985) (p. 157)

273. Wald, A.: Sequential Analysis. Wiley, New York (1947) (p. 316)

274. Wang, M.Z., Massey, J.L.: The characterisation of all binary sequences with a perfect linear complexity profile. Paper presented at the Eurocrypt '86 (1986) (p. 49)

275. Ward, J.B.: The Beal Papers. Virginian Book and Job Print (1885) (p. 2)

276. Warren, H.S. Jr.: Hacker's Delight. Addison-Wesley, Boston (2003). Revisions and additional material are on the homepage of the book. http://www.hackersdelight.org/ (pp. 53, 113)

277. Warren, H.S. Jr: Homepage of Hacker's Delight. http://www.hackersdelight.org/. Contains example programs, errata and additional material (pp. 263, 350)

278. Wegener, I.: Branching Programs and Binary Decision Diagrams. SIAM, Philadelphia (2000) (p. 125)

279. Wegner, P.: A technique for counting ones in a binary computer. Commun. ACM **3**, 322 (1960) (p. 358)

280. Wiedemann, D.H.: Solving sparse linear equations over finite fields. IEEE Trans. Inf. Theory **IT-32**(1), 54–62 (1986) (pp. 291, 292)

281. Wilkes, M.V., Wheeler, D.J., Gill, S.: The Preparation of Programs for an Electronic Digital Computer. 2nd edn. Addison-Wesley, Reading (1957) (p. 263)

282. Winograd, S.: A new algorithm for inner product. IEEE Trans. Comput. **C-18**, 693–694 (1968) (p. 279)

283. Wozencraft, J.M., Rieffen, B.: Sequential Decoding. MIT Press/Wiley, Cambridge (1961) (p. 109)

284. Wu, H.: Cryptanalysis of a 32-bit RC4-like Stream Cipher. Technical report, Katholieke Universiteit Leuven, Dept. ESAT/COSIC (2005). http://eprint.iacr.org/2005/219.pdf (p. 224)
285. Xiao, G.Z., Massey, M.L.: A spectral characterisation of correlation immune combining functions. IEEE Trans. Inf. Theory **34**, 569–571 (1988) (p. 77)
286. Yang, B.Y., Chen, O.C.H., Bernstein, D.J., Chen, J.M.: Analysis of QUAD. In: Biryukov, A. (ed.) Fast Software Encryption: 14th International Workshop, FSE 2007. Lecture Notes in Computer Science, vol. 4593, pp. 290–308. Springer, Berlin (2007) (p. 256)
287. Yang, Y.X., Guo, B.: Further enumerating Boolean functions of cryptographic significance. J. Cryptol. **8**, 115–122 (1995) (p. 81)
288. Zhang, J.Z., You, Z.S., Li, Z.L.: Enumeration of binary orthogonal arrays of strength 1. Discrete Math. **239**, 191–198 (2001) (p. 81)
289. Zigangirov, K.S.: Some sequential decoding procedures. Probl. Inf. Transm. **2**, 13–25 (1966) (p. 109)
290. Zsigmondy, K.: Zur Theorie der Potenzreste. Monatshefte Math. Phys. **3**, 265–284 (1892) (p. 297)

Index

0–9
2-adic numbers, 261

A
A5/1, 176
A5/2, 170–176
Accessible predicate, 243
Adjacency matrix, 62
Aggregating table, 284
Aircrack, 185
Algebraic attacks, 65
Algebraic normal form, 309
Algorithm
    Berlekamp-Massey ~, 33
Almost bent function, 30
Alternating step generator, 157, 158
Alternative, 313
Ascending chain condition, 137
Asynchronous stream cipher, 5
Attack
    ~based on Golić's correlation, 202–209
    algebraic ~, 65
    CJS ~, 91–105
    correlation ~, 66
    second round ~against RC4, 207–209
    state recovering ~against RC4, 209–212
Auto key cipher, 5
Auto-correlation, 25
    sequences with two level ~, 27
Auto-correlation test, 25

B
Balanced, 76
Balanced colorings of a hypercube, 81
Basis
    Gröbner~, 137
BDD, 117

Beale cipher, 2
Bent function, 30
Berlekamp-Massey algorithm, 33
Berlekamp's algorithm, 146
Bernoulli numbers, 329
Binary decision diagram, 117
Birthday paradox, 185
Black box linear algebra, 292
Bluetooth, 126
Blum-Blum-Shub generator, 244–247
Blum-Micali generator, 243
Borel σ-algebra, 311
Brownian motion, 322
Buchberger, Bruno, 137
Buchberger's Algorithm, 140

C
Caesar cipher, 1
Carry save adder, 263
Characterization
    ~of m-sequences, 27
Chinese remainder theorem, 293
Cipher
    asynchronous stream ~, 5
    auto key ~, 5
    homophone ~, 2
    polyalphabetic ~, 2
    polygraphic ~, 2
    RC4 ~, 183
    self-synchronizing stream ~, 5, 6
    synchronous stream ~, 5
    Vigenère ~, 3
CJS-attacks, 91–105
Code
    cyclic redundancy ~(CRC), 189
Code of LFSR, 24
Column distance, 106

Combiner with memory, 149
Companion matrix, 18
Complete problem, 277
Complexity class
    DSpace, 275
    DTime, 275
    NSpace, 275
    NTime, 275
Conditional expectation, 316
Constancy on cyclotomic cosets, 26
Continued fraction, 46
Control graph, 124
Convolution, 101
Convolutional codes, 105–111
Coppersmith's method, 303
Correlation
    Golić's ~, 203
Correlation attack, 66
Correlation immune function, 75
Correlation-attack, 91–115
Correlation-attack on the shrinking generator,
    161
CRC-code, 189
Cross-correlation, 29
Cyclic redundancy check code, 189
Cyclotomic polynomial, 297

D
De Bruijn graph, 59
De Bruijn sequence, 59, 157
Decoding
    sequential ~, 109–111
    twice step ~, 101
    Viterbi ~, 107–109
Degree
    ω-~, 141
Degree reduction, 147
Deterministic test, 313
Dickson's lemma, 136
Difference set, 27
Differential power analysis, 164
Digraph probabilities, 213
Distance profile, 106, 111
Distinguisher, 242, 312
Distribution test, 25
Doob's theorem, 319
DPA, 164
DSpace, 275

E
E_0, 126, 153
Enigma, 8
Error probability
    ~of the first kind, 313

~of the second kind, 314
Eulerian function, 294
Euler's constant, 331
Euler's summation formula, 330
Extended linearization, 145

F
F_4 algorithm, 147
Fano metric, 110
Fast algebraic attacks, 149
Feedback polynomial, 19
    irreducible ~, 20
    primitive ~, 21
    reducible ~, 21
Feedback shift register, 17
Fibonacci implementation, 51
Filter
    non-linear ~, 72
Filtration, 317
Fisher Stern generator, 255
FMS-attack, 199–202
Formal language, 275
Fortuitous states, 218
Fourier transform, 101
Fourier transformation, 104, 105
Free binary decision diagrams, 124
Free distance, 106
Frobenius automorphism, 305
Function
    almost bent ~, 30
    bent ~, 30
Functional central limit theorem, 326

G
Galois group of a finite field, 305
Galois implementation, 51
Geffe generator, 64, 143
Generating function, 19
Generator matrix, 24
GGHN generator, 222
Golić correlation, 203
Golomb axioms, 24
Graded lexicographic ordering, 132
Graded reverse lexicographic ordering, 132
Gram-Schmidt orthogonalization, 302
Gröbner basis, 137
Gröbner walk, 140
GSM protocol, 169

H
Hadamard difference set, 28
Hadamard matrix, 28
Hamming weight, *see* sideway addition
Hard problem, 277
Harmonic numbers, 331

Hilbert's basis theorem, 137
Homogeneous polynomial, 141
Homophone cipher, 2
Hypothesis, 313

I
Ideal
  monomial ~, 135
Initial form, 141
Input hard predicate, 243
Involution, 332

J
Jacobi symbol, 300

K
Karatsuba algorithm, 247
Kerckhoffs' principle, 1, 169
Key scheduling
  ~of RC4, 190–199
Knapsack generator, 255
Kullback-Leibner information, 320

L
Labeled tree, 334
Language
  formal ~, 275
Laplacian matrix, 62
Lattice, 301
Laurent series, 46
Leading coefficient, 134
Leading form, 140
Leading monomial, 134
Leading term, 134
Legendre symbol, 299
Lexicographic ordering, 132
LFSR, 17
  closed formula of an ~sequence, 22
  Fibonacci implementation of a ~, 51
  Galois implementation of a ~, 51
  linear complexity of an ~sequence, 31
  non-linear combination of ~, 66
  periods of ~, 22
  product of two ~sequences, 67
  sum of two ~sequences, 32
  trace representation of an ~sequence, 20
Likelihood quotient test, 315
LILI-128, 151
Linear complexity, 31
  ~of the stop-and-go generator, 156
Linear complexity profile, 45
  good ~, 46
  perfect ~, 45
Linear feedback shift register, *see* LFSR

Linearization, 143

M
M-sequence, 18, 24–30
  characterization of a ~by the shift-and-add
    property, 27
Martingale, 317
Martingale convergence theorem, 319
Massey, J.L., 77
Matrix
  Hadamard ~, 28
  Toeplitz ~, 31
Matrix order, 133
Measurable function, 311
Measurable space, 311
Measure, 311
Metric
  Fano ~, 110
Möbius function, 306
Modular shift register generator, *see* Galois
    implementation
Monic polynomial, 305
Monoalphabetic cipher, 2
Monomial ideal, 135
Monomial order, 132
Montgomery multiplication, 250
Mosquito, 235
Most powerful test, 314
Moustique, 235
MRSRG, *see* Galois implementation
MSRG, *see* Galois implementation
Multidegree, 134
Multiple-return shift register generator, *see*
    Galois implementation
MXOR, 288

N
NESSIE-Project, 151
Next bit predictor, 242
Neyman-Pearson test, 314
Noetherian ring, 137
Non-linear filter, 72
Normal form
  algebraic ~, 309
NSpace, 275
Ntime, 275

O
O-Notation, 272
ω-degree, 141
ω-homogeneous, 141
One-time pad, 4
Optimal non-linearity, 79
Optimal sampling theorem, 318

Ordering on monomials, 132
Orthogonal array, 76

P

Pan-normal-form, 284
Parameter space, 313
Parity check matrix, 24
Pattern
    ~of RC4, 212
Perfect local randomizer, 256
Permanent, 74
Permutation, 332
Playfair cipher, 2
Polyalphabetic cipher, 2
Polygraphic cipher, 2
Polynomial
    cyclotomic ~, 297
    monic ~, 305
Population count, see sideway addition
Power generator, 253
Predictor
    next bit ~, 242
    previous bit ~, 242
Preferred pair, 30
Previous bit predictor, 242
Prime number theorem, 295
Primitive polynomial, 21
Protocol
    WEP ~, 184
    WPA ~, 185–187

Q

QUAD, 256
Quadratic reciprocity law, 300

R

Rabbit, 232
Randomized test, 313
RC4, 183
    key scheduling of ~, 190–199
RC4A, 224
Reduction, 277
Reflection principle, 324
Relinearization, 144
Resilient, 76
Resultant, 69
Rotor machines, 8
RSA generator, 253

S

S-polynomial, 138
Self-shrinking generator, 342
Self-synchronizing stream cipher, 5, 6
Sequential decoding, 109–111, 113

Sequential likelihood ratio test, 319
Sequential test, 100
Serial test, 25
Shannon's coding theorem, 93
Shift-and-add property, 26
Shrinking generator, 158–163
Side channel attacks, 163
Sideway addition, 53, 103, 262, 263, 343
Siegenthaler's inequality, 79
σ-algebra, 311
Simple power analysis, 163
Simple shift register generator, see Fibonacci
    implementation
SPA, 163
Sparse feedback polynomials, 114
Sparse linear algebra, 292
SSRG, see Fibonacci implementation
Step-once-twice generator, 157
Stop-and-go generator, 155–157
Stopping time, 318
Strassen's algorithm, 280
Stream cipher
    Mosquito, 235
    Moustique, 235
    Rabbit, 232
    Trivium, 229
Strong Markov property, 324
Submartingale, 317
Subset sum generator, see knapsack generator
Supermartingale, 317
Symmetric group, 332
Synchronous stream cipher, 5
Syzygien polynomial, 138

T

Temporal key hash, 187
Term, 134
Test
    auto-correlation ~, 25
    distribution ~, 25
    likelihood quotient ~, 315
    most powerful ~, 314
    Neyman-Pearson ~, 314
    sequential likelihood ratio ~, 319
    serial ~, 25
Test problem, 313
Test (statistical), 313
Toeplitz matrix, 31
Transform
    Walsh ~, 30, 77
Tree diagram, 109
Trilinear form, 282
Trinomial, 308
Trivium, 229

Turing machine, 273
Twice step decoding, 101

V
Variation distance, 312
Vertical bit count algorithm, 113
Vigenère cipher, 3
Viterbi decoding, 107–109, 112
Von-Neumann-generator, 234

W
Wald's test, 319
Walsh spectrum, 30

Walsh transform, 30, 77
Well-ordering, 132
WEP, 184
Wiener process, 322
Wireless LAN, 184–190
WPA, 185–187

X
XL-Algorithm, 145

Z
Zsigmondy's theorem, 71, 297